ADVANCES IN INSECT CHEMICAL ECOLOGY

Chemical signals mediate all aspects of insects' lives and their ecological interactions. The discipline of chemical ecology seeks to unravel these interactions by identifying and defining the chemicals involved, and by documenting how perception of these chemical mediators modifies behavior and, ultimately, reproductive success.

Chapters in this volume consider how plants use chemicals to defend themselves from insect herbivores; the complexity of floral odors that mediate insect pollination; tritrophic interactions of plants, herbivores, and parasitoids, and the chemical cues that parasitoids use to find their herbivore hosts; the semiochemically mediated behaviors of mites; pheromone communication in spiders and cockroaches; the ecological dependence of tiger moths on the chemistry of their host plants; and the selective forces that shape the pheromone communication channel of moths.

Each review is written by an internationally recognized expert and presents descriptions of the chemicals involved, the effects of semiochemically mediated interactions on reproductive success, and the evolutionary pathways that have shaped the chemical ecology of arthropods.

Professors RING CARDÉ and JOCELYN MILLAR are both based in the University of California at Riverside. Between them, they have written over 300 articles on chemical ecology and have co-edited six books.

ADVANCES IN INSECT CHEMICAL ECOLOGY

Edited by

RING T. CARDÉ AND JOCELYN G. MILLAR
University of California at Riverside, USA

CAMBRIDGE
UNIVERSITY PRESS

PUBLISHED BY THE PRESS SYNDICATE OF THE UNIVERSITY OF CAMBRIDGE
The Pitt Building, Trumpington Street, Cambridge, United Kingdom

CAMBRIDGE UNIVERSITY PRESS
The Edinburgh Building, Cambridge, CB2 2RU, UK
40 West 20th Street, New York, NY 10011–4211, USA
477 Williamstown Road, Port Melbourne, VIC 3207, Australia
Ruiz de Alarcón 13, 28014 Madrid, Spain
Dock House, The Waterfront, Cape Town 8001, South Africa

http://www.cambridge.org

First published 2004

Printed in the United Kingdom at the University Press, Cambridge

Typeface Times 11/14 pt. *System* LaTeX 2_ε [TB]

A catalogue record for this book is available from the British Library

Library of Congress Cataloguing in Publication data

Advances in insect chemical ecology / edited by Ring T. Cardé and Jocelyn G. Millar.
p. cm.
Includes bibliographical references and index.
ISBN 0 521 79275 4
1. Insects – Ecophysiology. 2. Animal chemical ecology. I. Cardé, Ring T. II. Millar, Jocelyn G., 1954–
QL495.A18 2004
573.8′77157 – dc22 2003055731

ISBN 0 521 79275 4 hardback

Contents

Contributors

John T. Arnason
Faculty of Science, University of Ottawa, ON KIN 6N5, Canada

Ring T. Cardé
Department of Entomology, University of California, Riverside,
CA 92521 USA

William E. Conner
Department of Biology, Wake Forest University, Winston-Salem,
NC 27109, USA

Tony Durst
Faculty of Science, University of Ottawa, ON KIN 6N5, Canada

César Gemeno
Department of Entomology, North Carolina State University, Raleigh,
NC 27695, USA
Current address: Departamento de Producción Animal y Ciencia Vegetal,
Universidad de Lleida, Lleida 25198, Spain

Gabriel Guillet
Faculty of Science, University of Ottawa, ON KIN 6N5, Canada

Kenneth F. Haynes
Department of Entomology, University of Kentucky, Lexington,
KY 40546, USA

Yasumasa Kuwahara
Division of Applied life Sciences, Graduate School of Agriculture,
Kyoto University, Sakyo-ku, Kyoto 606-8502, Japan

Jocelyn G. Millar
Department of Entomology, University of California, Riverside,
CA 92521, USA

Robert A. Raguso
Department of Biological Sciences, University of South Carolina, Columbia,
SC 29208, USA

Coby Schal
Department of Entomology, North Carolina State University, Raleigh,
NC 27695, USA

Stefan Schulz
Institut für Organische Chemie, Technische Universität Braunschweig,
Hagenring 30, D-38106 Germany

Ted C. J. Turlings
Institute of Zoology, University of Neuchatel, 2007 Neuchatel, Switzerland

Felix Wäckers
Netherlands Institute of Ecology, N100-CTO Heteren, the Netherlands

Susan J. Weller
Department of Entomology, J. F. Bell Museum of Natural History, University
of Minnesota, St Paul, MN 55108, USA

Preface

In contrast to other animals, humans sense their world chiefly by vision, sound, and touch. We have, in general, a remarkably undeveloped sense of smell, and so it is not surprising that we fail to appreciate how important chemical signals are in the lives of other organisms. Chemical signals and cues serve insects in numerous ways, including sexual advertisement, social organization, defense, and finding and recognizing resources. Chemical ecology seeks to identify these chemicals and to establish how they affect an organism's behavior, physiology, and interactions with other organisms. As the techniques to identify fully the structures of natural products have become increasingly sophisticated and powerful, the amounts of natural products needed for characterization have diminished, and the number of identified compounds that mediate behavioral and physiological interactions has proliferated. Our understanding of precisely how organisms employ such chemical information, however, continues to lag behind our ability to characterize the chemicals involved. It is also clear that the discoveries to date represent a miniscule sampling of the multitude of insect species that use information conveyed by chemical signals and cues.

These reviews are designed to provide in-depth overviews and syntheses of defined areas in the chemical ecology of insects and their closely related arthropods. The topics covered in this volume include: chemical defenses of plants against insect herbivores; floral odors mediating insect pollination; how parasitic wasps use odors emitted by herbivores and the plants on which they are feeding to find their herbivore hosts; semiochemicals of mites; pheromones of spiders; pheromones of cockroaches; the intricate defensive and pheromonal relationships between arctiid moths and chemicals from their host plants; and the selective forces that structure moth communication by pheromones. The perspective of these reviews ranges from proximate issues – what are the chemicals and their functions – to ultimate questions – how might evolutionary processes influence these chemical messages?

We also intend that this volume will represent the first in a series focussed on the chemical ecology of insects, both from the point of view of career practitioners of chemical ecology such as ourselves, and for those scientists whose studies lead them into some aspect of chemical ecology as a part of a broader study. By far the largest part of what we know of chemical communication between organisms has come about as the result of studies on insects, and there is every reason to believe that this trend will continue for the foreseeable future. Therefore, in addition to being intrinsically interesting and fascinating systems to study, semiochemical research with insects has played a fundamental role in the development of knowledge and understanding of olfaction and taste. Insects have been key players in all aspects, from the deciphering of the chemical messages, through the integration and processing of chemical signals in the brain, to the neural outputs resulting in a particular behavioral or physiological event, and, finally, to the development of hypotheses and theories of how these communication systems have evolved. Rapid progress is being made in all of these areas, and so it seems fitting that selected topic areas be reviewed and summarized in an ongoing series of volumes.

We thank our colleagues for their contributions to this volume, and Ward Cooper of Cambridge University Press for his advice and counsel on this project.

1

Phytochemical diversity of insect defenses in tropical and temperate plant families

John T. Arnason, Gabriel Guillet and Tony Durst
Faculty of Science, University of Ottawa, Canada

Phytochemical diversity and redundancy

One of the most intriguing features of the chemical ecology of plant–insect interactions is the remarkable number of different phytochemical defenses found in plants. A single plant may contain five or six biosynthetic groups of secondary metabolites and within each group these defenses may include many structurally related analogs and derivatives. Across the different species of higher plants, there is a bewildering array of different substances and modes of actions of substances. During the course of our research on phytochemical defenses in specific plant families, we have become interested in the raison d'être for this diversity, as have many other researchers (Romeo *et al.*, 1996). How much do we know about different types of defenses in plants? How did they arise? Are some of these secondary metabolites "redundant," with no function? How do they interact with one another? In this chapter, we will address some of these issues with observations on the defenses of several plant families against insects, using results from our own research and the published literature.

In general, we do not have a comprehensive picture of the different types of defenses in plants. Over a decade ago, Soejarto and Farnsworth (1989) estimated that of the 250 000 species of flowering plants, only 5000 species had been thoroughly investigated according to the Natural Product Alert (NAPRALERT) database, leaving 98% of species with potential for phytochemical discovery. Taking a more focussed view of the potential for discovering drugs of phytochemical origin in tropical forests, Mendelsohn and Balick (1995) estimated that current prescription drugs represent only 12% of what might be there to discover. Although steady progress in identifying the phytochemical defenses of plants is being made (thousands of compounds per year), the percentage of higher plants studied is

Advances in Insect Chemical Ecology, ed. R. T. Cardé and J. G. Millar. Published by Cambridge University Press.

increasing only slowly. About 70% of angiosperm species are tropical or subtropical and, according to Wilson (2002), we may lose 20–50% of these within the coming decades because of habitat loss and fragmentation. For this reason, there is a great urgency to learn more about their phytochemistry and chemical ecology before they are lost forever.

Progress in evaluating the diversity of defenses in the neotropical Meliaceae

The family Meliaceae (the mahogany family) provides an example of the high level of phytochemical diversity that can be found in tropical trees. The neem tree, *Azadirachta indica*, is a member of this family and its insect antifeedant and growth-reducing effects on insects have generated considerable research interest in the defenses of the family against insects. Intensive study of neem has revealed it to contain more than 50 limonoids and related compounds in its various tissues (Isman *et al.*, 1996). Many of these substances have feeding deterrent effects on herbivorous insects, and the specific mode of action of the most well-known compound, azadirachtin, has been studied at the electrophysiological and biochemical level. The neem tree is one of the most extreme examples of phytochemical redundancy in a single species. Characterization of limonoids from other species has been of considerable interest to the phytochemical research community and a large number of compounds have been identified in Meliaceae and related families of the Rutales, especially from Asian and African forests (Waterman and Grundon, 1983; Champagne *et al.*, 1992, 1996).

Neotropical forests are somewhat less well studied (Table 1.1). Approximately 20% of species have been investigated, although some of the more important timber genera in the family, especially mahogany (*Swietenia*) and tropical cedar (*Cedrela*), have received considerable attention. The insecticidal properties of these trees were known to indigenous peoples, such as the Maya, who valued cedar for its insect resistance and traditionally used pressed oils from its seeds as a treatment for lice (Pennington *et al.*, 1981). Our collaborative group (including R. Mata's laboratory in Mexico, the tropical dendrology group in Costa Rica (L. Poveda and P. Sanchez) and M. Isman's insect chemical ecology group in Vancouver) recently has investigated the insect-deterrent properties and substances in a number of these species. From *Swietenia humilis*, Jimenez *et al.* (1996, 1998) recovered more than 10 limonoids, including six novel humilinolides (Fig. 1.1) that were found in concentrations of 0.021–0.29% in seeds. Several of these were isolated in sufficient quantity for the assessment of insect growth-reducing activity against European corn borer, *Ostrinia nubilalis*, a highly polyphagous insect. These limonoids incorporated into insect diets at 0.005% caused significant reduction in growth, comparable to that seen with the positive control toosendanin (Fig. 1.2), a limonoid that is commercially available

Table 1.1. *Number of species and state of phytochemical investigation in genera of neotropical Meliaceae*

Genus	Number of species	Number of species with phytochemical identifications
Trichilia	70	14
Guarea	35	9
Swietenia	3	3
Ruagea	5	1
Cabralea	1	1
Cedrela	8	4
Carapa	3	3
Schmardia	1	0

Data from Pennington *et al.* (1981) and recent literature.

in China. They also caused delays in time to pupation and adult emergence as well as elevated mortality compared with controls. A recent study (Omar, 2000) with a model phytophagous weevil, *Sitophilus zeamais*, showed that the same compounds caused significant reduction in consumption of treated diets by these insects. Clearly these compounds are present in seeds at 5–50 times the concentration necessary to produce significant effects on generalist phytophagous insects. The presence of more than 10 compounds shows that the multiple defenses found in neem are not unique.

Other recent studies have demonstrated the insect growth-reducing and antifeedant properties of limonoids isolated from seeds. A similar study of *Guarea macrophylla* led to the isolation of six limonoids and protolimonoids, that were active in the *Ostrinia* bioassays (Jimenez *et al.*, 1998). Hirtin, a limonoid isolated from the seeds of *Trichilia hirta*, was active as a growth reducer to dark-sided cutworm, *Peridroma saucia* (Xie *et al.*, 1994). Recently, we isolated two other insect growth-reducing limonoids from the seeds of *Trichilia maritana* (MacKinnon *et al.*, 1997a) (Fig. 1.2). These results and studies by other groups, for example work on the genus *Raugea* (Mootoo *et al.*, 1996), show continuing potential for identification of insecticidal limonoids in seeds.

Plant parts other than seeds have been less well studied and little attention has been placed on insecticidal modes of action of the defenses of the neotropical Meliaceae. Our study of the insect growth-reducing activity of 50 extracts of bark, leaf and wood of Central American Meliaceae showed the potent effect of these extracts against lepidopteran larvae and the potential for isolation of bioactive compounds from a large number of these species. Extracts from the genera *Trichilia* and *Cedrela* (Xie *et al.*, 1994; Ewete *et al.*, 1996a; Wheeler *et al.*, 2001) show exceptional activity. *Trichilia americana* extracts have strong antifeedant activity to *Spodoptera*

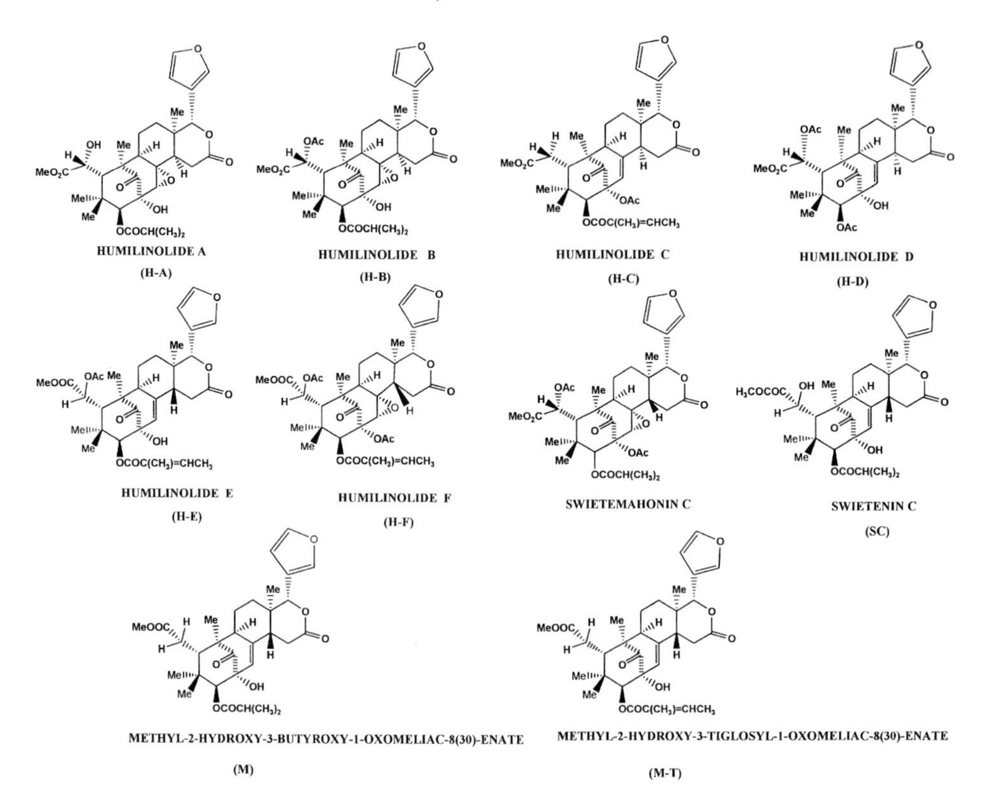

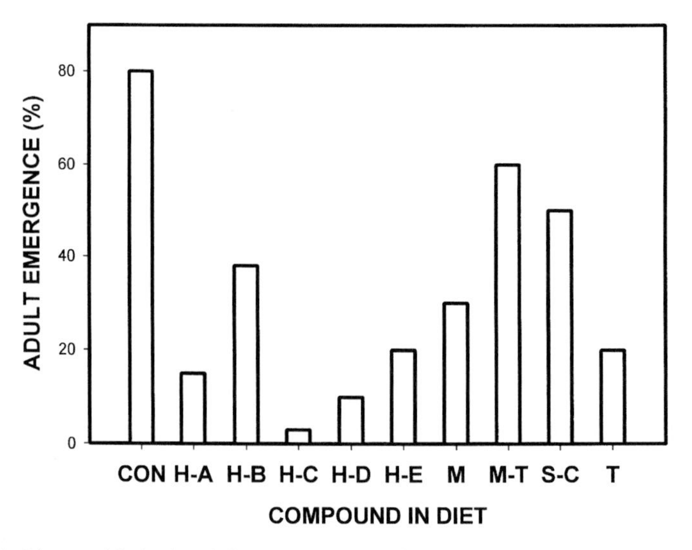

Fig. 1.1. Limonoids isolated from the seeds of *Swietenia humilis* and their effect on adult emergence of European corn borer following dietary administration at 50 ppm. CON, control; T, tenulin (toosendanin; Fig. 1.2); $n = 30$.

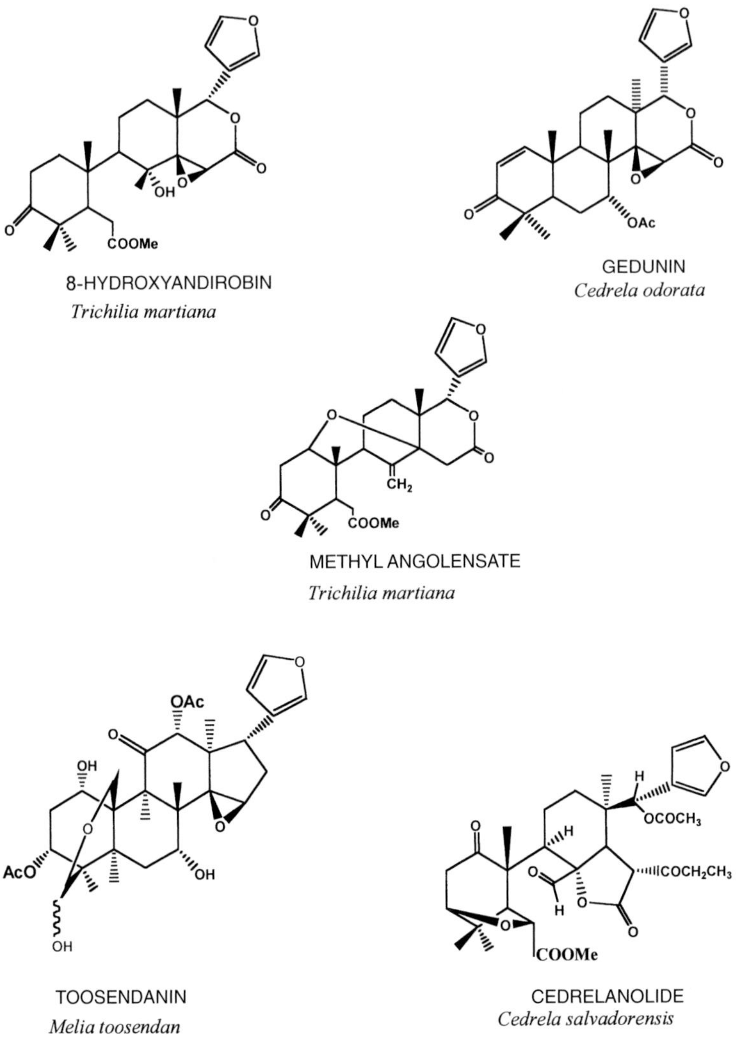

8-HYDROXYANDIROBIN
Trichilia martiana

GEDUNIN
Cedrela odorata

METHYL ANGOLENSATE
Trichilia martiana

TOOSENDANIN
Melia toosendan

CEDRELANOLIDE
Cedrela salvadorensis

Fig. 1.2. Limonoids isolated from the Meliaceae used in recent studies with insects.

litura, reducing growth by lowered food intake and lowered efficiency of conversion of ingested and digested food (Wheeler *et al*., 2001). The common tropical cedar tree of the Americas, *Cedrela odorata*, contains the limonoid gedunin (Fig. 1.2), and closely related derivatives in substantial amounts in the bark and wood. We previously had found gedunin and *C. odorata* extract to have antifeedant properties, and significantly to reduce growth and delay development of lepidopteran larvae (Arnason *et al*., 1987; Ewete *et al*., 1996a). Gedunin is also a potent antimalarial compound (MacKinnon *et al*., 1997b). Little was known about the specific mode of action of this (and other neotropical limonoids), but its relatively easy isolation

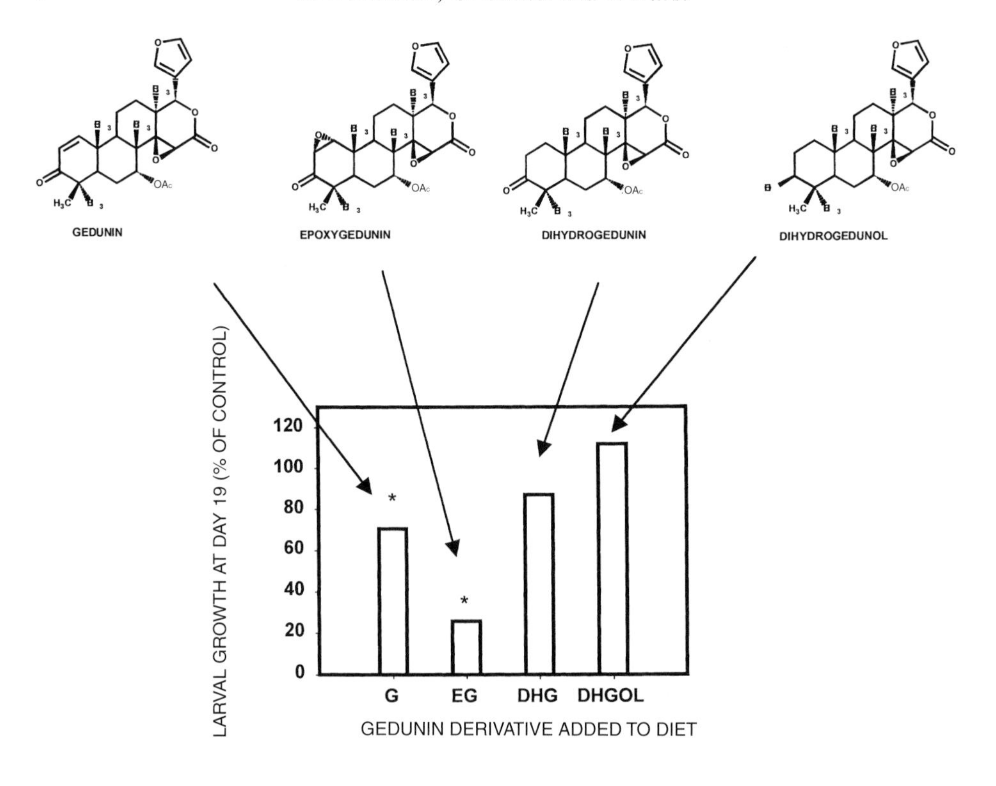

GEDUNIN + R-SH →

Fig. 1.3. Effect of semi-synthetic derivatives of gedunin on European corn borer growth and proposed mechanism of action. Significant effects.

and purification from wood provided enough starting material for semi-synthesis of derivatives (Fig. 1.3) (MacKinnon *et al.*, 1997b). Modification of the A ring by reduction of the double bond or both the keto group and the double bond eliminated insect growth-reducing activity. In contrast, the epoxy derivative produced an even greater growth-reducing effect than gedunin. These results are consistent with a mechanism of action in which gedunin or its epoxide acts as an alkylating agent, as shown in Fig. 1.3, whereas reduction of the keto group or double bond eliminates the

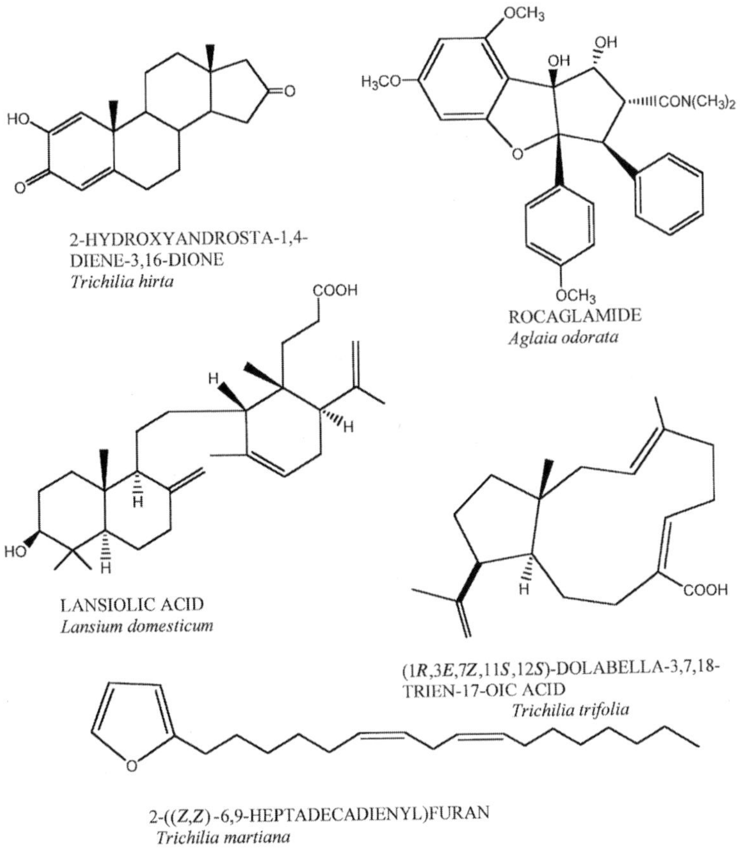

2-HYDROXYANDROSTA-1,4-DIENE-3,16-DIONE
Trichilia hirta

ROCAGLAMIDE
Aglaia odorata

LANSIOLIC ACID
Lansium domesticum

(1*R*,3*E*,7*Z*,11*S*,12*S*)-DOLABELLA-3,7,18-TRIEN-17-OIC ACID
Trichilia trifolia

2-((Z,Z)-6,9-HEPTADECADIENYL)FURAN
Trichilia martiana

Fig. 1.4. Phytochemical diversity in the Meliaceae, illustrating bioactive compounds other than limonoids.

potential for this reaction. Recent work on a related species of tropical cedar, *Cedrela salvadorensis* yielded the limonoid cedrelanolide (Fig. 1.2), which is comparable in activity to gedunin (Jimenez *et al.*, 1996) but which must have a different mode of activity.

Other studies show the presence of a diversity of compounds other than limonoids as defenses in the tissues of Meliaceae. Woody tissues of *Trichilia trifolia* afforded three novel dolabellanes with flexible C_{11} ring structures. These substances were very active antifeedants in the *Sitophilus* bioassay (Ramirez *et al.*, 2000). *T. martiana* seeds yielded large amounts of 2-((Z,Z)-6,9-heptadecadienyl)furan. *T. hirta* and *T. americana* bark have yielded novel steroids by insect bioassay-guided isolation and application of a nanoprobe nuclear magnetic resonance (NMR) technique for structure elucidation (Chaurest *et al.*, 1996). Compounds isolated included hydroxyandrosta-1,4-diene-3,16-dione (Fig. 1.4) and derivatives. However, studies by Wheeler *et al.* (2001) suggest that other unidentified compounds may also be involved in the activity of these species. Asian species have also provided a number of interesting non-limonoids. *Aglaia odorata* contains several benzofuran derivatives including rocaglamide (Fig. 1.4), which is the only compound that we have tested that has insecticide activity comparable to azadirachtin when incorporated in *O. nubilalis* diets (Ewete *et al.*, 1996b). We have also isolated nine triterpenes from bark of *Lansium domesticum*, collected in Borneo (Omar, 2000). These compounds, called lansiolides, including lansiolic acid (Fig. 1.4), have insect feeding-deterrent activity as well as antimalarial activity.

Clearly there is still much to discover about the insect chemical ecology of this tropical family. More information is also needed about the chemical ecology of generalist and specialist fauna on these plants, although economically important insect pests, such as the mahogany shoot borer (*Hipsipyla grandella*), have been studied. This pest attacks and damages the shoots of *Cedrela* and *Swietenia* but does not attack the closely related Australian genus *Toona*. Grafts of *C. odorata* shoots onto root stock of *Toona ciliata* are resistant because of the translocation of *Toona* compounds into the graft (da Silva *et al.*, 1999).

Overall, the published studies reveal that the defenses of the Meliaceae involve use of multiple defenses in one species and a broad diversity of biosynthetic types of phytochemical defense other than the well-studied limonoids, especially in plant parts other than seeds. Evolution of these diverse defenses was predicted in Ehrlich and Raven's (1964) stepwise co-evolution theory, later described as the "chemical arms race hypothesis" (Berenbaum, 1983), which was developed in part by observation of tropical families, including the Meliaceae, and their associated insect fauna. Although the concept that chance mutation and elaboration of novel defensive phytochemicals has led to the escape of plants from their herbivores continues to provide a plausible explanation for the evolution of phytochemical diversity, many modern chemical ecologists suggest that aspects of the original co-evolution theory requires reworking because of inconsistencies, such as a lack of observed reciprocal effects and ecophysiological constraints on co-evolution (Thompson, 1994).

Analog synergism in the Lepidobotryaceae and Piperaceae

Although plant–herbivore co-evolutionary theory provides a model for the evolution of the phytochemical diversity between species described above, the diversity of closely related compounds or analogs observed in any single species is not readily explained. Jones and Firn (1991), Berenbaum (1985) and Feng and Isman (1995) provided several hypotheses that have looked specifically at the evolution of diversity of phytochemical substances in a single species and the possibility of "analog redundancy" (a duplication of effort or production of compounds with no specific function). Jones and Firn (1991) proposed a "screening hypothesis" in which mechanisms exist to produce and retain a large number of compounds, many of which may have low potency but at least one or a few of which are molecules of high potency that will provide a defensive role. In our research, another example of extreme analog synergism similar to that of *S. humilis* is found in *Ruptiliocarpon caracolito*, a species that produces a remarkable number of unusual triterpene analogs (MacKinnon *et al.*, 1997c). This species was originally thought to belong to the Meliaceae, based on similarities in wood structure, but was more recently identified as the only neotropical member of the family Lepidobotryaceae, which contains two other species in Africa. Extracts of the wood, bark, and leaves were the most active insect growth reducers tested in a survey of *Rutales* species assayed against corn borer and cutworm. From this tree, our group has isolated more than 14 triterpenoids with a unique spiro C–D structure. Published structures are shown in Fig. 1.5. Because of the large number of analogs and derivatives, this is a useful species to test the Jones and Firn (1991) screening hypothesis, which is that these compounds are being produced in large numbers because a few may be highly effective insect deterrents. We isolated and screened 12 of these compounds at 5 and 50 ppm in lepidopteran diets, concentrations well below their naturally occurring levels. Assessed over the larval period, the results (Fig. 1.5) showed that most of the compounds increased larval mortality and significantly reduced the growth of survivors (not shown). They also caused delays in development and mortality at pupal and adult stages. Such a high level of activity suggests screening for a rare effective defense molecule is not occurring.

An alternative hypothesis for this phytochemical redundancy is analog synergism (Berenbaum, 1986). We isolated six of the compounds in sufficient quantity for studies of their activity alone and in mixtures. When tested alone, not all of the compounds reduced growth significantly and the calculated mean effect of the six compounds was to reduce growth of larvae by 50% compared with controls. In the mixture, each of the six individually tested compounds was added at one-sixth of the individual compounds. The growth of insects fed the mix was reduced significantly (80%; $P = 0.01$) suggesting synergism between the analogs (data not shown).

We were able to test the analog synergism in another setting where the analogs could be produced in sufficient quantity by synthesis. The wild pepper, *Piper tuberculatum*, was the most active of 16 pepper plants from Costa Rica assessed in our study (Bernard *et al.*, 1995). Four insecticidal piperamides from *P. tuberculatum* were produced by synthesis (Scott *et al.*, 2002) and the lethal concentrations of the compounds were assessed alone and in binary, tertiary and quaternary mixtures. Although binary mixtures were no more toxic than individual compounds, toxicity increased with three and four compounds in the mixture, while keeping the total

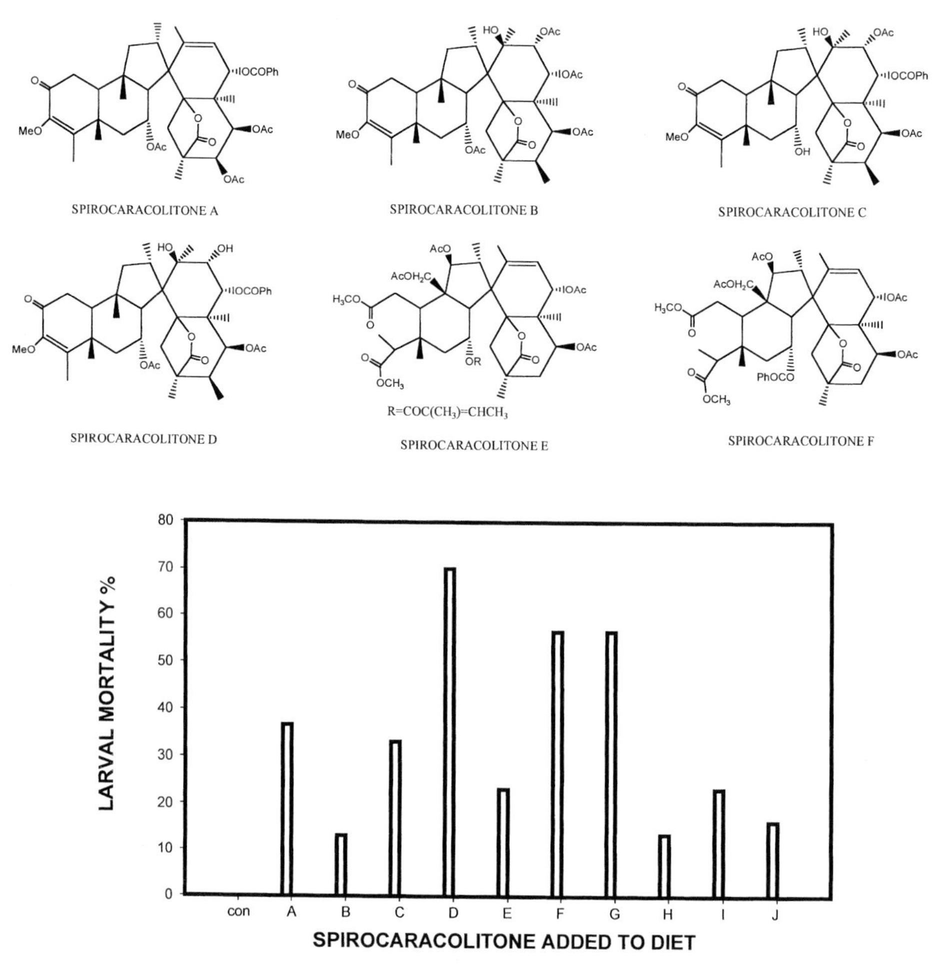

Fig. 1.5. Six of the more than 12 spirocaracolitones isolated from the neotropical tree *Ruptiliocarpon caracolito*. The effect of spirocaracolitones A–J on European corn borer larval mortality is illustrated when added to diets at 50 ppm.

the same. These results provide some support for the analog synergism hypothesis (Fig. 1.6).

Other benefits of large numbers of analogs to plants may include slower evolution of tolerance or lower rates of metabolism of mixtures compared with that of single compounds in herbivorous insects. Feng and Isman (1995) investigated the possibility of adaptation in herbivores by repeated selection of peach aphid colonies with either pure azadirachtin or neem seed extracts containing a large number of limonoids. The colonies treated with azadirachtin soon showed evidence of tolerance of this pure compound, whereas no evidence of tolerance was

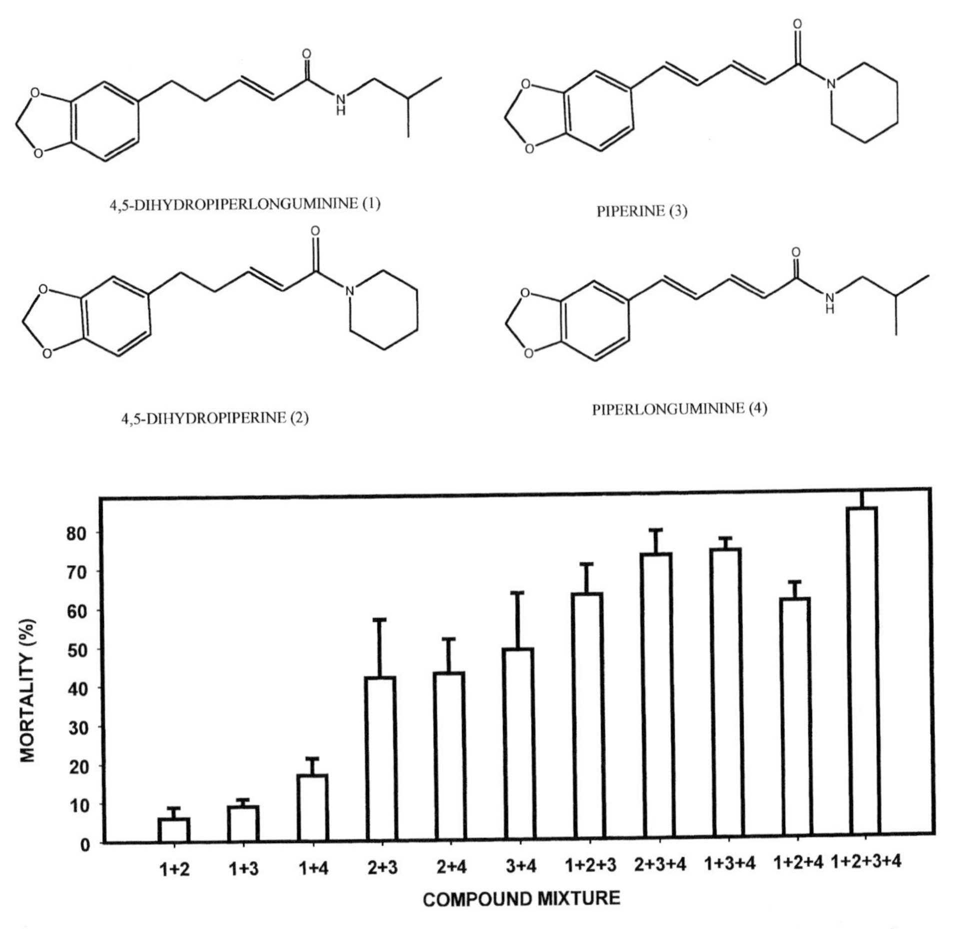

Fig. 1.6. Insecticidal synergism of binary, ternary, and quaternary mixtures of piperamides from *Piper tuberculatum* tested against *Aedes atropalpus* mosquito larvae.

observed with the mixture. In their study of wild parsnip defenses, Berenbaum and Zangerl (1996) showed another facet of multiple defenses. Parsnip webworms can metabolize parsnip furanocoumarins almost twice as fast when they are fed diets containing a single compound compared with diets contain mixtures of six furanocoumarins.

Interaction of insect defense metabolites of the Asteraceae

Many plant families have different biosynthetic groups of secondary metabolite defenses, and their possible interaction in plant defense has received little attention. The Asteraceae (daisy family) presents several characteristic groups of widely

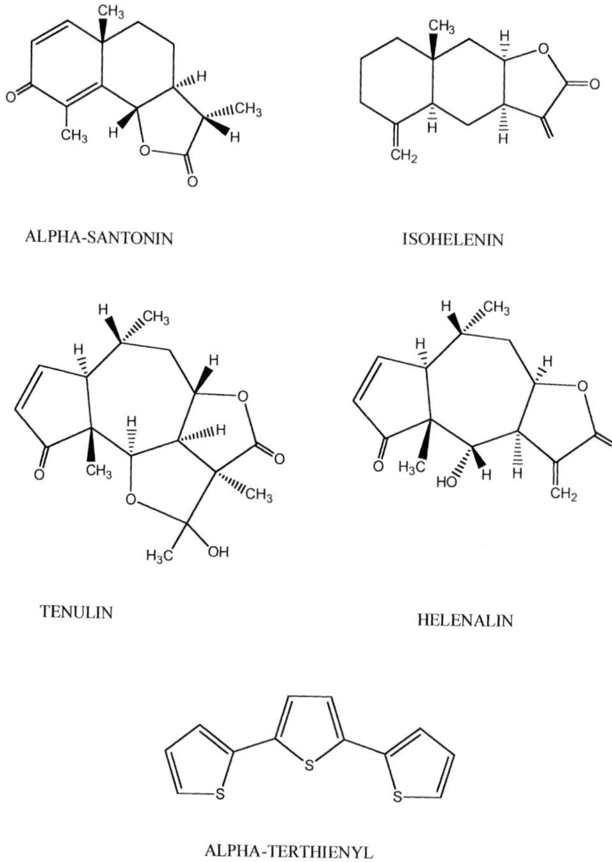

ALPHA-SANTONIN

ISOHELENIN

TENULIN

HELENALIN

ALPHA-TERTHIENYL

Fig. 1.7. Sesquiterpene lactones and thiophenes isolated from species of the Asteraceae used in synergism study.

distributed and co-occurring plant defensive compounds. These include thiophene derivatives and related polyacetylenes, sesquiterpene lactones (Fig. 1.7), and leaf volatiles (monoterpenes and fatty acid derivatives), and we have examined their interactions.

We investigated the interaction of leaf volatiles and the thiophene derivative α-terthienyl, from *Porophyllum ruderale*, a subtropical composite in which these compounds co-occur in leaves (Guillet *et al.*, 1998, 2000). Each larva was exposed in a closed Petri dish to a single leaf with five to seven essential oil glands (containing mainly limonene and 7-tetradecene) but made inaccessible to the larva by a fine wire mesh. Meridic diet offered to the insects was treated with 50 ppm α-terthienyl, a well-known photosensitizer. Exposure to the leaf volatiles alone did not significantly reduce larval growth, but α-terthienyl treatments (alone) did, as expected (Fig. 1.8). The combination of volatiles amplified the α-terthienyl effect on growth reduction

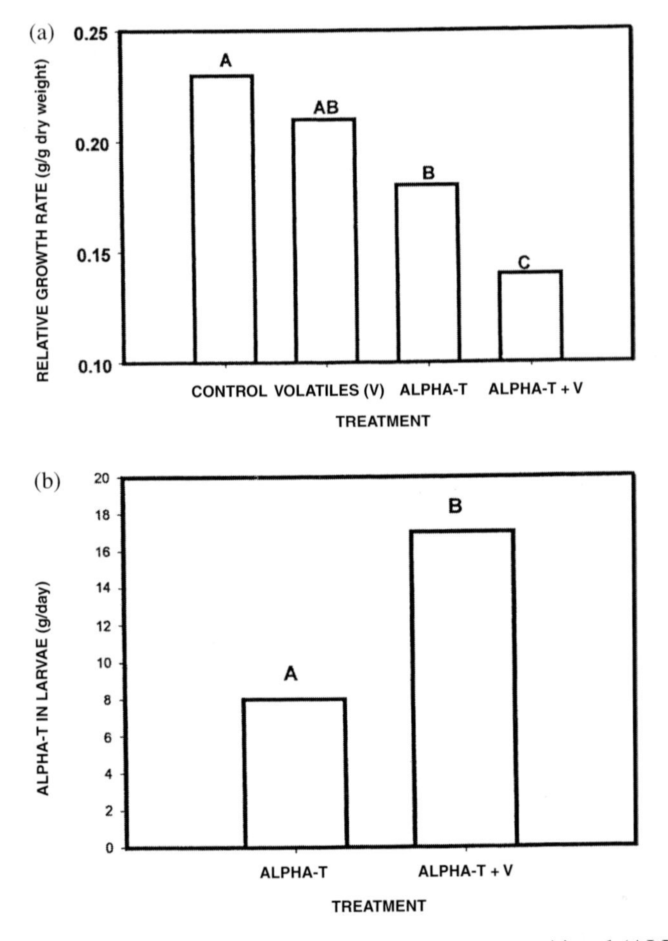

Fig. 1.8. The interaction between leaf volatiles (V) and α-terthienyl (ALPHA–T) of *Porophylum ruderale*. (a) Synergistic effects on European corn borer growth. (b) The concentration of α-terthienyl in larvae. Means followed by the same letter are not significantly different in Tukey's test ($P < 0.05$).

substantially. The mechanism of this synergism was an observed increase in tissue levels of α-terthienyl in insects exposed to the volatiles, compared with tissue levels in insects treated with only α-terthienyl.

In a second experiment, we investigated the co-administration of four sesquiterpene lactones, santonin, isohelenalin, tenulin, and helenalin, with α-terthienyl (Figs. 1.7 and 1.9). The combination of all these sesquiterpene lactones except santonin with α-terthienyl significantly reduced the level of glutathione in the insect larvae, increased the level of lipid peroxidation associated with α-terthienyl toxicity, and increased larval mortality. The synergism can be explained by the Michael addition reaction of the α-methylene-lactone or the cyclopentenone group in the sesquiterpene lactone with glutathione, which reduces antioxidant capacity of insect

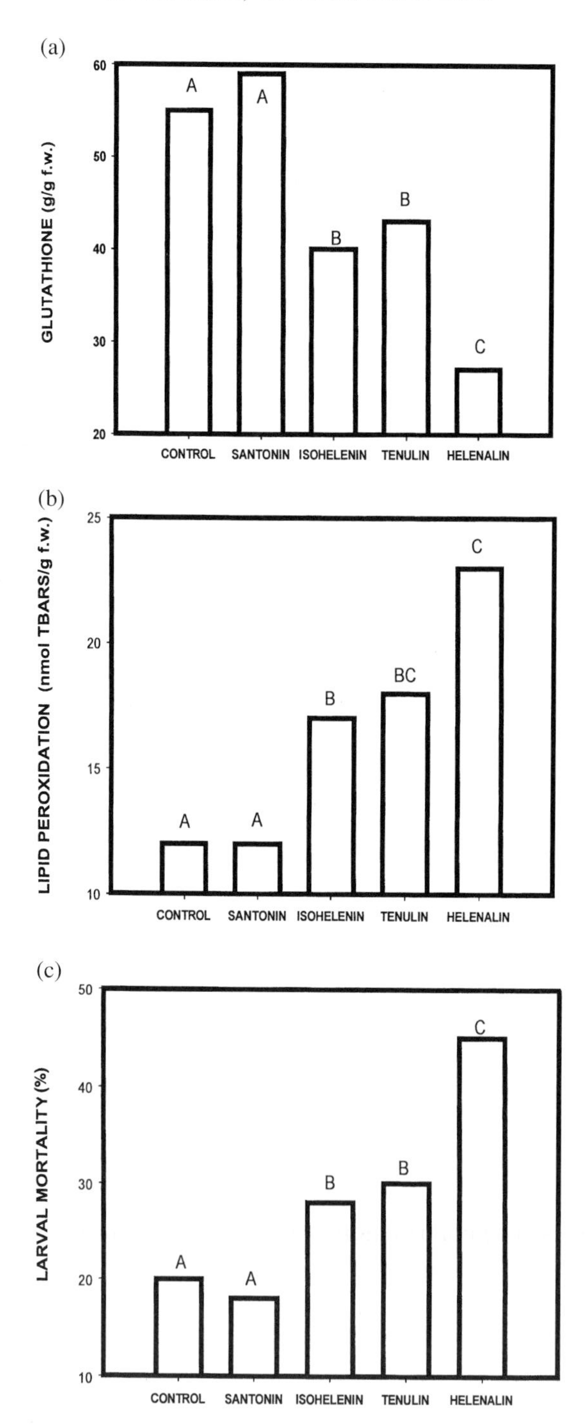

Fig. 1.9. Effects of sesquiterpene lactones on European corn borer. (a) Glutathione levels; (b) lipid peroxidation synergistic effects α-terthienyl; (c) with sesquiterpene lactones from the Asteraceae on larval mortality. Means followed by the same letter are not significantly different in Tukey's test ($P < 0.05$).

tissues and makes them more vulnerable to the photooxidation of lipids by α-terthienyl. Santonin did not act as a synergist because it does not contain either of the reactive groups required for Michael addition. These results suggest a mechanism-based interaction between these co-occurring substances.

Other interactions are likely in this phytochemically diverse family. Straight-chain polyacetylenes and sulfur heterocycles derived from them, including the thiophenes and thiarubrins, co-occur in several species, including *Rudbeckia hirta*, the common ox eye daisy of eastern North America. Each of these substances is highly insecticidal and has a different mode of action (Guillet *et al.*, 1997). Although we did not isolate enough of the compounds for interaction studies, it is likely that these occur.

Plant integrated chemical defense hypothesis

One of us (Guillet, 1997) developed an evolutionary hypothesis in chemical ecology called the plant integrated chemical defense hypothesis (PICD). The PICD hypothesis was developed to explain the observed toxic interaction of phototoxic defenses and other classes of secondary chemicals (SC) in the context of the ecophysiological constraints acting on plants. It can be extended to other chemical situations as well. The PICD hypothesis proposes that the toxic effect of a particular chemical may be increased by the presence of other chemicals, whether they are biosynthetically distinct types of metabolite or simply analogs. Both the synergism of volatiles and sesquiterpene lactones with phototoxic defenses and analog synergism of the Piperaceae and Lepidobotryaceae are consistent with the PICD hypothesis.

The establishment of plant integrated defenses involves the preferential evolutionary retention and production of those SCs exerting synergistic toxic effects and is possible only if a diversification of secondary metabolism in a given plant has previously occurred. This preliminary diversification of secondary metabolism could be mediated via the classical reciprocal co-evolutionary interactions between a host plant and its major pests, as predicted by the chemical arms race model (Berenbaum and Zangerl, 1996). The PICD hypothesis is consequently not an exclusive evolutionary hypothesis because it is compatible with and dependent on other evolutionary processes. The contribution of the PICD hypothesis is to provide both a functional explanation for the diversity of SCs within plants (Romeo *et al.*, 1996) and a reconciliation between different evolutionary models.

The integration step, which would be initiated only when a sufficient phytochemical diversity had been generated in a plant, would involve an evolutionary optimization of the plant genome in order to obtain a coordinated expression of the genes coding for the SCs having synergistic, beneficial effects. In other words, the PICD strategy would involve a genotypic "effort" to integrate and optimize the production of secondary metabolism to make it as profitable as possible to host plants. Although the "gene mechanics" purportedly involved in such fine-tuning processes remain an elusive notion, similar concepts have already been proposed. For example, it is accepted that oviposition preference by female insects on specific host plants relies on a dynamic fine tuning of co-adapted gene complexes (Charlesworth *et al.*, 1987; Thompson, 1994). Jarvis and Miller (1995) presented experimental evidence for ". . . self-organizing driving forces operating . . ." on genomes of plants to optimize their chemical defenses against fungi and insects. The review provided by these researchers focusses on the importance for plants to cluster the different genes involved in a given defensive pathway to optimize its general efficiency.

In summary, the PICD strategy involves two successive and possibly overlapping sequences: (i) a diversification of phytochemicals occurring in plants and (ii) a functional integration for a preferential biosynthesis of a few SCs with synergistic properties.

Plant integrated chemical defenses: comparisons with chemical arms race model

The original chemical arms race hypothesis indirectly predicts that the number of major pests is related to the number of classes of secondary metabolite occurring in a plant. In other words, "... the selection pressures exerted by any particular herbivore on a plant population is independent of the presence or absence of other herbivore species" (Hougen-Eitzman and Rausher, 1994). Because of the ecophysiological constraints limiting the ability of a plant to produce SCs (Zangerl and Bazzaz, 1992), it is unlikely that such independent reciprocal co-evolutionary processes could occur widely in nature, especially in plants that are attacked by a diversified range of pest organisms (Fox, 1981).

The PICD hypothesis specifically takes into consideration the trade-offs related to the costs required for the production of SCs. The integration step of the PICD strategy is assumed to be driven by the excessive costs required for biosynthesis of SCs. The PICD strategy can thus be seen as the ultimate outcome to dead-end reciprocal co-evolutionary interactions that have become too expensive for host plants to maintain, given that an increasing diversity of SCs requires successively more energy for their synthesis. The benefits of a plant relying on a PICD strategy would relate not only to lower costs associated with the production of synergist SCs (Berenbaum and Zangerl, 1996), but also to delaying the onset of insect resistance (Berenbaum, 1985; Isman *et al.*, 1996).

These considerations regarding the major differences between the chemical arms race model and the PICD hypothesis are summarized in Table 1.2. It is suggested that

Table 1.2. *Differences between the chemical arms race and the integrated chemical defenses hypotheses*

	Chemical arms race hypothesis	Integrated chemical defense hypothesis
Aspects related to herbivore pressure		
Major pest insects	Few specialist insects	Diversified suite of specialist and generalist insects
Selective pressure exerted on herbivorous insects	Hard, because of an oligogenic resistance	Soft, because of a polygenic resistance
Delay for insects to evolve adaptations	Low	High
Aspects related to phytochemical diversity		
Evolutionary effects of herbivore pressure	Independent diversification of few classes of secondary chemical	Interrelated diversification of different classes of secondary chemical
Costs of production of chemical defenses	High, because each major specialist pest involves an independent defensive strategy	Low because of synergistic interactions

these distinctions may simply originate from different "pictures" taken at different evolutionary steps of a common pattern of plant–insect interactions. The chemical arms race processes would constitute the initial responses of host plants to their major pests and thus provide the suitable phytochemical diversity that is required for PICD to eventually take place.

Predictions based on the plant integrated chemical defenses hypothesis

If the PICD hypothesis is ecologically relevant in nature, some specific patterns should be detectable. First, a crude phytochemical extract of a plant should possess a higher biological activity than the additive value of the different fractionated or purified extracts of the same plant. This appears to be the easiest way to test the PICD hypothesis, because the phytochemical identification of the different SCs present in plants may not be required.

Another prediction of the PICD hypothesis is that the different SCs acting in synergism should be produced in a synchronized way. In other words, for synergist integrated defenses to be functional, there should be a spatio-temporal convergence in the production of the different SCs acting in synergism. This implies that synergistic substances should be produced in plants at the same phenological stage of development and in the same parts. By considering an evolutionary time-scale, it is predicted that the phylogenetic distribution of synergist SCs should also converge. For instance, if the synergist insecticidal effects observed between phototoxic defenses and the sesquiterpene lactones having an α-methylene-γ-lactone and/or a cyclopentenone group (Fig. 1.7) are ecologically relevant in conferring an improved chemical protection to plants against insects and other pests, then a convergence in the phylogenetic distribution of both SCs should be detectable.

References

Arnason, J. T., Philogene, B. J. R., Donskov, N. and Kubo, I. (1987). Limonoids of the Meliaceae and Rutaceae reduce feeding, growth, and development of *Ostrinia nubilalis*. *Entomologia Experimentalis et Applicata* **43**: 221–226.

Berenbaum, M. (1983). Coumarins and caterpillars: the case for co-evolution. *Evolution* **37**: 163–179.

(1985). Brementown revisited: interactions among allelochemicals in plants. *Recent Advances in Phytochemistry* **19**: 139–169.

Berenbaum, M. and Zangerl, A. R. (1996). Phytochemical diversity: adaptation or random variation. *Recent Advances in Phytochemistry* **30**: 1–24.

Bernard, C. B., Krishnamurty, H. G., Chauret, D. *et al.* (1995). Insecticidal Piperaceae of the neotropics. *Journal of Chemical Ecology* **21**: 801–814.

Champagne, D., Koul, O., Isman, M. B., Scudder, J. and Towers, G. H. N. (1992). Biological activity of limonoids from the Rutales. *Phytochemistry* **31**: 377–394.

Charlesworth, B., Coyne, J. A. and Barton, N. H. (1987). The relative rates of evolution of sex chromosomes and autosomes. *American Naturalist* **130**: 113–146.

Chaurest, D., Durst, T., Arnason, J. T. *et al.* (1996). Novel steroids from *Trichilia hirta* identified by nanoprobe. *Tetrahedron Letters* **37**: 7875–7878.

da Silva, M. F., Agostinho, S., de Paula, J. *et al.* (1999). Chemistry of *Toona ciliata* and *Cedrela odorata* graft (Meliaceae). *Pure and Applied Chemistry* **71**: 1083–1087.

Ehrlich, P. R. and Raven, P. H. (1964). Butterflies and plants: a study in co-evolution. *Evolution* **18**: 586–608.

Ewete, F. K., Arnason, J. T., Larson, J. and Philogene, B. J. R. (1996a). Biological activity of extracts from traditionally used Nigerian plants against the European corn borer. *Entomologia Experimentalis et Applicata* **80**: 531–537.

Ewete, F. K., Nicol, R. W., Hengsawad, V. *et al.* (1996b). Inseciticdal activity of *Aglaia odorata* extract and its insecticidal principle rocaglamide to the European corn borer. *Journal of Applied Entomology* **120**: 483–488.

Feng, R. and Isman, M. B. (1995). Selection for resistance in green peach aphid. *Experientia* **51**: 831–833.

Fox, L. R. (1981). Defense and dynamics in plant-herbivore systems. *American Zoologist* **21**: 853–864.

Guillet, G. (1997). Ecophysiological Importance of Phototoxins in Plant–Insect Relationships. Ph.D. Thesis, University of Ottawa.

Guillet, G., Philogene, B. J. R., Meara, J., Durst, T. and Arnason, J. T. (1997). Multiple modes of insecticidal action of three classes of polyacetylene derivatives. *Phytochemistry* **46**: 495–498.

Guillet, G., Belanger, A. and Arnason, J. T. (1998). Volatile monoterpenes in *Porophyllum gracile* and *P. ruderale*. *Phytochemistry* **49**: 423–429.

Guillet, G., Harmatha, J., Wadell, T. G., Philogene, B. J. R. and Arnason, J. T. (2000). Synergism between sesquiterpene lactones and alpha-terthienyl. *Photochemistry and Photobiology* **71**: 111–115.

Hougen-Eitzman, D. and Rausher, M. D. (1994). Interactions between herbivorous insects and plant–insect coevolution. *American Naturalist* **143**: 677–697.

Isman, M. B., Matsuura, H., MacKinnon, S., Durst, T., Towers, G. H. N. and Arnason, J. T. (1996). Phytochemistry of the Meliaceae. *Recent Advances in Phytochemistry* **30**: 155–178.

Jarvis, B. B. and Miller, J. P. (1995). Phytochemical diversity and redundancy in ecological interactions. *Recent Advances in Phytochemistry* **30**: 265–293.

Jimenez, A., Mata, R., Pereda-Miranda, R. *et al.* (1996). Insecticidal limonoids from *Swietenia humilis* and *Cedrela salvadorensis*. *Journal of Chemical Ecology* **23**: 1225–1234.

Jimenez, A., Villarreal, C., Toscano, R. *et al.* (1998). Limonoids from *Swietenia humulis* and *Guarea grandiflora*. *Phytochemistry* **49**: 1981–1998.

Jones, C. and Firn, R. D. (1991). On the evolution of phytochemical diversity. *Philosophical Transactions of the Royal Society, London* **333**: 273–280.

MacKinnon, S., Chauret, D., Wang, M. *et al.* (1997a). Botanicals from the Piperaceae and Meliaceae of the American Neotropics. *American Chemical Society Symposium Series* **658**: 49–57.

MacKinnon, S., Angerhoffer, C., Pezzutto, J. *et al.* (1997b). Antimalarial activity of neotropical Meliaceae extracts and gedunin derivatives. *Journal of Natural Products* **60**: 336–341.

MacKinnon, S., Bensimon, C., Arnason, J. T., Sanchez-Vindas, P. E. and Durst, T. (1997c). Spirocaracolitones, CD-spiro triterpenoids from *Ruptiliocarpon caracolito*. *Journal of Organic Chemistry* **62**: 840–845.

Mendelsohn, R. T. and Balick, M. J. (1995). The value of undiscovered pharmaceuticals in tropical forests. *Economic Botany* **49**: 223–238.

Mootoo, B. S., Ransawan, R., Khan, A. *et al.* (1996). Tritetranorterpenoids from *Raugea glabra*. *Journal of Natural Products* **59**: 544–547.

Omar, S. (2000). Antifeedant and Antimalarial Activity of Tropical Meliaceae. Ph.D. Thesis, University of Ottawa.

Pennington, T., Styles, B. T. and Taylor, D. A. H. (1981). *Flora Neotropica Monograph 28*. New York: New York Botanical Garden.

Ramirez, M-C., Toscano, R., Arnason, J., Omar, S., Cerda-Garcia-Rojas, C. M. and Mata, R. (2000). Structure, conformation and absolute configuration of new antifeedant dolabellanes from *Trichilia trifolia*. *Tetrahedron* **56**: 5085–5091.

Romeo, J., Saunders, J. A. and Barbosa, P. (eds.) (1996). *Phytochemical Diversity and Redundancy in Ecological Interactions. Recent Advances in Phytochemistry*, vol. 30. New York: Plenum Press.

Soejarto, D. D. and Farnsworth, N. R. (1989). Tropical rainforests: potential source of new drugs. *Perspectives in Biology and Medicine* **32**: 244–256.

Thompson, J. N. (1994). *The Coevolutionary Process*. Chicago, IL: University of Chicago Press.

Waterman, P. G. and Grundon, M. F. (1983) (eds.). *Chemistry and Chemical Taxonomy of the Rutales*. New York: Academic Press.

Wheeler, D. A., Isman, M. B., Sanchez-Vindas, P. E. and Arnason, J. T. (2001). Screening of Costa Rican *Trichilia* species for biological activity against the larvae of *Spodoptera litura* (Lepidoptera: Noctuidae). *Biochemical Systematics and Ecology* **24**: 347–335.

Wilson, E. O. (2002). *The Future of Life*. New York: Alfred A. Knopf Press.
Xie, S.; Gunning, P., Mackinnon, S. *et al.* (1994). Biological activity of extracts of *Trichilia* species and the limonoid hirtin against lepidopteran larvae. *Biochemical Systematics and Ecology* **22**: 129–136.
Zangerl, A. R. and Bazzaz, F. A. (1992). Theory and pattern in plant defense allocation. In *Plant Resistance to Herbivores and Pathogens: Ecology, Evolution, and Genetics*, eds. R. S. Fritz and E. L. Simms, pp. 363–391. Chicago: University of Chicago Press.

2

Recruitment of predators and parasitoids by herbivore-injured plants

Ted C. J. Turlings

Institute of Zoology, University of Neuchatel, Switzerland

Felix Wäckers

Netherlands Institute of Ecology, Heteren, the Netherlands

Introduction

In recent years, induced plant defenses have received widespread attention from biologists in a variety of disciplines. The mechanisms underlying these defenses and the interactions that mediate them appeal not only to plant physiologists, ecologists, and evolutionary biologists but also to those scientists that search for novel strategies in plant protection. Several recent books (Karban and Baldwin, 1997; Agrawal *et al.*, 1999) and reviews (Baldwin, 1994; Karban *et al.*, 1997; Agrawal and Rutter, 1998; Agrawal and Karban, 1999; Baldwin and Preston, 1999; Dicke *et al.*, 2003) have been devoted entirely to the subject of induced plant defenses. Various forces, ranging from abiotic stresses to biotic factors such as pathogens, arthropods, or higher organisms, may trigger different plant defense responses. Yet, the biochemical pathways that are involved appear to show considerable similarities. This is also true for the so-called indirect defenses.

The term indirect defense refers to those adaptations that result in the recruitment and sustenance of organisms that protect the plants against herbivorous attackers. The early published examples of indirect defenses involved intimate plant–ant interactions, in which myrmecophilous plants were shown to have evolved a range of adaptations providing ants with shelter (domatia) and various food sources (Belt, 1874; Janzen, 1966). In return, these plants may obtain a range of benefits because ants can provide nutrition (Thomson, 1981) or more commonly, protection against herbivores, pathogens, and competing plants (e.g. Koptur, 1992; Oliveira, 1997). The well-documented fitness benefits of ant attendance in myrmecophilous plants (Rico-Gray and Thien, 1989; Oliveira, 1997), combined with the fact that domatia and food supplements are difficult to reconcile with other functions, are convincing arguments for the interpretation that these adaptations represent examples

Advances in Insect Chemical Ecology, ed. R. T. Cardé and J. G. Millar. Published by Cambridge University Press.

Fig. 2.1. Extrafloral nectar droplets on *Ricinus communis* (castor bean).

of indirect defense. The above-mentioned studies have all focussed on intimate examples of plant–ant mutualisms. However, similar adaptations are also found in non-myrmecophilous plants. Acarodomatia have been recorded from so-called "mite plants." These preexisting structures facilitate symbiotic interactions with predatory or fungivorous mites (Bakker and Klein, 1992; Whitman, 1994).

Extrafloral nectaries (Fig. 2.1) are probably the most frequently described adaptations believed to serve as indirect defenses. They have been described in approximately 1000 species from 93 plant families including numerous dicotyledonous species, ferns, and such diverse monocotyledonous taxa as lilies, orchids, sedges, and grasses (Koptur, 1992). They are found in virtually all plant types including herbs, vines, shrubs and trees, annuals as well as perennials, and successional as well as climax species.

Often extrafloral nectaries show prominent colorations (primarily black and red), which set them off against the (green) background. In contrast to their floral counterparts, extrafloral nectaries are generally exposed (Zimmerman, 1932), giving insects easy access to the nectar. The nectaries are often situated on leaves or petioles (Fig. 2.1), where they are ideally situated for crawling insects or flying insects that land on the leaf surface (Fig. 2.2). In other plants, they are found on petioles or the (leaf) stem, which is an effective placement for ants and other natural enemies crawling up the plant.

Less evident is the primary function of plant odor emissions. Although it is clear that plant odors are used by parasitoids (Fig. 2.3) and predators to locate potential prey (Vinson *et al.*, 1987; Nordlund *et al.*, 1988; Whitman, 1988), they are likely to have other functions as well (Harrewijn *et al.*, 1995; Turlings and Benrey, 1998).

Fig. 2.2. A female of the parasitoid *Cotesia glomerata* feeding on extrafloral nectar of *Vicia faba*.

Yet, the notion that plant volatiles may serve as signals to recruit members of the third trophic level has been reinforced by the fact that they are inducible. So far, evidence for plant-produced signals has been limited to interactions between plants and arthropods, but a recent study showed that plants may also recruit nematodes that can infect beetle larvae feeding on the roots of these plants (van Tol *et al.*, 2001). The accumulating evidence strongly suggests that herbivore-induced plant signals play a very important role in the indirect protection of plants against herbivory.

The increasing number of studies on the interactions between plants and the natural enemies of herbivores attacking these plants is revealing an astonishing sophistication. This is most apparent in the specificity of the interactions; plants may respond differently to different herbivores and the natural enemies are able to distinguish among these differences (Sabelis and van de Baan, 1983; Takabayashi *et al.*,

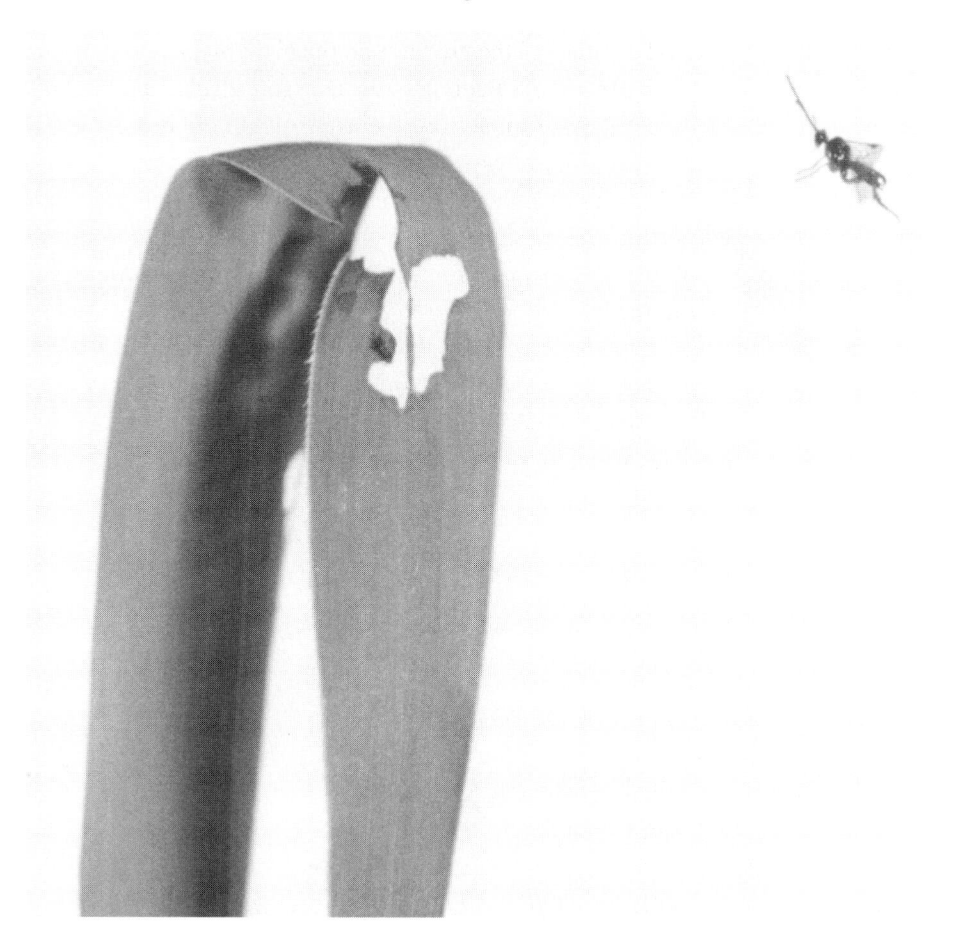

Fig. 2.3. A female of the parasitoid *Cotesia marginiventris* attracted to the odor emitted by a maize leaf that has been damaged by a *Spodoptera exigua* larva.

1995; De Moraes *et al.*, 1998; Powell *et al.*, 1998). There is even evidence to suggest that plants selectively employ direct and indirect defenses depending on which herbivore feeds on them (Kahl *et al.*, 2000). An additional twist to the refinement of the interactions is that there is now clear evidence for information transfer among plants mediated by volatile signals (Arimura *et al.*, 2000a; Dolch and Tscharntke, 2000; Karban *et al.*, 2000). These potent plant signals can be expected to affect multiple interactions within entire food webs (Janssen *et al.*, 1998; Sabelis *et al.*, 1999) and many more interchanges are likely to be discovered.

This chapter gives an overview of the developments in research on induced indirect defenses. We discuss both the ecological aspects as well as our knowledge of the mechanisms that are involved. This chapter differs from most other reviews in that it includes both attraction by means of induced volatiles and the plant's strategy to keep the natural enemies of the predator on the plant by increased nectar production in response to herbivory. We compare these two strategies, particularly in terms of timing and specificity of induction. We argue that there is a danger of overinterpreting results if we do not always recognize the fact that plants need to benefit from the proposed function of the induced responses. Hence, our discussion of how natural selection may have shaped the various interactions emphasizes the role of the plant and to what extent its interests are in tune with those of the third trophic level. Some recent studies provide evidence for the adaptiveness of inducible indirect defenses, but it is concluded that field experiments, preferentially with natural systems, are needed to establish truly if plants do benefit from these inducible responses. Field data are also still lacking for a conclusive appreciation of the full potential of exploiting indirect plant defenses in the protection of crop plants.

Inducible volatile signals

The role of plant volatiles as prey and host location cues

The evolutionary "cat-and-mouse game" between entomophagous insects and their prey has led to various refined adaptations on both sides. Potential prey may minimize the encounter rates with their natural enemies by being cryptic, visually as well as chemically. Although various natural enemies make use of prey-derived cues (for review see Tumlinson *et al.*, 1992), others have evolved to rely primarily on indirect cues that may be less reliable but are more readily available and detectable (Vet *et al.*, 1991; Vet and Dicke, 1992). For the many natural enemies of herbivores, the plants on which the herbivores feed play a key role in providing useful cues.

That predators and especially parasitoids of herbivores are attracted to plants has long been known. In their review of this topic, Nordlund *et al.* (1988) suggest that Picard and Rabaud (1914) were the first to realize the importance of plants for foraging entomophagous insects. In numerous studies since (e.g., Taylor, 1932; Zwölfer and Kraus, 1957; Salt, 1958; Harrington and Barbosa, 1978), predators and parasitoids were found more on one plant species than another. That plant volatiles may be responsible for such differential attractiveness was apparent from studies by, among others, Monteith (1955), Arthur (1962), Flint *et al.* (1979), and Elzen *et al.* (1984). These studies considered the importance of the plant only at the level of habitat locations, as defined by Vinson (1981). It was not until papers by Price *et al.* (1980), Vinson (1981), and Barbosa and Saunders (1985) that the more direct role of plants in mediating the step of host/prey location was considered. Initial studies suggested only that insect-derived attractants (kairomones) were affected by the plant diet of the host or prey (e.g., Roth *et al.*, 1978; Sauls *et al.*, 1979;

Loke *et al.*, 1983). For instance, parasitoid females tend to respond more strongly to feces from host larvae that have fed on their customary host plant than to feces from larvae fed on an artificial diet (Roth *et al.*, 1978; Sauls *et al.*, 1979; Mohyuddin *et al.*, 1981; Nordlund and Sauls, 1981).

The first series of studies to provide complete behavioral as well as chemical evidence for the ability of herbivore-injured plants to actively attract natural enemies of their predators was obtained with studies on plant–mite interactions (e.g., Sabelis and van de Baan, 1983; Dicke and Sabelis, 1988; Dicke *et al.*, 1990a,b). First, it was found that plants infested with spider mites were far more attractive to predators than were uninfested plants (Sabelis and van de Baan, 1983). Subsequently, Dicke and Sabelis (1988) showed the presence of several unique volatile substances in the headspace collections of lima bean leaves infested with the spider mite *Tetranychus urticae*. These volatiles were not emitted by the spider mites but by the infested plant. Synthetic versions of some of these substances, which were not present in the collections from uninfested plants, were attractive to the predatory mite *Phytoseiulus persimilis*. These first studies and many since (e.g., Dicke *et al.*, 1990a,b; Takabayashi *et al.*, 1991a, 1994; Scutareanu *et al.*, 1997) show that *P. persimilis* and various other predators that use phytophagous mites as prey make effective use of a specific blend of mite-induced compounds to locate plants with prey. This well-studied model system has proven very valuable in revealing the intricacies and complexity of a multitude of interactions that can be affected by the herbivore-induced plant volatiles (Janssen *et al.*, 1998; Sabelis *et al.*, 1999).

The majority of other studies show conclusively that herbivory induces plants to emit volatiles that may serve as an indirect defense involving the attraction of parasitoids (Table 2.1). In particular, studies with caterpillars on cotton (McCall *et al.*, 1993, 1994; Loughrin *et al.*, 1994, 1995a; Röse *et al.*, 1996), *Brassica* spp. (Steinberg *et al.*, 1992; Agelopoulos and Keller, 1994; Geervliet *et al.*, 1994; Mattiacci *et al.*, 1994; Benrey *et al.*, 1997), and maize (Turlings *et al.*, 1990, 1991a,b, 1995; Potting *et al.*, 1995; Takabayashi *et al.*, 1995; Alborn *et al.*, 1997) have revealed that caterpillar damage results in the release of parasitoid attractants. In most cases, injury by caterpillars was found to cause a much stronger reaction in terms of odor emissions than mechanical damage. Relatively new is the finding that egg deposition by herbivores on plants can cause plants to release volatiles that are attractive to egg parasitoids (Meiners and Hilker, 1997, 2000; Meiners *et al.*, 2000). The plant response to egg deposition also appears to be systemic (Wegener *et al.*, 2001).

Many additional studies show that herbivore-induced emissions of volatiles are very common and are found in a wide range of plant taxa, induced by numerous herbivorous arthropods, and affect many different natural enemies (Table 2.1). The list in Table 2.1 is not meant to be complete but rather to illustrate how common these

Table 2.1. *Examples of predators and parasitoids that use induced plant odors to locate their prey or hosts*

Natural enemy	Herbivore(s)[a]	Plant(s)	Evidence	Selected references
Predators				
Acari: Phytoseiidae				
Phytoseiulus persimilis	*Tetranychus urticae* (Acari: Tetranychidae)	Apple, cucumber, gerbera, lima bean	B + C	Dicke *et al.*, 1990a,b; Takabayashi *et al.*, 1991a,b; Shimoda *et al.*, 1997; Krips *et al.*, 2001
Amblyseius andersoni	*Tetranychus urticae* (Acari: Tetranychidae)	Lima bean	B + C	Dicke *et al.*, 1990a; Dicke and Groeneveld, 1986
Amblyseius finlandicus	*Tetranychus urticae* (Acari: Tetranychidae) *Panonychus ulmi* (Acari: Tetranychidae)	Apple	B + C	Sabelis and van de Baan, 1983; Takabayashi *et al.*, 1991a
Coleoptera: Cleridae				
Thanasimus dubius	*Ips pini* (Coleoptera: Scolytidae)	Pine	B	Aukema *et al.*, 2000
Thysanoptera: Thripidae				
Scolothrips takahasshi	*Tetranychus urticae* (Acari: Tetranychidae)	Lima bean	B + C	Shimoda *et al.*, 1997
Hemiptera: Anthocoridae				
Orius laevigatus	*Frankliniella occidentalis* (Thysanoptera: Thripidae)	Cucumber	B	Venzon *et al.*, 1999
Hemiptera: Miridae	*Tetranychus urticae* (Acari: Tetranychidae)			
Cyrthorinus lividipennis	*Nilaparvata lugens* (Homoptera: Delphacidae)	Rice	B	Rapusas *et al.*, 1996
Hemiptera: Pentatomidae				
Perillus bioculatus	*Leptinotarsa decemlineata* (Coleoptera; Chrysomelidae)	Potato	B + C	van Loon *et al.*, 2000a; Weissbecker *et al.*, 2000
Parasitoids				
Hymenoptera: Braconidae				
Apanteles (Cotesia) sp.	*Ectropis obliqia*	Tea	B/C	Xu and Chen, 1999
Apanteles (Cotesia) chilonis	*Chilo suppressalis*	Rice	B	Chen *et al.*, 2002
Coeloides bostrichorum	*Ips typographus* (Coleoptera: Scolytidae)	Spruce	B + C	Pettersson *et al.*, 2001

(*cont.*)

Table 2.1. (*cont.*)

Natural enemy	Herbivore(s)[a]	Plant(s)	Evidence	Selected references
Cotesia glomerata, Cotesia rubecula	*Pieris* spp.	Cabbage and related subspecies	B + C	Steinberg *et al.*, 1993; Angelopoulos and Keller, 1994; Mattiacci *et al.*, 1994, 1995
Cotesia kariyai	*Pseudaletia separata Mythimna separata*	Maize, kidney bean, Japanese radish	B + C	Takabayashi *et al.*, 1995; Fujiwara *et al.*, 2000
Cotesia marginiventris	*Spodoptera* spp.	Maize, cotton, cowpea	B + C	Turlings *et al.*, 1990, 1991a, 2002
Diachasmimorpha longicaudata	*Anastrepha spp.* (Diptera: Tephritidae)	Mango, grapefruit	B	Eben *et al.*, 2000
Macrocentrus grandii	*Ostrinia nubilalis*	Maize	B + C	Udayagiri and Jones, 1992a,b
Microplitis croceipes	*Helicoverpa* and *Heliothis* spp.	Cotton, cowpea, maize	B + C	McCall *et al.*, 1993; Turlings *et al.*, 1993a; Röse *et al.*, 1996
Microplitis demolitor	*Pseudoplusia includens*	Plant volatiles	C	Ramachandran and Norris, 1991
Hymenoptera: Mymaridae				
Anagrus nilaparvatae	*Nilaparvata lugens* (Homoptera: Delphacidae)	Rice	B	Lou and Cheng, 1996
Hymenoptera: Aphidiidae				
Aphidius ervi	*Acyrthosiphon pisum* (Homoptera: Aphididae)	Broad bean	B + C	Du *et al.*, 1996; Powell *et al.*, 1998
Hymenoptera: Eulophidae				
Diglyphus isaea	*Liriomyza trifolii* (Diptera: Agromyzidae)	Bean	B + C	Finidori-Logli *et al.*, 1996
Oomyzus gallerucae	*Xanthogaleruca luteola* (Coleoptera: Chrysomelidae)	Elm	B + C	Meiners and Hilker, 1997, 2000; Wegener *et al.*, 2001
Hymenoptera: Pteromalidae				
Rhopalicus tutela	*Ips typographus* (Coleoptera: Scolytidae)	Pine	B + C	Pettersson, 2001
Hymenoptera: Scelionidae				
Trissolcus basalis	*Nezara viridula* (Hemiptera: Pentatomidae)	Soybean	B	Loch and Walter, 1999

[a]If not otherwise indicated the hosts or prey are lepidopteran larvae.
B, behavioral evidence; C, chemical evidence.

tritrophic interactions are. The evidence is not equally conclusive in all cases, but behavioral observations and/or chemical analyses strongly indicate that induced plant volatiles play a key role in host or prey location. Current research in this area focusses on the mechanisms of induction and on questions concerning the ecological significance and evolutionary history of these interactions, as well as on the possibility of exploiting this indirect plant defense for crop protection. We attempt to give an overview of the most significant findings of these research efforts.

Elicitors and induction mechanisms

Mere mechanical damage of leaves may result in the temporary emission of some volatiles, but in most cases these emissions can be greatly enhanced and prolonged by eliciting factors that come directly from a feeding insect. These factors also elicit odor emissions when undamaged leaves take them up via the petiole, and the response to these factors has been shown to be systemic (Dicke *et al.*, 1990a; Turlings *et al.*, 1993a). After the isolation and identification of a β-glucosidase (Mattiacci *et al.*, 1995) and the fatty acid derivative volicitin (Alborn *et al.*, 1997) as elicitors from caterpillar oral secretion (regurgitant), a multitude of studies have revealed details about the mechanisms that may mediate the formation and action of these compounds. Notably the groups of Tumlinson (Gainesville, Florida) and Boland (Jena, Germany) have made considerable progress in these areas, as summarized below.

Beta-glucosidase

A β-glucosidase in the regurgitant of *Pieris brassicae* larvae causes a release of volatiles in brassica plants that is similar to the release observed after feeding by these larvae (Mattiaci *et al.*, 1995). Beta-glucosidases are present in many organisms (e.g., Robinson *et al.*, 1967; Sano *et al.*, 1975; Wertz and Downing, 1989; Yu, 1989) and may function in catalyzing biochemical pathways that involve glycoside cleavage. The emissions by Brassicaceae are characterized by glucosinolate breakdown products, which have not been observed in other plant families that have been studied in this context (Agelopoulos and Keller, 1994; Mattiaci *et al.*, 1994; Geervliet *et al.*, 1996). It is expected that enzymes like β-glucosidase facilitate this glucosinolate breakdown. This notion is reinforced by the fact that a systemic emission after caterpillar feeding is only observed if the distant leaves are mechanically damaged (L. Mattiaci, personal communication), suggesting that the substrate comes in contact with enzymes only when it is released from ruptured cells. The key enzyme involved in the hydrolysis of glucosinolates is myrosinase, which causes the release of volatile defensive compounds such as isothiocyanates (Bones and Rossiter, 1996). The bioactivity of enzymes in the oral secretions of insects as inducers of volatile emissions may be limited to specific plant taxa.

Volicitin

Incubation of young maize plants in the regurgitant of several lepidopteran larvae and a grasshopper was found to induce the release of a blend of terpenoids and indole that is typical for plants with caterpillar damage (Turlings *et al.*, 1993a). Similarly, Potting *et al.* (1995) found that stem borer regurgitant applied to mechanically damaged sites caused an increase in induced odor emissions in maize plants. In both cases, induced plants were more attractive to parasitoids than plants that were not induced.

Volicitin (*N*-(17-hydroxylinolenoyl)-L-glutamine) was identified by Alborn *et al.* (1997) as the active substance in the regurgitant of *Spodoptera exigua* larvae. Low concentrations of this elicitor alone cause the same reaction in maize plants as pure regurgitant and render the plants equally attractive to the parasitoids (Turlings *et al.*, 2000). So far, volicitin has only been shown to be active in maize (Turlings *et al.*, 2002) and does not induce a reaction in, for instance, lima bean (Koch *et al.*, 1999). Studies of the source and biosynthesis of volicitin revealed that this fatty acid derivative is formed in the bucal cavity of the insect (Paré *et al.*, 1998). Linolenic acid, the fatty acid part of volicitin, is ingested with plant material and is then 17-hydroxylated and conjugated with insect-derived glutamine (Paré *et al.*, 1998). Spiteller *et al.* (2000) showed that bacteria isolated from caterpillar gut contents are able to synthesize volicitin and other *N*-acylglutamine conjugates from externally added precursors. Hence, it is not necessarily the plant or the insect that controls the biosynthesis of volicitin. It remains surprising, however, that the insects "allow" the synthesis of elicitors that trigger plant reactions with such negative consequences for the insect. It is, therefore, expected that these metabolites play an essential role in the insects' physiology. Perhaps they serve as surfactants that facilitate the transport and digestion of food, or they may neutralize the effects of plant toxins (Alborn *et al.*, 2000; Spiteller *et al.*, 2000). Numerous *N*-acylglutamates that may show elicitor activity (Halitschke *et al.*, 2001) occur in the oral secretions of various insects (Pohnert *et al.*, 1999; Spiteller *et al.*, 2000).

Not surprisingly, some factors in the oral secretions of caterpillars may suppress induced plant defenses (Felton and Eichenseer, 2000). Musser *et al.* (2002) elegantly showed that the enzyme glucose oxidase in the saliva of *Helicoverpa zea* counteracts the induced production of nicotine. The presence of such suppressing agents would explain the fact that the isolated active fraction containing volicitin showed more activity than pure caterpillar regurgitant (Turlings *et al.*, 2000).

Elicitors from plants

Plant hormones with various functions have been identified over the years and an increasing number of studies show that they may also affect volatile emissions

(Farmer, 2001). Even nectar production may be effected by such hormones (Heil *et al.*, 2001). The gaseous hormone ethylene plays an important role in plant development, but also in defense (Mattoo and Suttle, 1991). Upon perception of a pathogen, plants show enhanced ethylene production, which has been shown to be involved in the induction of defense reactions (Boller, 1991). Wild tobacco plants engineered with an *Arabidopsis* sp. ethylene-insensitive gene do not show typical leaf development arrestment in the presence of leaves of other tobacco plants, demonstrating the importance of ethylene in plant development (Knoester *et al.*, 1998). The ethylene-insensitive plants also showed reduced defense protein synthesis and were susceptible to soil pathogens to which they were normally fully resistant. In connection with the third trophic level, Kahl *et al.* (2000) found that attack by *Manduca* caterpillars on wild tobacco plants causes an ethylene burst that suppressed induced nicotine production but stimulated volatile emissions. They argued that the plant "chooses" to employ an indirect defense (the attraction of natural enemies) rather than a direct defense to which the attacker could adapt (Kahl *et al.*, 2000; Winz and Baldwin, 2001). This implies that the plant is capable of identifying its attacker. We discuss this possibility in more detail in the discussion of specificity.

Studies into the effects and mechanisms of induced resistance against pathogens and insects have revealed the role of salicylic acid (SA) and jasmonic acid (JA). These compounds are seen as the key signals for defense gene expression (Reymond and Farmer, 1998). It was generally thought that SA regulates resistance to fungal, bacterial, and viral pathogens (Enyedi *et al.*, 1992; Ryals *et al.*, 1996), whereas JA induces the production of various proteins via the octadecanoid pathway that provides plants with resistance against insects (Broadway *et al.*, 1986; Farmer *et al.*, 1992). However, this distinction between the two pathways is not that clear and pathogens and arthropods may sometimes trigger both (Farmer *et al.*, 1998; Reymond and Farmer, 1998; Walling, 2000). SA and JA, as well as synthetic mimics, can be applied exogenously to plants to induce the same metabolic changes that lead to resistance as induced by pathogens and insects (Ryals *et al.*, 1992; Kessmann *et al.*, 1994; Görlach *et al.*, 1996; Thaler *et al.*, 1996). The two different pathways that the elicitors stimulate can compromise each other (Doherty *et al.*, 1988). Thaler (1999) demonstrated this in a field situation, where tomato plants stimulated with a SA mimic reduced resistance to *S. exigua*, while JA treatment rendered plants more vulnerable to the bacterial speck pathogen *Pseudomonas syringae* pv. tomato.

Treatment with SA, JA, or their mimics can also induce the release of volatiles in plants, but the blends produced are somewhat different for the two elicitors (Dicke *et al.*, 1993, 1999; Hopke *et al.*, 1994; Ozawa *et al.*, 2000; Rodriguez-Saona *et al.*, 2001; Wegener *et al.*, 2001). In a rare field experiment (see below), Thaler (1999) showed that treatment of tomato plants with JA increased the parasitism rate of caterpillars on the plants, which was most likely the result of JA-induced increases in odor emissions. The overall evidence clearly indicates that these inducers of general defense reactions also play a role in volatile signaling.

Arimura *et al.* (2000a) found that several of the induced volatiles themselves can serve as elicitors by triggering gene activation in neighboring leaves that leads to further emissions. In this context, (*Z*)-jasmone was shown to be a potent plant-derived volatile elicitor that triggers the release of (*E*)-β-ocimene in the bean plant, *Vicia faba* (Birkett *et al.*, 2000). These examples of plant odors inducing plant defense pathways have important implications for plant–plant communication (see below).

Pathogen-derived elicitors

Cellulysin is a fungus-derived enzyme mixture of exo- and endoglucanases that is an extremely potent elicitor of plant volatile biosynthesis through the upregulation of the octadecanoid pathway (Piel *et al.*, 1997). The low-molecular-weight phytotoxin coronatin, which is produced by certain bacteria (Bender *et al.*, 1996; Ichihara *et al.*, 1977), is also a strong elicitor of volatile emissions and mimics specific compounds within the pathway (Weiler *et al.*, 1994; Boland *et al.*, 1995; Schüler *et al.*, 2001). More recently, a mixture containing the ion channel-forming peptide of the peptaibol class (alamethicin), isolated from the plant parasitic fungus *Tricoderma viride*, has also been implicated in volatile induction via the octadecanoid-signaling pathway (Engelberth *et al.*, 2000, 2001). It should be noted that the induced volatile blends show considerable differences for the different elicitors. In lima bean, alamethicin only induces the production of the two common homoterpenes (3*E*)-4,8-dimethyl-1,3,7-nonatriene (DMNT) and (3*E*,7*E*)-4,8,12-trimethyl-1,3,7,11-tridecatetraene (TMTT), and of methyl salicilate. These compounds are barely induced after treatment with JA or cellulysin, which stimulate the production of other inducible volatiles (Engelberth *et al.*, 2001). Coronatin and its synthetic mimic coronalon induce the production of a complete blend of all these compounds (Schüler *et al.*, 2001). The common elicitation of volatile synthesis by pathogens is likely to affect insect induction if simultaneous infections occur and should be considered in future studies on variability and specificity of plant-provided signals (see below).

The genetic basis for induction

Common elicitors like JA and SA and knowledge about the biochemical pathways that they induce are used to identify the plant genes that are involved in the induction process (Reymond *et al.*, 2000). Various genes that are induced by JA and related compounds have been identified (Reymond and Farmer, 1998; Stinzi *et al.*, 2001) and several of these genes can also be activated by some of the induced volatiles themselves (Arimura *et al.*, 2000a). However, very little is known about the genes that code for the enzymes involved in the direct synthesis of specific induced volatiles.

One of the main substances induced in maize by volicitin is indole, an intermediate in at least two biosynthetic pathways. Frey *et al.* (2000) identified a new enzyme, indole-3-glycerol phosphate lyase, which converts indole-3-glycerol phosphate to free indole. They found that the corresponding gene *igl* is selectively activated by volicitin. This differs from previously known enzymes like BX1, which catalyzes the conversion of indole-3-glycerol phosphate to indole to form the secondary defense compounds DIBOA (2,4-dihydroxy-2*H*-1,4-benzoxazin-3(4*H*)-one) and DIMBOA (2,4-dihydroxy-7-methoxy-2*H*-1,4-benzoxazin-3(4*H*)-one), or tryptophan synthase, which produces the amino acid trytophan (Frey *et al.*, 1997). The selective activation of the evolutionarily similar genes *igl* and *bx1* strongly suggests that the plants are capable of selecting which induced defense to use depending on the attacking species.

Volicitin has also been shown to activate a specific maize sesquiterpene cyclase gene, *stc1*, which is also activated in response to caterpillar feeding or regurgitant treatment (Shen *et al.*, 2000). The transcription of *stc1* results in the production of a naphthalene-based sesquiterpenoid, which we have not yet detected from the many maize lines we have studied (e.g., Gouinguené *et al.*, 2001). It would be interesting to see if this volicitin-induced substance shows attractiveness to natural enemies of the caterpillars that induce its production.

One of the terpenoids that is almost always found in induced odor blends of many plants species is the acyclic C_{11} homoterpene DMNT (Boland *et al.*, 1992; Dicke, 1994). Biosynthesis of DMNT proceeds via (*E*)-nerolidol, a sesquiterpene alcohol (Boland and Gäbler, 1989; Donath and Boland, 1994; Degenhardt and Gershenzon, 2000). Degenhardt and Gershenzon (2000) demonstrated the activity of a (*E*)-nerolidol synthase that converts farnesyl bisphosphate, a common precursor of sesquiterpenes, to (3*S*)-(*E*)-nerolidol in maize leaves after the leaves had been damaged by *Spodoptera littoralis* larvae. Activity of (*E*)-nerolidol synthase has also been shown in lima bean and cucumber leaves in response to spider mite feeding on these leaves (Bouwmeester *et al.*, 1999). (*E*)-Nerolidol synthase appears to be specifically committed to the formation of DMNT and could play a key role in determining the attractiveness of herbivore-injured plants to natural enemies (Degenhardt and Gershenzon, 2000).

The apparent selective activation of genes responsible for induced odor production and the committed function of the resulting enzymes may allow for a precise fine tuning between insect-derived elicitors and the responses of the plant. Thus, plants have the potential to adapt their signals specifically to the insect that feeds on a plant. Several studies present evidence for such specificity.

Specificity

If plants respond differentially to different herbivores, producing a distinct blend of volatiles in each case, the signals may provide the natural enemies with specific information on the identity and perhaps even stage of the herbivores present on a plant. Evidence for and against such specificity is accumulating. Dicke (1999) has listed various examples that indicate specificity as well as those that suggest a lack of specificity. For instance, Takabayashi *et al.* (1995) found that only the 1st through the 4th instar larvae of *Pseudaletia separata* (Lepidoptera: Noctuidae) induced a significant production of volatiles in maize. In accordance, the parasitoid *Cotesia kariyi* is attracted primarily to maize plants eaten by early instar larvae, which are suitable for parasitization (Takabayashi *et al.*, 1995). However, no such specificity was found for the interaction between maize plants, larvae of *S. littoralis* (Lepidoptera: Noctuidae), and the parasitoid *Microplitis rufiventris*, which also attacks only the early stages of this preferred host (Gouinguené, 2000; Gouinguené *et al.*, 2003).

Other examples of specificity show that different herbivore species cause different reactions in a plant. These differences can be in the total quantity of volatiles released (Turlings *et al.*, 1998) or in actual differences in the composition of the odor blend (Turlings *et al.*, 1993a; Du *et al.*, 1998; De Moraes *et al.*, 1998). A very distinct difference occurs in the ratios among typical green leaf volatiles released by plants damaged by either *Spodoptera frugiperda* or *S. exigua* (Turlings *et al.*, 1993a). Maize damaged by *S. frugiperda* emitted far more (*E*)-2-hexenal than maize damaged by *S. exigua*. The parasitoid was able to learn to distinguish between the two types of damage (Turlings *et al.*, 1993a), but it remains unclear whether (*E*)-2-hexenal played a role in this. It should be noted that the release of green leaf volatiles in maize does not appear to be enhanced by elicitors; these volatiles "bleed" instantaneously from damaged sites.

Learning is not required for the aphid parasitoid *Aphidius ervi* to recognize pea plants that are damaged by its specific host, the pea aphid *Acyrthosiphon pisum* (Du *et al.*, 1998; Powell *et al.*, 1998). This parasitoid is far more attracted by pea plants infested by this host than by pea plants infested by a non-host, *Aphis fabae*. Implicated in the specificity of the signal is 6-methyl-5-hepten-2-one, a substance that was only detected in the odor profile of plants infested by *A. pisum* (Wadhams *et al.*, 1999); the pure compound was found to be highly attractive to *A. ervi* (Du *et al.*, 1998).

Behavioral and chemical evidence for signal specificity was also obtained by De Moraes *et al.* (1998). They found that *Cardiochiles nigriceps*, a parasitoid that specializes on *Heliothis virescens*, is much more attracted to plants attacked by its host than by plants attacked by the closely related non-host *H. zea*. Volatile collections from maize and tobacco plants that had been subjected to feeding by these noctuids showed differences in the relative ratios of some of the major compounds. It remains to be determined whether these observed differences allow *C. nigriceps* to recognize plants with hosts.

Two novel studies (Kahl *et al.*, 2000; De Moraes *et al.*, 2001) have reached some spectacular conclusions concerning specific responses to insect feeding. Kahl *et al.* (2000) showed that wild tobacco, *Nicotiana attenuata*, does not increase its production of nicotine after it has been damaged by nicotine-tolerant *Manduca sexta* caterpillars. Any other form of damage is known to result in the accumulation of nicotine in this plant, through stimulation of the JA signal cascade. It was subsequently confirmed that an ethylene burst resulting from *M. sexta* feeding suppressed nicotine production (Winz and Baldwin, 2001). The authors suggested that the plant chooses not to use its direct defense against this well-adapted adversary but instead mobilizes a strong indirect defense with the release of considerable amounts of volatiles that were shown to attract natural enemies (Kessler and Baldwin, 2001). They also point out that ingested nicotine probably has not much effect on *M. sexta* but may negatively affect its natural enemies. Equally interesting is the finding by De Moraes *et al.* (2001) that in *Nicotiana tabacum* the odor emitted after caterpillar feeding is different during the night than during the day. The day-time volatiles are known to attract parasitoids (De Moraes *et al.*, 1998), whereas the night time volatiles repelled female *H. viresens* moths and kept them from laying eggs on the emitting plants (De Moraes *et al.*, 2001). Again the plant appears to choose what and when to emit.

These examples suggest great sophistication in how the plants "choose" to respond to herbivore attack. However, the ability of natural enemies to take advantage of this specificity may be hampered by the great variability that can be observed among different genotypes of a plant species in the release of the major volatile compounds. Possibly, subtle differences in some of the minor compounds play an important role in determining signal specificity. Below we discuss this genotypic variation and its implication for the reliability and specificity of herbivore-induced signals.

Variability

It seems that plant-provided signals are limited in the specific information that they can provide because of the high variability that is found among plant genotypes (Takabayashi *et al.*, 1991b; Loughrin *et al.*, 1995a; Gouinguené *et al.*, 2001; Krips *et al.*, 2001). Variation can also be found between plant parts (Turlings *et al.*, 1993b), between different growth stages of a plant (Gouinguené, 2000; Turlings *et al.*, 2002) and between plants grown under different conditions (Gouinguené and Turlings,

2002). Moreover, many parasitoids and predators, whether they are generalists or not, can find their hosts or prey on a variety of plant species and each of these has its own characteristic basic odor blend. Therefore, natural enemies that use plant odors to locate their prey will need to determine which odors are most reliably associated with a certain prey at a certain time.

The variation in odor emissions that can be found among plant species is illustrated in Fig. 2.4. The chromatograms depict the volatile blends released by four crop plant species (maize, cotton, cowpea, and alfalfa) at different times after an attack by the common lepidopteran pest *S. littoralis*. For this experiment, 2- to 4-week-old plants that had been grown in pots in a climate chamber were transplanted into a glass vessel (a cylinder 10 cm in diameter and 45 cm high) the day before odor collections started. Pure humidified air was pumped into each vessel just above soil level, while close to the open top of the vessel most of the air was pulled out through a collection trap. With this technique, which is similar to the one described by Turlings *et al.* (1998), the volatiles emitted by each plant were collected for periods of 3 h. The volatiles were extracted from the traps, two internal standards were added, and each sample was analyzed on a gas chromatograph coupled to a mass spectrometer. The top chromatograms in Fig. 2.4 show the odor emissions before caterpillar attack. Most plants are virtually odorless when undamaged, but some, like the maize line used here, constitutively release a few substances (e.g., linalool and (*E*,*E*)-α-farnesene).

After the first 3 h collection, 20 starved 3-day-old *S. littoralis* larvae were placed on each plant. A new 3 h collection was started immediately after. As can be seen in the second chromatogram for each plant, there was considerable variation in the types of substance that were released by each plant, but all of them released the highly volatile green leaf odors (e.g., (*E*)-2-hexenal, (*Z*)-3-hexenol, and (*Z*)-3-hexenyl acetate). These volatiles are characteristic for fresh damage and may play a common role in the initial attraction of naïve natural enemies to damaged plants (Fritzsche Hoballah *et al.*, 2002) as well as in the location of recently damaged sites on a plant. Of the plants tested, cotton was the only one that showed an immediate release of significant amounts of several of terpenoids (e.g., α-pinene, β-pinene, and caryophyllene). These terpenoids are stored in glands located near the surface of cotton leaves and are released when the glands are ruptured (Elzen *et al.*, 1985; Loughrin *et al.*, 1994). In maize, only small amounts of induced terpenoids were collected during the first hours of attack. We had previously shown (Turlings *et al.*, 1998) that these are compounds induced after caterpillar damage and that the reaction in this plant can be observed within hours.

That maize is faster in the production of induced substances than the other plants is clear from the remaining chromatograms. On the second day of the experiment, the maize plants showed a full release of all induced substances, while for the other plants the release takes more time. After 2 days, cotton plants also released induced

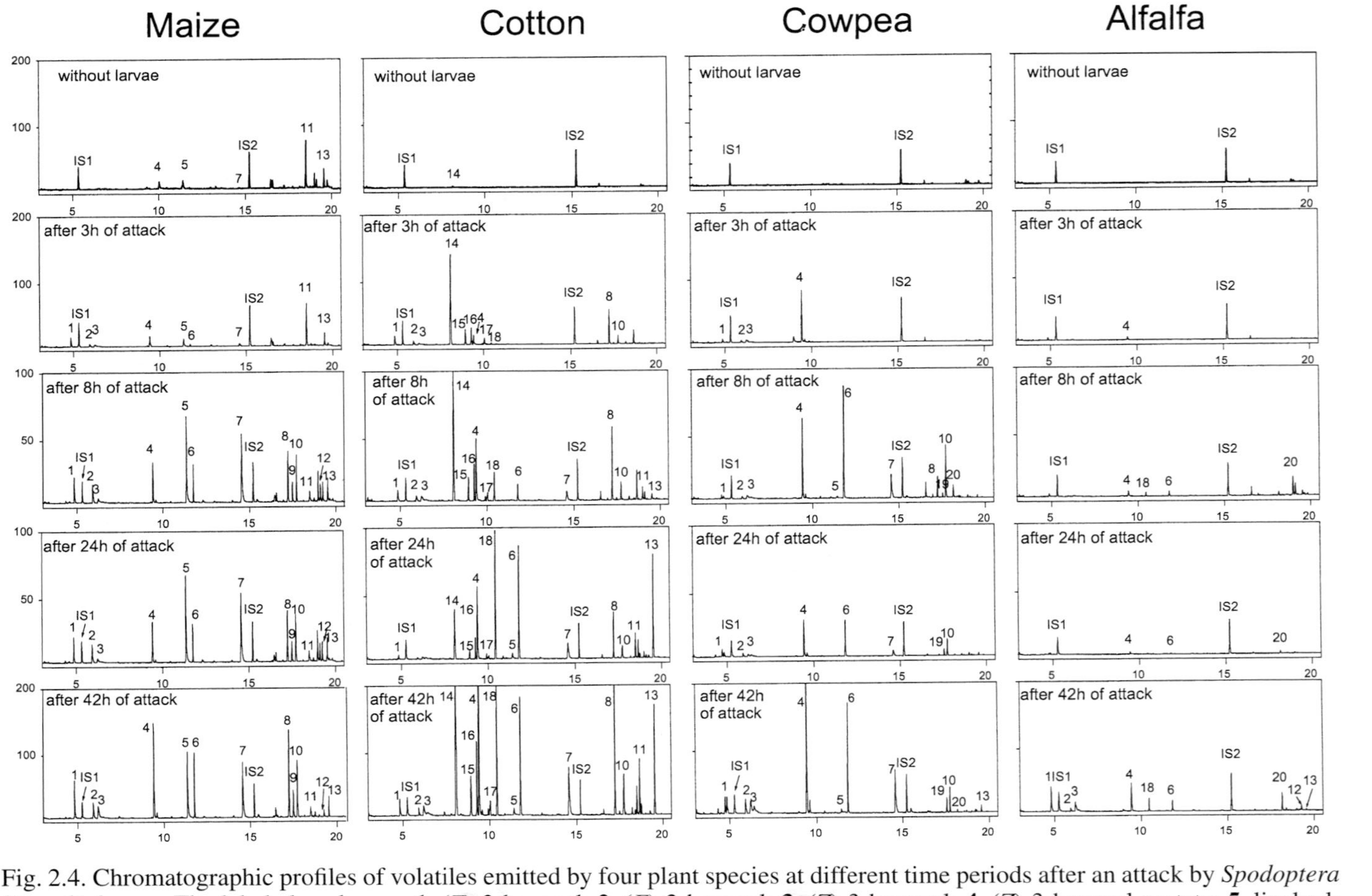

Fig. 2.4. Chromatographic profiles of volatiles emitted by four plant species at different time periods after an attack by *Spodoptera littoralis* larvae. The labeled peaks are: **1**, (*Z*)-3-hexenal; **2**, (*E*)-2-hexenal; **3**, (*Z*)-3-hexenol; **4**, (*Z*)-3-hexenyl acetate; **5**, linalool; **6**, (*E*)-4,8-dimethyl-1,3,7-nonatriene; **7**, indole; **8**, (*E*)-β-caryophyllene; **9**, (*E*)-α-bergamotene; **10**, (*E*)-β-farnesene; **11**, (*E,E*)-α-farnesene; **12**, nerolidol; **13**, (*E,E*)-4,8,12-trimethyl-1,3,7,11-tridecatetraene; **14**, α-pinene; **15**, β-pinene; **16**, β-myrcene; **17**, D-limonene; **18**, (*E*)-β-ocimene; **19**, β-sesquiphellandrene; **20**, germacrene D. Two internal standards, *n*-octane and nonyl acetate, are labeled with **IS1** and **IS2**, respectively.

terpenoids (e.g., (*E*)-β-ocimene, DMNT, (*E*)-β-farnesene) alongside the ones that are released constitutively from the glands (see also McCall *et al.*, 1993; Loughrin *et al.*, 1994; Röse *et al.*, 1996). The late reaction in this plant may be a strategy in which it first relies on its stored defenses and then, when an attack continues, switches to an induced defense.

The cowpea and especially alfalfa plants released relatively few substances and in lesser amounts. Parasitoids and predators that can find their victims on all of these plants will have to deal with all this variability and are likely to show differences in their preferences for these odors based on their interactions with certain plant species over evolutionary time. One behavioral characteristic that has been frequently shown for parasitoids, and which may help them to deal with this tremendous variation, is the ability to learn by association. This ability allows parasitoid females to change their responses in accordance with the odor cues that they experience to be most reliably associated with the presence of hosts (Turlings *et al.*, 1993b; Vet *et al.*, 1995). This associative learning is expected to be important for generalist parasitoids, which are unlikely to rely on innate preferences for specific cues but rather need to establish what cues are most reliably associated with the presences of suitable hosts at a given time. This may be different for highly specialized parasitoids such as *C. nigriceps*, which only attacks *H. virescens* (De Moraes *et al.*, 1998). It too relies on plant volatiles for host location but apparently has adapted to respond to subtle differences in the plants' responses to damage by different insects. It still has to be determined what these differences are. Studies on the host-locating behavior of the aphid parasitoid *A. ervi* suggest that it distinguishes between plants that carry host and non-host aphids with the use of a single compound, 6-methyl-5-hepten-2-one. So far, this compound has only been found in the odor blend emitted by host-infested plants (Du *et al.*, 1998; Powell *et al.*, 1998). Further studies with additional plant genotypes and plant species will reveal if such specific indicators are indeed provided by the plants.

Benefits

Among others, Faeth (1994), van der Meijden and Klinkhamer (2000), and Hare (2002) have stressed the need for ecological evidence that plants benefit from recruiting natural enemies of herbivores. Van der Meijden and Klinkhamer (2000), who focus on parasitoids, criticized the studies that imply mutualistic interactions between the first and third trophic level. They cite several papers on parasitoids that may not reduce herbivory in their hosts. Indeed, there are examples where plants do not benefit from the action of parasitoids (e.g., Coleman *et al.*, 1999), but the authors overlooked most of the papers that found such a reduction in herbivory after parasitization (see Beckage, 1985). In fact, van Loon *et al.* (2000b) pointed out that all studied solitary parasitoids cause their hosts to feed less, whereas for gregarious parasitoids this can vary. However, a reduction of herbivory does not necessarily imply a fitness gain for the plant. That plant fitness can indeed increase as a result of parasitization was convincingly demonstrated by Gómez and Zamora (1994). They studied the effects of chalcid parasitoids that attack a seed weevil (*Ceutorhynchus* sp.) on the fitness of a woody crucifer, *Hormathophylla spinosa*. With exclusion experiments in the field, they were able to show that, in the presence of the parasitoids, plants that were attacked by the weevil produced more seeds per fruit than weevil-infested plants without parasitoids. The parasitoids reduced weevil-inflicted seed damage to such an extent that the plants produced almost three times as much seed (Gómez and Zamora, 1994).

For leaf-feeding insects, which have been most studied in the context of induced odors that are attractive to parasitoids, such evidence was missing until recently. In a first study, van Loon *et al.* (2000b) showed that *Arabidopsis thaliana* plants attacked by larvae of *Pieris rapae* (Lepidoptera: Pieridae) produced considerably more seed when the larvae were parasitized by the solitary endoparasitoid *Cotesia rubecula*. We obtained similar results with the maize–*Spodoptera* system and found that plants infested with larvae parasitized by *Cotesia marginiventris* yielded more seed than those attacked by healthy *Spodoptera* larvae (Fritzsche Hoballah and Turlings, 2001). This evidence clearly shows the potential of plant signals indirectly to reduce herbivory and enhance plant fitness, but it remains to be seen what the consequences of these interactions are for wild plants in their natural environment.

Other ecological consequences of induced odor emissions

Attraction or repellence of herbivores by induced plant odors

A limited number of studies have looked at how induced plant volatiles affect the attractiveness of herbivores. It was found that different herbivores are affected differently. Landolt (1993) showed that adult females of the cabbage looper, *Trichoplusia ni* (Lepidoptera: Noctuidae), may be more attracted to cotton plants that have already been damaged by its larvae, but they prefer to oviposit on healthy plants rather than damaged plants. In the case of cabbage plants, cabbage looper females avoided previously damaged plants altogether (Landolt, 1993). Repellence of plants that emit herbivore-induced volatiles was also observed for the corn-leaf aphid *Rhopalosiphum maidis*. This was demonstrated under laboratory as well as field conditions (Bernasconi *et al.*, 1998).

Interestingly, Lepidoptera and aphids seem to avoid already infested plants, whereas Coleoptera are in general attracted to volatiles emitted by plants that are under attack by conspecifics. This has been shown for scarabaeid (Domek and Johnson, 1988; Harari *et al.*, 1994; Loughrin *et al.*, 1995b) and chrysomelid beetles (Peng and Weiss, 1992; Bolter *et al.*, 1997; Kalberer *et al.*, 2001). The Colorado potato beetle, a chrysomelid, is more attracted not only to potato plants damaged by conspecifics rather than undamaged plants (Bolter *et al.*, 1997; Landolt *et al.*, 1999) but also to plants treated with insect regurgitant or the synthetic elicitors of odor emissions volicitin and methyl jasmonate (Landolt *et al.*, 1999), as well as to plants exposed to damaging ozone levels (Schutz *et al.*, 1995). These beetles specialize on specific host plants and are well adapted to, and may even exploit, their hosts' chemical defenses. Increases in these defensive chemicals in response to damage or elicitors may not be harmful to these insects. Moreover, beetles may be less vulnerable to natural enemies, especially if they rely on plant-derived chemicals for their defense. They may, therefore, be under less or no pressure to avoid plants that emit attractants for natural enemies. It has been proposed that the beetles visiting already attacked plants increase their chances of finding a suitable mate (Loughrin *et al.*, 1995b; Kalberer *et al.*, 2001) and a mass attack may weaken the plants' chemical defense potential.

For some herbivores, the responses to herbivore-induced plant odors differ under different circumstances. For instance, the spider mite *T. urticae* is more attracted to healthy lima bean leaves than leaves that emit volatiles induced by spider mite infestation (Dicke, 1986; Dicke and Dijkman, 1992). However, Pallini *et al.* (1997) found that the same mite is attracted to cucumber plants that are already infested by conspecifics. In contrast, *T. urticae* avoids the odor of cucumber plants under attack by the western flower thrips, *Frankliniella occidentalis*, which is a herbivore but also feeds on spider mites. Bark beetles can cause strong reactions in their host trees, resulting in the emission of a blend of volatile terpenoids that, in combination with aggregation pheromenes, is used in mass attacks. These same substances may attract predators (Byers, 1989) and parasitoids (Sullivan *et al.*, 2000; Pettersson, 2001; Pettersson *et al.*, 2001) to infested trees.

As yet, there is no specific pattern in how induced volatiles affect the attractiveness of plants to herbivores. Obviously, the responses will be correlated with fitness consequences. Insects vulnerable to natural enemies and induced plant toxins are, therefore, expected to avoid induced plants, whereas those that are adapted to plant defenses and/or benefit from aggregating are likely to be attracted. Comparative studies could test such hypotheses.

Plant–plant "communication"

Evidence for interactions among plants mediated by airborne chemicals was first obtained some 20 years ago (Baldwin and Schultz, 1983; Haukioja *et al.*, 1985; Rhoades, 1983, 1985), but skepticism and criticism of methodology and statistical procedures (Fowler and Lawton, 1985) initially prevented general acceptance by biologists. Evidence obtained since then has changed this. Ethylene was shown to activate defense genes (Ecker and Davis, 1987) and, in the seminal paper by Farmer and Ryan (1990), it was shown that methyl jasmonate induced the synthesis of proteinase inhibitors in tomato plants.

In the spider mite–lima bean system, it has now been shown that mite infestation activates defense genes in the plants and, in addition, several of these genes can also be activated when a lima bean plant is exposed to some of the induced volatiles of neighboring conspecifics (Arimura *et al.*, 2000a,b). Clearly, the genetic basis for plant–plant communication is in place. That it can actually take place in the field has now also been confirmed.

Dolch and Tscharntke (2000) studied the effects of artificial defoliation of alder trees on subsequent herbivory by alder leaf beetle (*Agelastica alni*). After defoliation, herbivory by *A. alni* was significantly lower in the defoliated trees and its neighbors compared with trees distant from the manipulated trees. Laboratory studies confirmed that resistance was induced not only in defoliated alders but also in their undamaged neighbors (Dolch and Tscharntke, 2000). Follow-up work showed that alder leaves respond to herbivory by *A. alni* with the release of ethylene and of a blend of volatile mono-, sesqui-, and homoterpenes. This herbivory also increased the activity of oxidative enzymes and proteinase inhibitors (Tscharntke *et al.*, 2001).

Additional convincing evidence for odor-mediated interactions between plants comes from a field study by Karban *et al.* (2000). They showed, over several years, that clipping sagebrush caused neighboring wild tobacco plants to become more resistant to herbivores. Preventing root contact between the plants did not change this effect, but preventing the exchange of volatiles between the plants by enclosing the clipped shoots in plastic bags did mitigate the effect. The explanation is that the release of methyl jasmonate by the damaged sagebrush caused an increase in phenol oxidase in the tobacco plants, rendering them more toxic to herbivores (Karban *et al.*, 2000). The relevance of such interactions in natural interactions remains to be elucidated for odor emissions resulting from natural herbivory.

In the context of tritrophic interactions, plant–plant communication has been subject to only few studies (Bruin *et al.*, 1995). In one such study, Bruin *et al.* (1992) demonstrated that healthy cotton plants that were exposed to spider mite-induced volatiles from conspecific plants increased in their attractiveness to predatory mites. This increased attraction was probably not simply the result of adsorbence and re-release of these volatiles from the healthy plants, because there is now clear evidence that volatiles from spider mite-infested plants can induce odor releases in neighboring plants (Arimura *et al.*, 2000a).

Inducible nutrition

Although insect predators and parasitoids are carnivorous by definition, they often also feed on plant-derived foods. This vegetarian side to their diet includes various plant substrates, such as nectar, food bodies, and pollen. In addition, they often utilize foods indirectly derived from plants (e.g. honeydew, or pycnial fluid of fungi). In some cases, predators may also feed on plant productive tissue, in which case they have to be classified as potential herbivores. The level in which predators or parasitoids depend on primary consumption varies.

Nutritional requirements of natural enemies

Ants display a broad variation in lifestyles, which is reflected in an equally broad dietary diversity, ranging from species that are primarily predators to species that rely almost entirely on honeydew and extrafloral nectar. Although it has long been held that the majority of ant species are predominantly carnivorous (Sudd and Franks, 1987; Hölldobler and Wilson, 1990), Tobin (1994) argued that the dominant species are largely primary consumers, for which the bulk of their diet consists of plant-derived carbohydrates. An important dichotomy might occur between the nutrition of immature and mature stages. Ants tend to feed protein-rich food preferentially to their larvae, whereas the adults survive mostly on a diet of plant-derived carbohydrates (Haskins and Haskins, 1950; Vinson, 1968). Further differentiation takes place among the adult castes, as it is believed that certain activities such as foraging, killing, and dismembering of prey, as well as the transporting of food items or building material, require most energy (Beattie, 1985). Foraging workers retain the majority of sugar-rich foods, while passing the bulk of protein-rich food to castes remaining in the nest (Markin, 1970; Schneider, 1972). The important role of carbohydrates to ant nutrition was also demonstrated by Porter (1989). He showed that fire ant colonies kept on insect prey only had a retarded growth and reproduction rate in comparison with colonies fed both prey and sugar water. It has been argued that displacement of the native fire ant *Solenopsis geminata* by the imported fire ant *Solenopsis invicta* is partly based on the latter species' higher efficacy in collecting liquids such as nectar (Tennant and Porter, 1991). The main carbohydrate sources exploited by ants are extrafloral nectar (Fisher *et al.*, 1990) and honeydew, the sugar-rich excretions from sap-feeding insects (Retana *et al.*, 1987). Interestingly, the use of floral nectar appears to be relatively uncommon (Tobin, 1994).

While sugar solutions can be a significant item in the diet of ants, parasitoids are often entirely dependent on carbohydrates as an adult food source (Jervis *et al.*, 1993). The parasitoids' longevity and fecundity are usually subject to energetic constraints (Leatemia *et al.*, 1995; Stapel *et al.*, 1997; Wäckers, 2001), whereas the parasitoids' behavior can also be strongly affected by their nutritional state (Wäckers, 1994; Takasu and Lewis, 1995). There is strong evidence that the availability of suitable sugar sources can play a key role in parasitoid host dynamics (Krivan and Sirot, 1997; Wäckers, 2003).

Plant-provided nutrition and its functions

Plants employ nutritional supplements in a range of mutualistic interactions. Best known are the floral rewards targeted at pollinators (Faegri and van der Pijl, 1971), and the fleshy fruit tissue promoting seed dispersal by vertebrates. Ants as well can play an important role in the dissemination of seeds. Their tendency to harvest seeds and to transport them to their (underground) nests makes ants efficient seed dispersers (Horvitz and Schemske, 1986; Jolivet, 1998). Some plant species stimulate this interaction by producing protein- and lipid-rich seed appendages, the so-called elaiosomes (Milewski and Bond, 1982). Ants collect these seeds preferentially, consume the nutrient-rich elaiosomes and may subsequently discard the hard seeds in underground waste dumps. The scarring of the seeds, the moist and nutrient-rich surroundings, as well as the clustering of seeds, might be factors benefiting germination and seedling growth (Beattie, 1985).

Defense is a further category in which plants employ food rewards to acquire protection by arthropod mutualists. The provision of food sources allows plants to recruit or sustain predators or parasitoids, which, in turn, can provide protection against herbivory. The plant-derived food structures involved in indirect defensive interaction can be divided in two main groups: food bodies and extrafloral nectaries.

Food bodies are protein- and/or lipid-rich epidermal structures, including Beltian bodies, Müllerian bodies and pearl bodies (Rickson, 1980). Food bodies can be harvested by ants for consumption by either larvae or adults. However, in some of the examples that have been described as "food bodies," actual collecting by ants has not yet been observed (Beattie, 1985). Unlike extrafloral nectar, food bodies can serve as an alternative to insect protein. This facilitates intimate interactions with ants, as it allows ants to remain on the plant (nesting) during times in which the availability of insect protein is low. However, it incurs the risk that ants become protein satiated, which may hamper carnivory. Some ant species rely entirely on food bodies of their particular host plant for their protein supply (Carroll and Janzen, 1973). Even though food bodies are collected by some non-mutualists (Letourneau, 1990), the range of potential consumers is not as broad as in the case of the easily accessible and digestible extrafloral nectar (Whitman, 1994). This makes food bodies less vulnerable to consumption by unintended consumers.

Extrafloral nectaries include a wide range of nectar-excreting structures, which are distinguished from their floral counterparts by the fact that they are not involved in pollination. Extrafloral nectar is typically dominated by sucrose and its hexose components glucose and fructose. The fact that these common sugars are acceptable to the majority of insects, combined with the exposed nature of extrafloral nectaries, makes them suitable food sources for a broad range of insects. Compared with floral nectar, extrafloral nectar often has increased fructose and glucose levels (Tanowitz and Koehler, 1986; Koptur, 1994), as well as a higher overall sugar concentration (Koptur, 1994; Wäckers *et al.*, 2001). These characteristics can be explained by the exposed nature of most extrafloral nectaries, which result in faster microbial breakdown of sucrose and increased evaporation. The high sugar concentration may also serve an ecological function, as high sugar concentrations reduce intake by visiting ants and increase durations of ant visits (Josens *et al.*, 1998). A further benefit of highly concentrated extrafloral nectar may lie in the fact that it prevents nectar use by a range of non-intended visitors (Wäckers *et al.*, 2001). This applies especially to Lepidoptera, whose mouthpart morphologies restrict them to feeding on nectar with relatively low sugar concentrations.

In addition to carbohydrates, extrafloral nectar may contain variable amounts of proteins, amino acids, and lipids (Baker *et al.*, 1978; Smith *et al.*, 1990). The particular amino acid composition can increase the attractiveness of extrafloral nectar as a food source (Lanza, 1988). Nevertheless, extrafloral nectar by itself falls short from providing a well-balanced diet. Low amino acid levels or the absence of certain essential amino acids forces nectar consumers to seek out supplementary protein sources, thereby stimulating predation.

Extrafloral nectar can make a significant contribution to the diet of ant species visiting these food sources. Fisher *et al.* (1990) reported that the six ant species investigated in their study derive between 11 and 48% of their diet from extrafloral nectar. Retana *et al.* (1987) found (extrafloral) nectar to be the main food source for *Camponotus foreli*. Extrafloral nectar can also be extensively used by other predators (Bakker and Klein, 1992; Ruhren and Handel, 1999), as well as parasitoids (Bugg *et al.*, 1989). In some instances, these carnivores, rather than ants, might represent the primary force protecting the plant (Ruhren and Handel, 1999; Cuautle and Rico-Gray, 2003).

Constitutive versus induced extrafloral nectar

Constitutive nectar production

Most plants excrete some extrafloral nectar irrespective of whether herbivores are present. Such constitutive nectar production may be synchronized with the most susceptible stages of plant growth (Bentley, 1977) or with the times during which damaging herbivores are usually active (Tilman, 1978). Furthermore, nectar production may coincide with the daily activity pattern of ants (Pascal and Belin-Depoux, 1991). Further synchronization is achieved through the ability of the plants to increase nectar secretion in response to herbivore presence (Koptur, 1989; Wäckers and Wunderlin, 1999; Heil *et al.*, 2001).

In general, constitutive nectar production may provide a degree of prophylactic protection, because it allows plants to accommodate some natural enemies before herbivores arrive (Wäckers *et al.*, 2001). Prophylactic protection by natural enemies may include, for example, the prevention of herbivore oviposition or removal of herbivore eggs. Maintaining some baseline nectar production in undamaged plants is also likely to assure some level of ant visitation, which expedites the defense response to herbivore attack.

Induction of food provision

In addition to the constitutive production of food supplements, some plants can actively adjust their food provision in response to their biotic environment (Table 2.2). Unlike other defense mechanisms, this induction can be elicited by two distinct mechanisms. Food provision can be raised both by food removal (Risch and Rickson, 1981; Koptur, 1992; Heil *et al.*, 2000) and by tissue damage (Koptur, 1989; Wäckers and Wunderlin, 1999; Heil *et al.*, 2001; Wäckers *et al.*, 2001). These mechanisms represent active responses by the plants to both ant attendance and herbivore feeding. This receptiveness toward the presence of both the second and the third trophic level represents a unique and highly dynamic type of plant response.

In the case of food bodies, the primary mechanism of induction might be food body removal. Risch and Rickson (1981) showed that the production of unicellular food bodies by *Piper cenocladum* is stimulated by the presence of the mutualist ant *Pheidole bicornis*. When ants are present, the plant produces 30 times as many food bodies as control plants. Similar effects had previously been reported for other types of food bodies (Carroll and Janzen, 1973). In *P. cenocladum*, a clerid beetle exploits this relationship. Their larvae are also able to stimulate food body production in the absence of the ants (Letourneau, 1990).

Extrafloral nectar production can be raised in response to both nectar removal (Koptur, 1992; Heil *et al.*, 2000) and tissue damage. Stephenson (1982), using *Catalpa speciosa*, was the first to investigate the latter mechanism. He diluted the nectar of individual nectaries with water and demonstrated that the diluted nectar collected from sphingid-damaged leaves was richer in solutes compared with nectar collected from undamaged leaves. Smith *et al.* (1990) point out that this does not resolve whether *C. speciosa* actually increased its nectar volume or whether it produced the same volume with an increased solute concentration.

Koptur (1989) reported that mechanical damage of *Vicia sativa* leaves increased the volume of extrafloral nectar production by a factor of 2.5. Heil *et al.* (2001) reported a two- to five-fold increase in volume of nectar secretion in *Macaranga tanarius* following leaf damage. In *Ricinus communis* and *Gossypium herbaceum*, the increase in extrafloral nectar production following *S. littoralis* herbivory was three-fold and ten-fold, respectively (Wäckers *et al.*, 2001). Through parallel high pressure liquid chromatographic analysis of sugars in the collected nectar, the

Table 2.2. *The effect of leaf damage on extrafloral nectar production*

Plant species	Damage type	Quantitative changes			Qualitative changes		References
		Nectar volume	Carbohydrates	Amino acids	Carbohydrates	Amino acids	
Catalpa speciosa	Herbivory	?	Increase in solutes	Increase in solutes	?	?	Stephenson, 1982
Ipomoea carnea	Mechanical	No	?	?	?	?	Koptur, 1989
Inga spp.	Mechanical	No	?	?	?	?	Koptur, 1989
Vicia sativa	Mechanical	2.5-fold increase[a]	?	?	?	?	Koptur, 1989
Impatiens sultani	Mechanical	No	No	5.6-fold increase	No	No	Smith *et al.*, 1990
Passiflora spp.	Mechanical	Yes	?	?	?	?	Swift and Lanza, 1993
Gossypium herbaceum	Herbivory (caterpillar) and mechanical	10-fold increase	No	?	No	?	Wäckers and Wunderlin, 1999; Wäckers *et al.*, 2001
Macaranga tanarius	Herbivory and mechanical	Yes	?	?	?	?	Heil *et al.*, 2000, 2001
Ricinus communis	Herbivory (caterpillar)	2.5-fold increase	No	?	No	?	Wäckers *et al.*, 2001
Vicia faba	Herbivory (aphids)	No	No	?	No	?	Engel *et al.*, 2001

[a]In one of the four defoliation levels tested.

latter study was the first to demonstrate that the increased nectar secretion actually represents a proportionate increase in carbohydrate secretion.

All these examples focus on the temporal aspect of nectar induction. In addition, extrafloral nectaries are also especially suited for the study of spatial dynamics following induction. This aspect can be easily assessed because of the discrete distribution of nectaries, the possibility of non-destructive sampling, as well as the ease of nectar collection. With respect to the spatial pattern of induction, Wäckers *et al.* (2001) showed that the impact of herbivory on extrafloral nectar induction is primarily localized (i.e., restricted to the damaged leaf). This local increase in nectar production can help in actively guiding ants to the site of attack. In addition, a weaker systemic response was found. This systemic induction was restricted to the younger leaves.

These examples show that several plants possess the ability to raise extrafloral nectar production in response to herbivory, but this induction is not necessarily universal and might vary depending on both plant and herbivore species (Table 2.2). Koptur (1989) could not demonstrate an effect of mechanical defoliation on extrafloral nectar production in *Ipomoea carnea*, *Inga brenesii* and *Inga punctata*. In *V. faba*, aphid feeding had no effect on the quantity of extrafloral nectar secretion (Engel *et al.*, 2001). A similar lack of induction was found following feeding by *S. littoralis* larvae (F. L. Wäckers, unpublished data).

Specificity of induction: elicitors and mechanisms

The few studies that have addressed the induction of extrafloral nectar production have examined either actual herbivory or mechanical damage. The fact that mechanical damage failed to elicit nectar induction in several plant systems (Koptur, 1989) could be interpreted as indicating that the method of mechanical damage is not a suitable mimic of herbivory.

Alternatively, induction of nectar secretion could require a herbivory-specific elicitor, similar to the induction of plant volatiles. To investigate this Wäckers and Wunderlin (1999) conducted a set of experiments analogous to those conducted by Turlings *et al.* (1990), in which cotton plants were subjected either to herbivory or to mechanical damage with and without caterpillar regurgitant. In contrast to the mechanism of herbivore-induced volatile emission, the induction of extrafloral nectar secretion was found to be elicited by tissue damage, irrespective of whether this damage was mechanical or caused by actual herbivory. The addition of *S. littoralis* regurgitant had no significant effect on the level, the timing, or the distribution of nectar secretion. These findings indicate that the induction of extrafloral nectar secretion constitutes a general response by the plant to tissue damage, rather than representing a herbivory-specific mechanism.

This rather unspecific induction of nectar secretion in cotton was surprising in light of the fact that the induction of volatile emission by this plant had been demonstrated to be specific. Herbivore-damaged plants show a higher rate of volatile emission compared with mechanically damaged plants (McCall *et al.*, 1994), and herbivore feeding induced *de novo* synthesis of various terpenoids (Paré and Tumlinson, 1997), which resulted in a quantitative as well as a qualitative response to herbivory. The specificity of the plant response is not restricted to the differentiation between mechanical damage and herbivory. The composition of the induced volatile blend also varies between (even closely related) herbivore species (De Moraes *et al.*, 1998).

The difference in induction specificity between the two categories of indirect defense indicates that the induction pathways involved are not entirely identical. It may also reflect differences in the costs and benefits of such specificity (Wäckers and Wunderlin, 1999). The use of volatiles as a signal to recruit natural enemies is dependent on induction, as this communication between plants and the third trophic level breaks down when the volatile signal is not reliably associated with herbivore presence. Extrafloral nectar, by comparison, constitutes a reward in itself rather than serving as a signal to indicate the location of a reward. The response by the third trophic level, as a result, is not dependent on the degree in which nectar secretion correlates with herbivore presence. Therefore, an increase in nectar production following mechanical damage entails the additional cost of nectar production but has no negative implications for the efficacy of this indirect defense mechanism.

Working with *M. tanarius*, Heil *et al.* (2001) also reported that mechanical damage is sufficient to induce nectar secretion. They were also able to achieve a similar response through exogenous application of JA to undamaged plants. This fact, combined with the finding that the response in damaged plants could be suppressed by phenidone, an inhibitor of JA synthesis, indicates that the induction of extrafloral nectar production is elicited via the octadecanoid signal cascade (Heil *et al.*, 2001), which is also involved in the production of various inducible plant volatiles (see above).

Costs and benefits

The benefit of extrafloral nectar production to plant fitness has been well established (Bentley, 1977; Inouye and Taylor, 1979; O'Dowd, 1979; Wagner, 1997; Koptur *et al.*, 1998). Whether induction further enhances plant fitness over constitutive nectar production remains an open issue. The fact that both inducible and constitutive nectar production occurs (Table 2.2) indicates that the costs and benefits of nectar induction vary among plants.

It is often believed that the primary benefit of defense induction is economical, as it restricts defensive investments to those periods in which plants are actually under attack (Rhoades, 1979; Zangerl and Rutledge, 1996). In addition to these economic benefits, induction of extrafloral nectar production may also enhance the effectiveness of natural enemy recruitment, because it results in an accumulation of natural enemies on the site of attack (F. L. Wäckers and F. Frei, unpublished data). However, these benefits of induction come at the price of increased vulnerability during the plant's non-induced state. To understand the pattern in which extrafloral nectar is produced, we need to identify and quantify the particular costs involved in the use of this indirect defense.

Costs of extrafloral nectar production

Pyke (1991) demonstrated a trade-off between floral nectar secretion and seed production in hand-pollinated *Brandfordia nobilis*. Comparable studies on fitness consequences of extrafloral nectar production have yet to be conducted. However, strong indirect evidence for the high cost of extrafloral nectar production is provided by the finding that some plant species have lost extrafloral nectaries in ecosystems void of mutualist ant species. Rickson (1977) was able to track the gradual regression of *Cecropia peltata* extrafloral nectaries from *Azteca* ant-inhabited mainland Central America over a range of Caribbean islands lacking the mutualist ant species. Bentley (1977) described a decline in sepal nectaries of *Bixa orellana* from ant-rich lowlands to higher altitudes where ant populations are scarce.

Direct costs

To the plant, the direct cost of producing extrafloral nectar can be relatively low. O'Dowd (1979) estimated that the energy invested in the lifetime petiolar nectar production of an individual *Ochroma pyramidale* leaf constitutes about 1% of the leaf's energy content. However, since leaf tissue makes up only part of the total plant mass, this figure does not reveal which fraction of the total assimilated energy is diverted to extrafloral nectar. A more accurate way to estimate allocational costs is to express the quantity of excreted sugars as a fraction of the daily production of assimilates. Wäckers *et al.* (2001) calculated that castor (*Ricinus communis*) diverts 1% of its daily assimilates to the production of extrafloral nectar. Even though this cost may seem unsubstantial, its cumulative nature could lead to rapid cost increments over the total period of plant growth. In addition to the loss of carbohydrates, nectar secretion also entails a loss of other compounds, in particular amino acids and water. Depending on the growth conditions of the plant, loss of these compounds may represent considerable additional cost factors.

Direct costs further include the costs involved in active nectar sequestration, as well as the cost involved in producing the nectary. This latter cost is probably low, as nectaries are often simple and small, showing little differentiation. In other types of defense, costs relating to biosynthesis, transport, and storage (i.e., autotoxicity) can be considerable (Karban and Baldwin, 1997). However, these costs do not apply in the case of extrafloral nectar as nectaries are usually vascularized and obtain non-toxic primary metabolites directly from the phloem or xylem (Frey-Wyssling, 1955; Beattie, 1985).

Ecological costs

In addition to the direct costs, the production of extrafloral nectar can also entail substantial indirect (ecological) costs. In insect-pollinated plants, extrafloral nectaries can have adverse effects on pollination efficacy. Interference with the pollination process can occur when extrafloral nectaries distract the pollinators away from the floral nectar (Koptur, 1989) or when nectary-attending ants attack flower visitors (F. L. Wäckers personal observation). Considerable ecological costs may arise when extrafloral nectaries are exploited by herbivores. Adult herbivores such as moths are often entirely or partly dependent on sugar solutions as an energy source. Nectar feeding frequently increases herbivore longevity as well as the number and size of matured eggs (Leahy and Andow, 1994; Binder and Robbins, 1996; Romeis and Wäckers, 2000, 2002). When herbivores are attracted or retained by extrafloral nectaries, this can severely increase herbivory levels on nectar-producing plants (Adjei-Maafo and Wilson, 1983; Rogers, 1985; McEwen and Liber, 1995). To reduce these ecological costs, plants may have adapted the extrafloral nectar composition to exclude unintended visitors and to cater selectively to those insects from which they benefit (Wäckers *et al.*, 2001).

How heavily these direct and indirect cost factors weigh on plant fitness depends on the plant species and its growing conditions. Induction of extrafloral nectar production, however, allows plants to minimize almost all of these cost factors simultaneously. In the absence of herbivory, nectar production and its associated costs may be all but eliminated, with the full costs only being assumed during periods of herbivory.

The cost-saving benefit of inducible defense is counterbalanced by the loss of preventative protection (Zangerl and Bazzaz, 1992). Any damage inflicted during the lag period between herbivore attack and the onset of the induced defense should be included in the costs of induction. It is our experience that the induced production of nectar takes about 24 h (Wäckers *et al.*, 2001). In the economic terms of the optimal defense theory, inducible defenses have a selective advantage over constitutive defenses when the savings in defensive costs during herbivore-free periods outweigh the loss in preventative protection during the lag time of induction. In comparison with direct defenses, lag time of indirect defense is extended because of the inherent delay in natural enemy response. In the case of ants responding to extrafloral nectaries, this delay includes the time for ant scouts to encounter the nectary, as well as the time required for nestmate recruitment (Wäckers *et al.*,

2001). This additional lag time likely reduces the economic benefits of induction of indirect defenses relative to those of direct defense. Plants may have developed various strategies to minimize the lag time of indirect defense induction. Maintaining some baseline nectar production in undamaged plants could be such a strategy. By accommodating at least a few natural enemies, the indirect defense can begin to operate quickly once the plant is attacked.

The need for more field data

To demonstrate that a plant trait has a defensive function, it is necessary to show that it has a negative effect on plant antagonists, reduces the damage done to the plants, and increases plant fitness under natural conditions (Hare, 2002). Attraction of natural enemies to herbivore-induced volatiles has mainly been demonstrated in laboratory studies, and the role of these volatiles for interactions in the field is still poorly understood (Sabelis *et al.*, 1999). Initial evidence comes from studies in which caged plants out in a field were found to attract more parasitoids or predators when damaged by herbivores than when undamaged. Drukker *et al.* (1995) showed that psyllid-infested pear trees attracted more predatory anthocorid bugs than trees without psyllids. In the laboratory, Scutareanu *et al.* (1997) demonstrated that infested trees release more and different volatiles than uninfested pear trees, and that the production of these volatiles was positively correlated with the density of the psyllids on the trees. Similar results were obtained by Shimoda *et al.* (1997), who found in a field experiment that the predator *Scolothrips takahashii* was attracted to cages that contained a lima bean plant infested with spider mites. Spider mite infestation is known to cause lima bean to emit a blend of specific terpenoids and methyl salicylate (Dicke and Sabelis, 1988; Dicke *et al.*, 1990a,b). However, as for the study with the psyllid-infested pear trees, it could not be excluded that the predators were directly attracted by the herbivores on the plants rather than the induced plant odor.

Conclusive field evidence has been obtained with the manipulation of odor emissions of free-standing plants. Thaler (1999), for example, observed an increase in parasitism of *S. exigua* larvae on tomato after the plants had been treated with JA. This treatment induces the octadecanoid pathway, which results in the production of various defense compounds, including volatiles. In an earlier study in a tobacco field, De Moraes *et al.* (1998) had already found that the specialist parasitoid *C. nigriceps* could distinguish between the odor of plants that have been damaged by its specific host *H. virescens* and the odor of plants damaged by a closely related non-host. In a natural, non-agricultural environment, Kessler and Baldwin (2001) supplemented the odor of wild tobacco plants with synthetic volatiles and found that (*Z*)-3-hexenol, linalool and (*Z*)-α-bergamotene all increased the predation rate of *M. sexta* eggs and neonate larvae by a generalist predator. Similar increases of predation were obtained by treating wild tobacco plants with methyl jasmonate (Kessler and Baldwin, 2001). In one of our own studies, we trapped considerably more parasitoids on sticky traps downwind from maize plants treated with caterpillar regurgitant than upwind from these plants or near untreated plants (Bernasconi Ockroy *et al.*, 2001). These studies provide good evidence for a role of induced plant volatiles in host and prey location. What is still missing, however, is field evidence from unmanipulated studies showing that plants actually benefit from these interactions.

The sophisticated equipment required for volatile identification has long confined the topic of herbivore-induced volatiles to the laboratory, but extrafloral nectaries have traditionally been studied in the field. Moreover, the work on extrafloral nectaries has mainly addressed wild plant species within their natural habitat, whereas the study of plant volatiles has long focussed on agricultural crops. As a result, we have a relative wealth of field evidence for the defensive function of extrafloral nectaries.

It has been well established that extrafloral nectaries are visited by a range of predators and parasitoids (Janzen, 1966; Bugg *et al.*, 1989; Koptur, 1992). Ants are by far the most common visitors to extrafloral nectaries. The facts that ants are social, show recruitment behavior, and have a strong tendency to defend lucrative sugar sources against competitors make them especially suitable as defensive agents. Nevertheless, not all ants are equally effective. Their aggressiveness ranges from species that attack large mammals (Bennett and Breed, 1985) to species that are passive or even tend to drop from the nectary when disturbed (O'Dowd, 1979).

In a number of cases, it has been demonstrated that increased levels of nectar production translates to higher levels of ant attendance (Passera *et al.*, 1994). The fact that the most aggressive ants monopolize the most productive nectar sources (Del-Claro and Oliveira, 1993) constitutes a further benefit to high levels of nectar production.

Using exclusion experiments, several studies were able to demonstrate that ants effectively protect the plant against herbivory (O'Dowd and Catchpole, 1983; Wagner, 1997; but see O'Dowd and Catchpole, 1983; Rico-Gray and Thien, 1989). In the same way, reduction of herbivory has recently been demonstrated in mutualisms between extrafloral nectaries and spiders (Ruhren and Handel, 1999), as well as predatory wasps (V. Rico-Gray, personal communication).

A number of studies have provided the ultimate proof for the defensive function of extrafloral nectaries by demonstrating that herbivory reduction by ants actually translates to an increased reproductive fitness of nectar-providing plants (Koptur, 1979; Rico-Gray and Thien, 1989; Oliveira, 1997; Wagner, 1997). In the most extreme cases, unattended plants die as result of herbivory in the absence of ants (Janzen, 1966).

In addition to these empirical studies, there is indirect ecological evidence for the defensive function of extrafloral nectaries. Several studies have reported correlations between the abundance of plants with extrafloral nectaries and ant abundance (Pemberton, 1998; Rico-Gray *et al.*, 1998). Bentley (1977) and Rickson (1977) showed that plants may lose extrafloral nectaries in ecosystems void of mutualist ant species.

Even though this evidence supports the defensive function of extrafloral nectaries, the evidence is largely based on myrmecophilous plants. In other plant species, the benefit of ant attendance is not always as clear (O'Dowd and Catchpole, 1983; Koptur and Lawton, 1988). In these species, the provision of extrafloral nectar may serve to enhance the effectiveness of other plant–predator (Ruhren and Handel, 1999) or plant–parasitoid interactions (Lingren and Lukefahr, 1977; Bugg *et al.*, 1989; Koptur, 1994), or serve other (non-defensive) functions.

Future directions

Although much is known about various intricacies of the active role of plants in tritrophic interactions, it is evident from the above review that numerous questions remain and several areas are virtually unexplored. We identify three areas that appear to us as particularly interesting and they can be expected to receive special attention in future research programs.

Cross-effects

Almost all studies on induced indirect defenses have looked at the effects of an attack by a single herbivore or pathogen species. In a natural situation, however, plants often suffer from simultaneous attacks by multiple adversaries. Many plants carry several herbivores and they can be infested by pathogens at the same time that they are eaten by herbivores. This should again contribute to the variability of reactions that plants exhibit. Plant infestations by multiple species and their cross-effects have been studied for direct defenses (Hatcher, 1995; Agrawal *et al.*, 1999; Rostàs *et al.*, 2003), but not yet in the context of indirect defenses.

Studies on the cross-effects of herbivore and pathogen infestation on direct defenses have yielded results that can be quite different for different systems (Karban and Kuc, 1999; Stout and Bostock, 1999; Rostàs *et al.*, 2003). The cucumber plant has been studied in detail with several pathogens and herbivores (Apriyanto and Potter, 1990; Ajlan and Potter, 1991; Moran, 1998). In most cases, infection with one pathogen caused a systemic resistance to other pathogens but had no systemic effect on insect herbivores, except for a positive effect on the striped cucumber beetle (Apriyanto and Potter, 1990). Moran (1998) reported that locally, at the site of pathogen infestation, both positive and negative effects on insects may occur. The most extensively studied system is that of *Rumex* spp. attacked by the leaf beetle *Gastrophysa viridula* and the biotrophic rust fungus *Uromyces rumicis*. Hatcher and co-workers (Hatcher *et al.*, 1994a,b, 1995; Hatcher and Paul, 2000) found that fungus infection made *Rumex* plants less preferred for oviposition and consumption by the beetle and, vice versa, that plants subjected to leaf beetle damage were less prone to rust infection. The studies reviewed by Rostàs *et al.* (2003) showed a general tendency of adverse effects of plant antagonists on each other. Very little information is available on how such cross-effects affect tritrophic interactions.

How does pathogen infestation affect odor emissions and does it interfere with emissions induced by insect herbivores? So far, only one study has specifically looked at this cross-effect (Cardoza *et al.*, 2002). It showed that insect feeding (beet armyworm, *S. exigua*) and fungus infection (white mold, *Sclerotium rolfsii*) resulted in distinctly different odor blends in peanut plants, whereas plants that were simultaneously infested by these two antagonists released a mix of both blends.

Shiojiri *et al.* (2001, 2002) revealed a fascinating cross-effect resulting from simultaneous feeding by larvae of two lepidopteran species. They showed that *Plutella xylostella* and *P. rapae* caused cabbage plants to release different odor blends that could be distinguished by *Cotesia plutella*. *Costesia glomerata* females were only attracted by plants damaged by *P. xylostella* and not by those damaged by *P. rapae*, which it cannot parasitize. Interestingly, the parasitoid is also less attracted to plants infested by both herbivores. This could explain why adult *P. xylostella* females show a preference to oviposit on plants that have already been infested by *P. rapae* (Shiojiri *et al.*, 2002),

How can multiple infestations affect each other? JA has typically been assumed to be involved in induced responses to herbivory and SA was assumed to be involved in most responses to pathogen infection. The interactions are not as straightforward and various insects and pathogens differ in the defense genes they activate (Walling, 2000). Ozawa *et al.* (2000) compared the induction of volatiles in lima bean leaves by caterpillars and spider mites with induction with JA and methyl salicylate. Their results suggest that response to caterpillar feeding involves the JA-related signaling pathway and that spider mite feeding triggers both the SA- and JA-related signaling pathways. Dicke *et al.* (1999) had already shown that JA-triggered emissions in lima bean showed some differences from mite-induced emissions and concluded that the induction involves more than just JA. This might indicate that the reaction to spider mite feeding is more similar to the reaction triggered by sucking insects such as whiteflies and aphids (Walling, 2000). A crucial issue in these types of study is how elicitors are applied (Schmelz *et al.*, 2001); ideally, the treatment should reflect natural conditions. The involvement of various pathways that can be triggered differently by different plant antagonists implies that infestation by multiple organisms will add to the variability in plant responses. In light of the likelihood that plants are subject to attack by more than one adversary, it seems pertinent to further study this so-called cross-talk and its ecological implications.

Exploitation of induced defenses for biological control

The above examples illustrate how plant attributes may contribute to successful prey location by natural enemies and it has been suggested that these attributes may be exploitable in pest control (Bottrell *et al.*, 1998; Lewis *et al.*, 1998; Cortesero *et al.*, 2000). It has been long recognized that efficacy of adult natural enemies as biological agents against insect pests may be increased by supplying them with food sources. Reviews on how plant-provided nutrition may aid in biological control are presented by Hagen (1986), Whitman (1994), Jervis and Kidd (1996), and Cortesero *et al.* (2000). Several examples show that predation and parasitism are higher on plants with extrafloral nectar than on plants without extrafloral nectar (Treacy *et al.*, 1986; Pemberton and Lee, 1996). Clearly, there is the potential that the production of extrafloral nectar could be optimized to increase the efficiency of biological control agents. Such selection or manipulation programs should also account for the risk that the nectar can be exploited by phytophagous insects (Rogers, 1985; Schuster and Calderon, 1986).

After it was recognized that plant volatiles play an essential role in host location by various parasitoids, it has been suggested that emission of these cues could be manipulated to facilitate prey finding and thus improve biological control (e.g., Nordlund *et al.*, 1988; Dicke *et al.*, 1990a; Turlings and Benrey, 1998; Cortesero *et al.*, 2000). The potential of such an approach remains unexplored, but two of the above-mentioned field studies suggest that it is feasible. Thaler's (1999) treatment of tomato plants with JA increased parasitism of an important pest. Equally promising is the increased predation of *M. sexta* eggs and neonate larvae that Kessler and Baldwin (2001) observed after supplementing the odor of wild tobacco plants with synthetic volatiles or by treating wild tobacco plants with methyl jasmonate. The possibilities offered by biotechnology will certainly make it possible to tailor the odor production of crop plants. The challenge will be to determine which odor blends are most effective in attracting the right control agents and again to avoid attracting herbivores at the same time.

Genetic transformation of plants for the purpose of enhancing biological control will still be some time away. These traits, however, could be well suited for the development of methods for the evaluation of other transgenic crops, as discussed in the next section.

Evaluation of transgenic crops

Current controversy over the use of transgenic crops places much emphasis on their potential effects on non-target insects. Various direct and indirect effects on natural enemies that are important in pest control are possible (Schüler *et al.*, 1999a) and a number of studies have addressed these potential effects. Most of these studies have involved *Bt* maize (maize plants producing a *Bacillus thuringiensis* toxin) and have shown little or no negative effect of *Bt* maize (e.g., Orr and Landis, 1997; Pilcher *et al.*, 1997). A possible exception is the reduced development and increased mortality of the predatory lacewing *Chrysoperla carnae* when it consumes *Bt* maize-fed prey (Hilbeck *et al.*, 1998).

We are aware of only one study that has looked at the attractiveness of transgenic plants to natural enemies. Schüler *et al.* (1999b) studied the attractiveness of *Bt* oilseed rape to the parasitoid *Cotesia plutellae*, which attacks the diamondback moth (*P. xylostella*), an important pest. As expected, feeding by the susceptible *P. xylostella* larvae was much reduced, resulting in fewer odors being emitted and a reduced attractiveness to the parasitoids. However, the outcome of this study was favorable in the sense that caterpillar-induced emissions of attractive volatiles was highest when *Bt*-resistant *P. xylostella* larvae were feeding on the plants. The authors argue that this could reduce the development of resistance to transgenic plants in field situations (Schüler *et al.*, 1999b).

For the evaluation of possible pleiotropic effects (side-effects resulting from genes effecting more than one phenotypic trait) in transgenic crops, the analyses of plant-produced odors and exudates (such as extrafloral nectar) may be ideal. Significant changes in biochemical pathways would likely result in alterations in the production of these secondary plant metabolites. Careful analyses of a large range of conventional varieties of a particular crop would also reveal the existing natural variation and would allow for a more realistic comparison than is commonly made. The significance of any observed changes for the interactions between the crop and beneficial insects could then be tested in appropriate bioassays. In such assays, conventional crops that exhibit clear differences in the trait under investigation (e.g., attractiveness to parasitoids or nutritional value of extrafloral nectar) could serve as realistic controls. For maize, we already have ample information on the considerable variation among genotypes in the emissions of volatiles (Gouinguené *et al.*, 2001 and unpublished data). In most cases, it is not to be expected that transgenesis has a major effect on the composition of odor blends or extrafloral nectar, as the production of, for instance, *Bt* involves entirely different biochemical pathways. Herbicide resistance in maize, however, does involve the shikimate pathway (Shah *et al.*, 1986; Padgette *et al.*, 1994), which is responsible for the production of several aromatic volatiles that are induced after insect attack (Paré and Tumlinson, 1999). It would be interesting to investigate if transgenic plants with herbicide resistance produce more or less of these substances and if this has any consequences for the attraction of parasitoids and predators. It would be equally interesting to study potential changes in the carbohydrate and amino acid composition of plant nectar and honeydew resulting from transgenesis. In all cases, it will be pertinent to compare such changes with the full spectrum of existing variability and to determine the ecological relevance of the changes.

Conclusions

Extrafloral nectar and plant odors play essential roles in the protection of plants from herbivores by natural enemies. Both these traits can be inducible, but plants without insect damage may have nectaries that produce significant amounts of nectar, whereas most undamaged plants are virtually odorless compared with damaged plants. The induction of plant odor emissions is relatively specific for insect feeding, and in some cases plants respond differentially to different herbivores. So far, studies into specificity have not addressed the considerable variation in signals emitted by different plant genotypes. There appears to be a danger of overinterpreting results from experiments conducted with just one insect–plant combination. Not only can different plant genotypes differ considerably in their induced responses (e.g., Gouinguené *et al.*, 2001; Krips *et al.*, 2001), but, in addition, the arthropods that make use of plant signals and food may show variation and rapid genetic changes in their responses (Margolies *et al.*, 1997; Maeda *et al.*, 1999; Dicke *et al.*, 2000). This variation needs to be considered and included in studies on specificity.

Both indirect defenses have now been shown to function in field situations, but further field studies are needed to confirm that plants do indeed benefit from emitting induced odors in natural settings. Moreover, nothing is yet known about the cross-effects of multiple infestations on plant indirect defenses. Also still lacking are appropriate field tests for the evaluation of plant-odor manipulation to enhance the effectiveness of biological control agents. However, various studies have indicated the potential of such approaches. Current biotechnology techniques offer ample opportunities for the manipulation of indirect plant defenses, which should largely facilitate the design of experiments that can help to answer the remaining questions on their function and exploitability.

Acknowledgements

We are grateful to Hans Alborn, Göran Birgersson, Cristina Faria, Monika Frey, Jonathan Gershenzon, Bernd Hägele, Yonggen Lou, Michael Rostàs, Goede Schüler and Cristina Tamò for advice and information on specific topics that are discussed in this chapter. Cristina Tamò also provide editorial assistance. This work is in part supported by funds from the Swiss National Science Foundation (grants 31-46237-95 and 31-58865-99) and the Swiss Center of Competence in Research on "Plant Survival."

References

Adjei-Maafo, I. K. and Wilson, L. T. (1983). Factors affecting the relative abundance of arthropods on nectaried and nectariless cotton. *Environmental Entomology* **12**: 349–352.

Agelopoulos, N. A. and Keller, M. A. (1994). Plant–natural enemy association in the tritrophic system *Cotesia rubecula–Pieris rapae*–Brassicaceae (Cruciferae). III: Collection and identification of plant and frass volatile. *Journal of Chemical Ecology* **20**: 1955–1967.

Agrawal, A. A. and Karban, R. (1999). Why induced defenses may be favored over constitutive strategies in plants. In *The Ecology and Evolution of Inducible Defenses*, eds. R. Tollrian and C. D. Harvell, pp. 45–61. Princeton: Princeton University Press.

Agrawal, A. A. and Rutter, M. T. (1998). Dynamic anti-herbivore defense in ant–plants: the role of induced responses. *Oikos* **83**: 227–236.

Agrawal, A. A., Tuzun, S. and Bent, E. (1999). *Induced Plant Defenses Against Pathogens and Herbivores*. St Paul, MO: APS Press.

Ajlan, A. M. and Potter, D. A. (1991). Does immunization of cucumber against anthracnose by *Colletotrichum lagenarium* affect host suitability for arthropods. *Entomologia Experimentalis et Applicata* **58**: 83–91.

Alborn, H. T., Turlings, T. C. J., Jones, T. H., Stenhagen, G., Loughrin, J. H. and Tumlinson, J. H. (1997). An elicitor of plant volatiles from beet armyworm oral secretion. *Science* **276**: 945–949.

Alborn, H. T., Jones, T. H., Stenhagen, G. S. and Tumlinson, J. H. (2000). Identification and synthesis of volicitin and related components from beet armyworm oral secretions. *Journal of Chemical Ecology* **26**: 203–220.

Apriyanto, D. and Potter, D. A. (1990). Pathogen-activated induced resistance of cucumber: response of arthropod herbivores to systemically protected leaves. *Oecologia* **85**: 25–31.

Arimura, G., Ozawa, R., Shimoda, T., Nishioka, T., Boland, W. and Takabyashi, J. (2000a). Herbivory-induced volatiles elicit defence genes in lima bean leaves. *Nature* **406**: 512–515.

Arimura, G., Tashiro, K., Kuhara, S., Nishioka, T., Ozawa, R. and Takabayashi, J. (2000b). Gene responses in bean leaves induced by herbivory and by herbivore-induced volatiles. *Biochemical and Biophysical Research Communications* **277**: 305–310.

Arthur, A. P. (1962). Influence of host tree on abundance of *Itoplectis conquistor* (Say) (Hymenoptera: Ichneumonidae), a polyphagous parasite of the European pine shoot moth, *Ryacionia buoliana* (Schiff) (Lepidoptera: Olethreutidae). *Canadian Entomologist* **94**: 337–347.

Aukema, B. H., Dahlsten, D. L. and Raffa, K. F. (2000). Improved population monitoring of bark beetles and predators by incorporating disparate behavioral responses to semiochemicals. *Environmental Entomology* **29**: 618–629.

Baker, D. A., Hall, J. L. and Thorpe, J. R. (1978). Study of extrafloral nectaries of *Ricinus communis*. *New Phytologist* **81**: 129–137.

Bakker, F. M. and Klein, M. E. (1992). Transtrophic interactions in Cassava. *Experimental and Applied Acarology* **14**: 293–311.
Baldwin, I. T. (1994). Chemical changes rapidly induced by folivory. In *Insect–Plant Interactions*, vol. V, ed. E. A. Bernays, pp. 1–23. Boca Raton, FL: CRC Press.
Baldwin, I. T. and Preston, C. A. (1999). The eco-physiological complexity of plant responses to insect herbivores. *Planta* **208**: 137–145.
Baldwin, I. T. and Schultz, J. C. (1983). Rapid changes in tree leaf chemistry induced by damage: evidence for communication between plants. *Science* **221**: 277–279.
Barbosa, P. and Saunders, J. A. (1985). Plant allelochemicals: linkage between herbivores and their natural enemies. In *Chemically Mediated Interactions Between Plants and Other Organisms*, eds. G. A. Cooper-Driver and T. Swain, pp. 197–137. New York: Plenum Press.
Beattie, A. J. (1985). *The Evolutionary Ecology of Ant–Plant Mutualisms*. Cambridge: Cambridge University Press.
Beckage, N. E. (1985). Endocrine interactions between endo-parasitic insects and their hosts. *Annual Review of Entomology* **30**: 371–413.
Belt, T. (1874). *The Naturalist in Nicaragua*. London: J. Murray.
Bender, C., Bailey, A. M., Jones, W. *et al.* (1996). Biosynthesis and regulation of the phytotoxin coronatine in *Pseudomonas syringae*. In *Molecular Aspects of Pathogenicity and Host Resistance Requirement for Signal Transduction*, eds. D. Mills, H. Kunoh, S. Mayama and N. Keen, pp. 233–244. St Paul, MO: APS Press.
Bennett, B. and Breed, M. D. (1985). On the association between *Pentaclethra macroloba* (Mimosacea) and *Paraponera clavata* (Hymenoptera: Formicidae) colonies. *Biotropica* **17**: 253–255.
Benrey, B., Denno, R. F. and Kaiser, L. (1997). The influence of plant species on attraction and host acceptance in *Cotesia glomerata* (Hymenoptera: Braconidae). *Journal of Insect Behavior* **10**: 619–630.
Bentley, B. L. (1977). Extra-floral nectaries and protection by pugnacious bodyguards. *Annual Review of Ecology and Systematics* **8**: 407–427.
Bernasconi, M. L., Turlings, T. C. J., Ambrosetti, L., Bassetti, P. and Dorn, S. (1998). Herbivore-induced emissions of maize volatiles repel the corn leaf aphid, *Rhopalosiphum maidis*. *Entomologia Experimentalis et Applicata* **87**: 133–142.
Bernasconi Ockroy, M. L., Turlings, T. J. C., Edwards, P. J. *et al.* (2001). Response of natural populations of predators and parasitoids to artificially induced volatile emissions in maize plants (*Zea mays* L.). *Agricultural and Forest Entomology* **3**: 1–10.
Binder, B. F. and Robbins, J. C. (1996). Age- and density-related oviposition behavior of the European corn borer, *Ostrinia nubilalis* (Lepidoptera: Pyralidae). *Journal of Insect Behavior* **9**: 755–769.
Boland, W. and Gäbler, A. (1989). Biosynthesis of homoterpenes in higher-plants. *Helvetica Chimica Acta* **72**: 247–253.
Boland, W., Feng, Z., Donath, J. and Gäbler, A. (1992). Are acyclic C-11 and C-16 homoterpenes plant volatiles indicating herbivory? *Naturwissenschaften* **79**: 368–371.
Boland, W., Hopke, J., Donath, J., Nuske, J. and Bublitz, F. (1995). Jasmonic acid and coronatin induce odor production in plants. *Angewandte Chemie: International Edition in English* **34**: 1600–1602.
Boller, T. (1991). Ethylene in pathogenesis and disease resistance. In *The Plant Hormone Ethylene*, eds. A. K. Matoo and J. C. Suttle, pp. 293–314. Boca Raton, FL: CRC Press.

Bolter, C. J., Dicke, M., Vanloon, J. J. A., Visser, J. H. and Posthumus, M. A. (1997). Attraction of Colorado potato beetle to herbivore-damaged plants during herbivory and after its termination. *Journal of Chemical Ecology* **23**: 1003–1023.
Bones, A. M. and Rossiter, J. T. (1996). The myrosinase–glucosinolate system, its organisation and biochemistry. *Physiologia Plantarum* **97**: 194–208.
Bottrell, D. G., Barbosa, P. and Gould, F. (1998). Manipulating natural enemies by plant variety selection and modification: a realistic strategy? *Annual Review of Entomology* **43**: 347–367.
Bouwmeester, H. J., Verstappen, F. W. A., Posthumus, M. A. and Dicke, M. (1999). Spider mite-induced (3S)-(*E*)-nerolidol synthase activity in cucumber and lima bean. The first dedicated step in acyclic C11-homoterpene biosynthesis. *Plant Physiology* **121**: 173–180.
Broadway, R. M., Duffey, S. S., Pearce, G. and Ryan, C. A. (1986). Plant proteinase-inhibitors: a defense against herbivorous insects. *Entomologia Experimentalis et Applicata* **41**: 33–38.
Bruin, J., Dicke, M. and Sabelis, M. W. (1992). Plants are better protected against spider-mites after exposure to volatiles from infested conspecifics. *Experientia* **48**: 525–529.
Bruin, J., Sabelis, M. W. and Dicke, M. (1995). Do plants tap Sos signals from their infested neighbors. *Trends in Ecology and Evolution* **10**: 167–170.
Bugg, R. L., Ellis, R. T. and Carlson, R. W. (1989). Ichneumonidae (Hymenoptera) using extrafloral nectar of faba bean (*Vicia faba* L., Fabaceae) in Massachusetts. *Biological Agriculture and Horticulture* **6**: 107–114.
Byers, J. A. (1989). Chemical ecology of bark beetles. *Experientia* **45**: 271–283.
Cardoza, Y. J., Alborn, H. T. and Tumlinson, J. H. (2002). In vivo volatile emissions from peanut plants induced by simultaneous fungal infection and insect damage. *Journal of Chemical Ecology* **28**: 161–174.
Carroll, C. R. and Janzen, D. H. (1973). Ecology of foraging by ants. *Annual Review of Ecology and Systematics* **4**: 231–257.
Chen, H., Lou, Y. and Cheng, J. (2002). Behavioral responses of the larval parasitoid *Cotesia chilonis* to the volatiles from its host and host plant. *Acta Entomologica Sinica* **45**: 617–622.
Coleman, R. A., Barker, A. M. and Fenner, M. (1999). Parasitism of the herbivore *Pieris brassicae* L. (Lep., Pieridae) by *Cotesia glomerata* L. (Hym., Braconidae) does not benefit the host plant by reduction of herbivory. *Journal of Applied Entomology (Zeitschrift für Angewandte Entomologie)* **123**: 171–177.
Cortesero, A. M., Stapel, J. O. and Lewis, W. J. (2000). Understanding and manipulating plant attributes to enhance biological control. *Biological Control* **17**: 35–49.
Cuautle, M. and Rico-Gray, V. (2003). The effect of wasps and ants on the reproductive success of the extrafloral nectaried plant *Turnera ulmifolia* (Turneraceae). *Functional Ecology* **17**: 417–423.
De Moraes, C. M., Lewis, W. J., Pare, P. W., Alborn, H. T. and Tumlinson, J. H. (1998). Herbivore-infested plants selectively attract parasitoids. *Nature* **393**: 570–573.
De Moraes, C. M., Mescher, M. C. and Tumlinson, J. H. (2001). Caterpillar-induced nocturnal plant volatiles repel nonspecific females. *Nature* **410**: 577–580.
Degenhardt, J. and Gershenzon, J. (2000). Demonstration and characterization of (*E*)-nerolidol synthase from maize: a herbivore-inducible terpene synthase participating in (3*E*)-4,8-dimethyl-1,3,7-nonatriene biosynthesis. *Planta* **210**: 815–822.
Del-Claro, K. and Oliveira, P. S. (1993). Ant–Homoptera interaction: do alternative sugar sources distract tending ants? *Oikos* **68**: 202–206.

Dicke, M. (1986). Volatile spider-mite pheromone and host-plant kairomone, involved in spaced-out gregariousness in the spider-mite *Tetranychus urticae*. *Physiological Entomology* **11**: 251–262.

(1994). Local and systemic production of volatile herbivore-induced terpenoids: their role in plant–carnivore mutualism. *Journal of Plant Physiology* **143**: 465–472.

(1999). Are herbivore-induced plant volatiles reliable indicators of herbivore identity to foraging carnivorous arthropods? *Entomologia Experimentalis et Applicata* **91**: 131–142.

Dicke, M. and Dijkman, H. (1992). Induced defense in detached uninfested plant-leaves: effects on behavior of herbivores and their predators. *Oecologia* **91**: 554–560.

Dicke, M. and Groeneveld, A. (1986). Hierarchical structure in kairomone preference of the predatory mite *Amblyseius potentillae*: dietary component indispensable for diapause induction affects prey location behavior. *Ecological Entomology* **11**: 131–138.

Dicke, M. and Sabelis, M. W. (1988). How plants obtain predatory mites as bodyguards. *Netherlands Journal of Zoology* **38**: 148–165.

Dicke, M., Sabelis, M. W., Takabayashi, J., Bruin, J. and Posthumus, M. A. (1990a). Plant strategies of manipulating predator–prey interactions through allelochemicals: prospects for application in pest-control. *Journal of Chemical Ecology* **16**: 3091–3118.

Dicke, M., Vanbeek, T. A., Posthumus, M. A., Bendom, N., Vanbokhoven, H. and Degroot, A. E. (1990b). Isolation and identification of volatile kairomone that affects acarine predator–prey interactions: involvement of host plant in its production. *Journal of Chemical Ecology* **16**: 381–396.

Dicke, M., Vanbaarlen, P., Wessels, R. and Dijkman, H. (1993). Herbivory induces systemic production of plant volatiles that attract predators of the herbivore – extraction of endogenous elicitor. *Journal of Chemical Ecology* **19**: 581–599.

Dicke, M., Gols, R., Ludeking, D. and Posthumus, M. A. (1999). Jasmonic acid and herbivory differentially induce carnivore-attracting plant volatiles in lima bean plants. *Journal of Chemical Ecology* **25**: 1907–1922.

Dicke, M., Schutte, C. and Dijkman, H. (2000). Change in behavioral response to herbivore-induced plant volatiles in a predatory mite population. *Journal of Chemical Ecology* **26**: 1497–1514.

Dicke, M., Van Poecke, R. M. P. and De Boer, J. G. (2003). Inducible indirect defence of plants: from mechanism to ecological functions. *Basic and Applied Ecology* **4**: 27–42.

Doherty, H. M., Selvendran, R. R. and Bowles, D. J. (1988). The wound response of tomato plants can be inhibited by aspirin and related hydroxybenzoic acids. *Physiological and Molecular Plant Pathology* **33**: 377–384.

Dolch, R. and Tscharntke, T. (2000). Defoliation of alders (*Alnus glutinosa*) affects herbivory by leaf beetles on undamaged neighbours. *Oecologia* **125**: 504–511.

Domek, J. M. and Johnson, D. T. (1988). Demonstration of semiochemically induced aggregation in the green june beetle, *Cotinis nitida* (L) (Coleoptera, Scarabaeidae). *Environmental Entomology* **17**: 147–149.

Donath, J. and Boland, W. (1994). Biosynthesis of acyclic homoterpenes in higher-plants parallels steroid-hormone metabolism. *Journal of Plant Physiology* **143**: 473–478.

Drukker, B., Scutareanu, P. and Sabelis, M. W. (1995). Do anthocorid predators respond to synomones from *Psylla*-infested pear trees under field conditions. *Entomologia Experimentalis et Applicata* **77**: 193–203.

Du, Y. J., Poppy, G. M. and Powell, W. (1996). Relative importance of semiochemicals from first and second trophic levels in host foraging behavior of *Aphidius ervi*. *Journal of Chemical Ecology* **22**: 1591–1605.

Du, Y. J., Poppy, G. M., Powell, W., Pickett, J. A., Wadhams, L. J. and Woodcock, C. M. (1998). Identification of semiochemicals released during aphid feeding that attract parasitoid *Aphidius ervi*. *Journal of Chemical Ecology* **24**: 1355–1368.

Eben, A., Benrey, B., Sivinski, J. and Aluja, M. (2000). Host species and host plant effects on preference and performance of *Diachasmimorpha longicaudata* (Hymenoptera: Braconidae). *Environmental Entomology* **29**: 87–94.

Ecker, J. R. and Davis, R. W. (1987). Plant defense genes are regulated by ethylene. *Proceedings of the National Academy of Sciences, USA* **84**: 5202–5206.

Elzen, G. W., Williams, H. J. and Vinson, S. B. (1984). Isolation and identification of cotton synomones mediating searching behavior by parasitoid *Campoletis sonorensis*. *Journal of Chemical Ecology* **10**: 1251–1264.

Elzen, G. W., Williams, H. J., Bell, A. A., Stipanovic, R. D. and Vinson, S. B. (1985). Quantification of volatile terpenes of glanded and glandless *Gossypium hirsutum* L. cultivars and lines by gas-chromatography. *Journal of Agricultural and Food Chemistry* **33**: 1079–1082.

Engel, V., Fischer, M. K., Wackers, F. L. and Volkl, W. (2001). Interactions between extrafloral nectaries, aphids and ants: are there competition effects between plant and homopteran sugar sources? *Oecologia* **129**: 577–584.

Engelberth, J., Koch, T., Kuhnemann, F. and Boland, W. (2000). Channel-forming peptaibols are potent elicitors of plant secondary metabolism and tendril coiling. *Angewandte Chemie: International Edition* **39**: 1860–1862.

Engelberth, J., Koch, T., Schuler, G., Bachmann, N., Rechtenbach, J. and Boland, W. (2001). Ion channel-forming alamethicin is a potent elicitor of volatile biosynthesis and tendril coiling. Cross talk between jasmonate and salicylate signaling in lima bean. *Plant Physiology* **125**: 369–377.

Enyedi, A. J., Yalpani, N., Silverman, P. and Raskin, I. (1992). Localization, conjugation, and function of salicylic-acid in tobacco during the hypersensitive reaction to tobacco mosaic-virus. *Proceedings of the National Academy of Sciences, USA* **89**: 2480–2484.

Faegri, K. and van der Pijl, L. (1971). *The Principles of Pollination Ecology*. Oxford: Pergamon Press.

Faeth, S. H. (1994). Induced plant responses: effects on parasitoids and other natural enemies of phytophagous insects. In *Parasitoid Community Ecology*, eds. B. A. Hawkins and W. Sheehan, pp. 245–260. Oxford: Oxford University Press.

Farmer, E. E. (2001). Surface-to-air signals. *Nature* **411**: 854–856.

Farmer, E. E. and Ryan, C. A. (1990). Interplant communication: airborne methyl jasmonate induces synthesis of proteinase-inhibitors in plant-leaves. *Proceedings of the National Academy of Sciences, USA* **87**: 7713–7716.

Farmer, E. E., Johnson, R. R. and Ryan, C. A. (1992). Regulation of expression of proteinase-inhibitor genes by methyl jasmonate and jasmonic acid. *Plant Physiology* **98**: 995–1002.

Farmer, E. E., Weber, H. and Vollenweider, S. (1998). Fatty acid signaling in *Arabidopsis*. *Planta* **206**: 167–174.

Felton, G. W. and Eichenseer, H. (2000). Herbivore saliva and its effects on plant defense against herbivores and pathogens. In *Induced Plant Defenses Against Pathogens and Herbivores: Biochemistry, Ecology, and Agriculture*, eds. A. A. Agrawal, S. Tuzan and E. Bent, pp. 19–36. St Paul, MO: APS Press.

Finidori-Logli, V., Bagneres, A. G. and Clement, J. L. (1996). Role of plant volatiles in the search for a host by parasitoid *Diglyphus isaea* (Hymenoptera: Eulophidae). *Journal of Chemical Ecology* **22**: 541–558.

Fisher, B. L., Sternberg, L. D. L. and Price, D. (1990). Variation in the use of orchid extrafloral nectar by ants. *Oecologia* **83**: 263–266.
Flint, H. M., Salter, S. S. and Walters, S. (1979). Caryophyllene: an attractant for the green lacewing. *Environmental Entomology* **8**: 1123–1125.
Fowler, S. V. and Lawton, J. H. (1985). Rapidly induced defenses and talking trees: the devil's advocate position. *American Naturalist* **126**: 181–195.
Frey, M., Chomet, P., Glawischnig, E. *et al.* (1997). Analysis of a chemical plant defense mechanism in grasses. *Science* **277**: 696–699.
Frey, M., Stettner, C., Pare, P. W., Schmelz, E. A., Tumlinson, J. H. and Gierl, A. (2000). An herbivore elicitor activates the gene for indole emission in maize. *Proceedings of the National Academy of Sciences, USA* **97**: 14801–14806.
Frey-Wyssling, A. (1955). The phloem supply to the nectaries. *Acta Botanica Neerlandica* **4**: 358–369.
Fritzsche Hoballah, M. E. F. and Turlings, T. C. J. (2001). Experimental evidence that plants under caterpillar attack may benefit from attracting parasitoids. *Evolutionary Ecology Research* **3**: 553–565.
Fritzsche Hoballah, M. E. F., Tamo, C. and Turlings, T. C. J. (2002). Differential attractiveness of induced odors emitted by eight maize varieties for the parasitoid *Cotesia marginiventris*: is quality or quantity important? *Journal of Chemical Ecology* **28**: 951–968.
Fujiwara, C., Takabayashi, J. and Yano, S. (2000). Effects of host-food plant species on parasitization rates of *Mythimna separata* (Lepidoptera: Noctuidae) by a parasitoid, *Cotesia kariyai* (Hymenoptera: Braconidae). *Applied Entomology and Zoology* **35**: 131–136.
Geervliet, J. B. F., Vet, L. E. M. and Dicke, M. (1994). Volatiles from damaged plants as major cues in long-range host-searching by the specialist parasitoid *Cotesia rubecula*. *Entomologia Experimentalis et Applicata* **73**: 289–297.
(1996). Innate responses of the parasitoids *Cotesia glomerata* and *C. rubecula* (Hymenoptera: Braconidae) to volatiles from different plant–herbivore complexes. *Journal of Insect Behavior* **9**: 525–538.
Gómez, J. M. and Zamora, R. (1994). Top-down effects in a tritrophic system: parasitoids enhance plant fitness. *Ecology* **75**: 1023–1030.
Gorlach, J., Volrath, S., Knaufbeiter, G. *et al.* (1996). Benzothiadiazole, a novel class of inducers of systemic acquired resistance, activates gene expression and disease resistance in wheat. *Plant Cell* **8**: 629–643.
Gouinguené, S. (2000). Specificity and variability in induced volatile signalling in maize plants, University of Neuchâtel, Switzerland.
Gouinguené, S. and Turlings, T. C. J. (2002). The effects of abiotic factors on induced volatile emissions in corn plants. *Plant Physiology* **129**: 1296–1307.
Gouinguené, S., Degen, T. and Turlings, T. C. J. (2001). Variability in herbivore-induced odour emissions among maize cultivars and their wild ancestors (teosinte). *Chemoecology* **11**: 9–16.
Gouinguené, S., Alborn, H. and Turlings, T. C. J. (2003). Induction of volatile emissions in maize by different larval instars of *Spodoptera littoralis*. *Journal of Chemical Ecology* **29**: 145–162.
Hagen, K. S. (1986). Ecosystem analysis: plant cultivar (HPR), entomophagous species and food supplements. In *Interactions of Plant Resistance and Parasitoids and Predators of Insects*, eds. D. J. Boethel and R. D. Eikenbary, pp. 153–197. New York: John Wiley & Sons.
Halitschke, R., Schittko, U., Pohnert, G., Boland, W. and Baldwin, I. T. (2001). Molecular interactions between the specialist herbivore *Manduca sexta*

(Lepidoptera, Sphingidae) and its natural host *Nicotiana attenuata*. III. Fatty acid–amino acid conjugates in herbivore oral secretions are necessary and sufficient for herbivore-specific plant responses. *Plant Physiology* **125**: 711–717.

Harari, A. R., Benyakir, D. and Rosen, D. (1994). Mechanism of aggregation behavior in *Maladera matrida* Argaman (Coleoptera, Scarabaeidae). *Journal of Chemical Ecology* **20**: 361–371.

Hare, J. D. (2002). Plant genetic variation in tritrophic interactions. In *Multitrophic Level Interactions*, eds. T. Tscharntke and B. A. Hawkins, pp. 8–43. Cambridge: Cambridge University Press.

Harrewijn, P., Minks, A. K. and Mollema, C. (1995). Evolution of plant volatile production in insect–plant relationships. *Chemoecology* **5/6**: 55–73.

Harrington, E. A. and Barbosa, P. (1978). Host habitat influences on oviposition by *Parasetigena silvestris* (R-D) (Diptera–Tachinidae), a larval parasite of gypsy moth (Lepidoptera–Lymantriidae). *Environmental Entomology* **7**: 466–468.

Haskins, C. P. and Haskins, E. F. (1950). Notes on the biology and social behavior of the archaic ponerine ants of the genera *Myrmeca* and *Promyrmeca*. *Annals of the Entomological Society of America* **43**: 461–491.

Hatcher, P. E. (1995). Three-way interactions between plant-pathogenic fungi, herbivorous insects and their host plants. *Biological Reviews of the Cambridge Philosophical Society* **70**: 639–694.

Hatcher, P. E. and Paul, N. D. (2000). Beetle grazing reduces natural infection of *Rumex obtusifolius* by fungal pathogens. *New Phytologist* **146**: 325–333.

Hatcher, P. E., Paul, N. D., Ayres, P. G. and Whittaker, J. B. (1994a). The effect of an insect herbivore and a rust fungus individually, and combined in sequence, on the growth of 2 *Rumex* species. *New Phytologist* **128**: 71–78.

(1994b). Interactions between *Rumex* spp., herbivores and a rust fungus: *Gastrophysa viridula* grazing reduces subsequent infection by *Uromyces rumicis*. *Functional Ecology* **8**: 265–272.

Hatcher, P. E., Ayres, P. G. and Paul, N. D. (1995). The effect of natural and simulated insect herbivory, and leaf age, on the process of infection of *Rumex crispus* L. and *R. obtusifolius* L. by *Uromyces rumicis* (Schum) Wint. *New Phytologist* **130**: 239–249.

Haukioja, E., Suomela, J. and Neuvonen, S. (1985). Long-term inducible resistance in birch foliage: triggering cues and efficacy on a defoliator. *Oecologia* **65**: 363–369.

Heil, M., Fiala, B., Baumann, B. and Linsenmair, K. E. (2000). Temporal, spatial and biotic variations in extrafloral nectar secretion by *Macaranga tanarius*. *Functional Ecology* **14**: 749–757.

Heil, M., Koch, T., Hilpert, A., Fiala, B., Boland, W. and Linsenmair, K. E. (2001). Extrafloral nectar production of the ant-associated plant, *Macaranga tanarius*, is an induced, indirect, defensive response elicited by jasmonic acid. *Proceedings of the National Academy of Sciences, USA* **98**: 1083–1088.

Hilbeck, A., Baumgartner, M., Fried, P. M. and Bigler, F. (1998). Effects of transgenic *Bacillus thuringiensis* corn-fed prey on mortality and development time of immature *Chrysoperla carnea* (Neuroptera: Chrysopidae). *Environmental Entomology* **27**: 480–487.

Hölldobler, B. and Wilson, E. O. (1990). *The Ants*. Cambridge, MA: Harvard University Press.

Hopke, J., Donath, J., Blechert, S. and Boland, W. (1994). Herbivore-induced volatiles: the emission of acyclic homoterpenes from leaves of *Phaseolus lunatus* and *Zea mays* can be triggered by a beta-glucosidase and jasmonic acid. *Febs Letters* **352**: 146–150.

Horvitz, C. C. and Schemske, D. W. (1986). Seed dispersal of a neotropical myrmecochore: variation in removal rates and dispersal distance. *Biotropica* **18**: 319–323.
Ichihara, A., Shiraishi, K., Sato, H. *et al.* (1977). Structure of coronatine. *Journal of the American Chemical Society* **99**: 636–637.
Inouye, D. W. and Taylor, O. R. (1979). Temperate region plant–ant–seed predator system: consequences of extra floral nectar secretion by *Helianthella quinquenervis*. *Ecology* **60**: 1–7.
Janssen, A., Pallini, A., Venzon, M. and Sabelis, M. W. (1998). Behaviour and indirect interactions in food webs of plant-inhabiting arthropods. *Experimental and Applied Acarology* **22**: 497–521.
Janzen, D. H. (1966). Coevolution of mutualism between ants and acacias in Central America. *Evolution* **20**: 249–275.
Jervis, M. A. and Kidd, N. A. C. (1996). Phytophagy. In *Insect Natural Enemies: Practical Approaches in Their Study and Avaluation*, eds. M. A. Jervis and N. A. C. Kidd, pp. 375–394. London: Chapman & Hall.
Jervis, M. A., Kidd, N. A. C., Fitton, M. G., Huddleston, T. and Dawah, H. A. (1993). Flower-visiting by hymenopteran parasitoids. *Journal of Natural History* **27**: 67–105.
Jolivet, P. (1998). *Myrmecophily and Ant–Plants*. Boca Raton, FL: CRC Press.
Josens, R. B., Farina, W. M. and Roces, F. (1998). Nectar feeding by the ant *Camponotus mus*: intake rate and crop filling as a function of sucrose concentration. *Journal of Insect Physiology* **44**: 579–585.
Kahl, J., Siemens, D. H., Aerts, R. J. *et al.* (2000). Herbivore-induced ethylene suppresses a direct defense but not a putative indirect defense against an adapted herbivore. *Planta* **210**: 336–342.
Kalberer, N. M., Turlings, T. C. J. and Rahier, M. (2001). Attraction of a leaf beetle (*Oreina cacaliae*) to damaged host plants. *Journal of Chemical Ecology* **27**: 647–661.
Karban, R. and Baldwin, I. T. (1997). *Induced Responses to Herbivory*. Chicago, IL: University Press of Chicago.
Karban, R. and Kuc, J. (1999). Induced resistance against pathogens and herbivore: an overview. In *Induced Plant Defenses Against Pathogens and Herbivores*, eds. A. A. Agrawal, S. Tuzun and E. Bent, pp. 1–15. St Paul, MO: APS Press.
Karban, R., Agrawal, A. A. and Mangel, M. (1997). The benefits of induced defenses against herbivores. *Ecology* **78**: 1351–1355.
Karban, R., Baldwin, I. T., Baxter, K. J., Laue, G. and Felton, G. W. (2000). Communication between plants: induced resistance in wild tobacco plants following clipping of neighboring sagebrush. *Oecologia* **125**: 66–71.
Kessler, A. and Baldwin, I. T. (2001). Defensive function of herbivore-induced plant volatile emissions in nature. *Science* **291**: 2141–2144.
Kessmann, H., Staub, T., Hofmann, C. *et al.* (1994). Induction of systemic acquired disease resistance in plants by chemicals. *Annual Review of Phytopathology* **32**: 439–459.
Knoester, M., van Loon, L. C., van den Heuvel, J., Hennig, J., Bol, J. F. and Linthorst, H. J. M. (1998). Ethylene-insensitive tobacco lacks nonhost resistance against soil-borne fungi. *Proceedings of the National Academy of Sciences, USA* **95**: 1933–1937.
Koch, T., Krumm, T., Jung, V., Engelberth, J. and Boland, W. (1999). Differential induction of plant volatile biosynthesis in the lima bean by early and late intermediates of the octadecanoid-signaling pathway. *Plant Physiology* **121**: 153–162.

Koptur, S. (1979). Facultative mutualism between weedy vetches bearing extrafloral nectaries and weedy ants in California. *American Journal of Botany* **66**: 1016–1020.
(1989). Is extrafloral nectar production an inducible defence? In *Evolutionary Ecology of Plants*, eds. J. Bock and Y. Linhart, pp. 323–339. Boulder, CO: Westview Press.
(1992). Extrafloral nectary-mediated interactions between insects and plants. In *Insect–Plant Interactions*, vol. IV, ed. E. A. Bernays, pp. 81–129. Boca Raton, FL: CRC Press.
(1994). Floral and extrafloral nectars of Costa Rican *Inga* trees: a comparison of their constituents and composition. *Biotropica* **26**: 276–284.
Koptur, S. and Lawton, J. H. (1988). Interactions among vetches bearing extrafloral nectaries, their biotic protective agents, and herbivores. *Ecology* **69**: 278–283.
Koptur, S., Rico-Gray, V. and Palacios-Rios, M. (1998). Ant protection of the nectaried fern *Polypodium plebeium* in central Mexico. *American Journal of Botany* **85**: 736–739.
Krips, O. E., Willems, P. E. L., Gols, R., Posthumus, M. A., Gort, G. and Dicke, M. (2001). Comparison of cultivars of ornamental crop *Gerbera jamesonii* on production of spider mite-induced volatiles, and their attractiveness to the predator *Phytoseiulus persimilis*. *Journal of Chemical Ecology* **27**: 1355–1372.
Krivan, V. and Sirot, E. (1997). Searching for food or hosts. The influence of parasitoids behavior on host–parasitoid dynamics. *Theoretical Population Biology* **51**: 201–209.
Landolt, P. J. (1993). Effects of host plant leaf damage on cabbage-looper moth attraction and oviposition. *Entomologia Experimentalis et Applicata* **67**: 79–85.
Landolt, P. J., Tumlinson, J. H. and Alborn, D. H. (1999). Attraction of Colorado potato beetle (Coleoptera: Chrysomelidae) to damaged and chemically induced potato plants. *Environmental Entomology* **28**: 973–978.
Lanza, J. (1988). Ant preferences for *Passiflora* nectar mimics that contain amino-acids. *Biotropica* **20**: 341–344.
Leahy, T. C. and Andow, D. A. (1994). Egg weight, fecundity, and longevity are increased by adult feeding in *Ostrinia nubilalis* (Lepidoptera, Pyralidae). *Annals of the Entomological Society of America* **87**: 342–349.
Leatemia, J. A., Laing, J. E. and Corrigan, J. E. (1995). Effects of adult nutrition on longevity, fecundity, and offspring sex-ratio of *Trichogramma minutum* Riley (Hymenoptera, Trichogrammatidae). *Canadian Entomologist* **127**: 245–254.
Letourneau, D. K. (1990). Code of ant–plant mutualism broken by parasite. *Science* **248**: 215–217.
Lewis, W. J., Stapel, J. O., Cortesero, A. M. and Takasu, K. (1998). Understanding how parasitoids balance food and host needs: importance to biological control. *Biological Control* **11**: 175–183.
Lingren, P. D. and Lukefahr, M. J. (1977). Effects of nectariless cotton on caged populations of *Campoletis sonorensis* (Hymenoptera–Ichneumonidae). *Environmental Entomology* **6**: 586–588.
Loch, A. D. and Walter, G. H. (1999). Multiple host use by egg parasitoid *Trissolcus basalis* (Wollaston) in a soyabean agricultural system: biological control and environmental implications. *Agricultural and Forest Entomology* **1**: 271–280.
Loke, W. H., Ashley, T. R. and Sailer, R. I. (1983). Influence of fall armyworm, *Spodoptera frugiperda* (Lepidoptera, Noctuidae) larvae and corn plant-damage on host finding in *Apanteles marginiventris* (Hymenoptera, Braconidae). *Environmental Entomology* **12**: 911–915.
Lou, Y. and Cheng, J. (1996). Behavioral responses of *Anagrus nilaparvatae* Pang et Wang to the volatiles of rice varieties. *Entomological Journal of East China* **5**: 60–64.

Loughrin, J. H., Manukian, A., Heath, R. R., Turlings, T. C. J. and Tumlinson, J. H. (1994). Diurnal cycle of emission of induced volatile terpenoids herbivore-injured cotton plants. *Proceedings of the National Academy of Sciences, USA* **91**: 11836–11840.

Loughrin, J. H., Manukian, A., Heath, R. R. and Tumlinson, J. H. (1995a). Volatiles emitted by different cotton varieties damaged by feeding beet armyworm larvae. *Journal of Chemical Ecology* **21**: 1217–1227.

Loughrin, J. H., Potter, D. A. and Hamiltonkemp, T. R. (1995b). Volative compounds induced by herbivory act as aggregation kairomones for the japanese-beetle (*Popillia japonica* Newman). *Journal of Chemical Ecology* **21**: 1457–1467.

Maeda, T., Takabayashi, J., Yano, S. and Takafuji, A. (1999). Response of the predatory mite, *Amblyseius womersleyi* (Acari: Phytoseiidae), toward herbivore-induced plant volatiles: variation in response between two local populations. *Applied Entomology and Zoology* **34**: 449–454.

Margolies, D. C., Sabelis, M. W. and Boyer, J. E. (1997). Response of a phytoseiid predator to herbivore-induced plant volatiles: selection on attraction and effect on prey exploitation. *Journal of Insect Behavior* **10**: 695–709.

Markin, G. P. (1970). Food distribution within laboratory colonies of argentine ant, *Iridomyrmex humilis* (Mayr). *Insectes Sociaux* **17**: 127–158.

Mattiacci, L., Dicke, M. and Posthumus, M. A. (1994). Induction of parasitoid attracting synomone in brussels-sprouts plants by feeding of *Pieris brassicae* larvae: role of mechanical damage and herbivore elicitor. *Journal of Chemical Ecology* **20**: 2229–2247.

(1995). Beta-glucosidase: an elicitor of herbivore-induced plant odor that attracts host-searching parasitic wasps. *Proceedings of the National Academy of Sciences, USA* **92**: 2036–2040.

Mattoo, A. K. and Suttle, J. C. (1991). *The Plant Hormone Ethylene*. Boca Raton, FL: CRC Press.

McCall, P. J., Turlings, T. C. J., Lewis, W. J. and Tumlinson, J. H. (1993). Role of plant volatiles in host location by the specialist parasitoid *Microplitis croceipes* Cresson (Braconidae, Hymenoptera). *Journal of Insect Behavior* **6**: 625–639.

McCall, P. J., Turlings, T. C. J., Loughrin, J., Proveaux, A. T. and Tumlinson, J. H. (1994). Herbivore-induced volatile emissions from cotton (*Gossypium hirsutum* L) seedlings. *Journal of Chemical Ecology* **20**: 3039–3050.

McEwen, P. K. and Liber, H. (1995). The effect of adult nutrition on the fecundity and longevity of the alive moth *Prays oleae* (Bern). *Journal of Applied Entomology (Zeitschrift für Angewandte Entomologie)* 119: 291–294.

Meiners, T. and Hilker, M. (1997). Host location in *Oomyzus gallerucae* (Hymenoptera: Eulophidae), an egg parasitoid of the elm leaf beetle *Xanthogaleruca luteola* (Coleoptera: Chrysomelidae). *Oecologia* **112**: 87–93.

(2000). Induction of plant synomones by oviposition of a phytophagous insect. *Journal of Chemical Ecology* **26**: 221–232.

Meiners, T., Westerhaus, C. and Hilker, M. (2000). Specificity of chemical cues used by a specialist egg parasitoid during host location. *Entomologia Experimentalis et Applicata* **95**: 151–159.

Milewski, A. V. and Bond, W. J. (1982). Convergence of myrmercochory in mediterranean Australia and South Africa. In *Ant–Plant Interactions in Australia*, ed. R. C. Buckley, pp. 89–98. The Hague: Junk.

Mohyuddin, A. I., Inayatullah, C. and King, E. G. (1981). Host selection and strain occurence in *Apalantes flavipes* (Cameron) (Hymenoptera: Braconidae) and its bearing on biological control of graminaceous stem-borers (Lepidoptera: Pyralidae). *Bulletin of Entomological Research* **71**: 575–581.

Monteith, L. G. (1955). Host preferences of *Drino bohemica* Messn. (Diptera: Tachinidae) with particular reference to olfactory responses. *Canadian Entomologist* **87**: 509–530.

Moran, P. (1998). Plant-mediated interactions between insects and a fungal plant pathogen and the role of plant chemical responses to infection. *Oecologia* **115**: 523–530.

Nordlund, D. A. and Sauls, C. E. (1981). Kairomones and their use for management of entomophagous insects. 11. Effect of host plants on kairomonal activity of frass from *Heliothis zea* (Lepidoptera, Noctuidae) larvae for the parasitoid *Microplitis croceipes* (Hymenoptera, Braconidae). *Journal of Chemical Ecology* **7**: 1057–1061.

Nordlund, D. A., Lewis, L. C. and Altieri, M. A. (1988). Influences of plant-produced allelochemicals on the host/prey selection behavior of entomophagous insects. In *Novel Aspects of Insect–Plant Interactions*, eds. P. Barbosa and D. Letourneau, pp. 65–90. New York: John Wiley & Sons.

O'Dowd, D. J. (1979). Foliar nectar production and ant activity on a neotropical tree, *Ochroma pyramidale*. *Oecologia* **43**: 233–248.

O'Dowd, D. J. and Catchpole, E. A. (1983). Ants and extrafloral nectaries: no evidence for plant-protection in *Helichrysum* spp. ant interactions. *Oecologia* **59**: 191–200.

Oliveira, P. S. (1997). The ecological function of extrafloral nectaries: herbivore deterrence by visiting ants and reproductive output in *Caryocar brasiliense* (Caryocaraceae). *Functional Ecology* **11**: 323–330.

Orr, D. B. and Landis, D. L. (1997). Oviposition of European corn borer (Lepidoptera: Pyralidae) and impact of natural enemy populations in transgenic versus isogenic corn. *Journal of Economic Entomology* **90**: 905–909.

Ozawa, R., Arimura, G., Takabayashi, J., Shimoda, T. and Nishioka, T. (2000). Involvement of jasmonate- and salicylate-related signaling pathways for the production of specific herbivore-induced volatiles in plants. *Plant and Cell Physiology* **41**: 391–398.

Padgette, S. R., Re, D. B., Barry, G. F. *et al.* (1994). New weed control opportunities: development of soybeans with a Roundup Ready^{TM} gene. In *Herbicide-resistant Crops: Agricultural, Economics, Environmental, Regulatory, and Technologycal Aspects*, ed. S. O. Duke. Boca Raton, FL: CRC Press.

Pallini, A., Janssen, A. and Sabelis, M. W. (1997). Odour-mediated responses of phytophagous mites to conspecific and heterospecific competitors. *Oecologia* **110**: 179–185.

Paré, P. W. and Tumlinson, J. H. (1997). *De novo* biosynthesis of volatiles induced by insect herbivory in cotton plants. *Plant Physiology* **114**: 1161–1167.

(1999). Plant volatiles as a defense against insect herbivores. *Plant Physiology* **121**: 325–331.

Paré, P. W., Alborn, H. T. and Tumlinson, J. H. (1998). Concerted biosynthesis of an insect elicitor of plant volatiles. *Proceedings of the National Academy of Sciences, USA* **95**: 13971–13975.

Pascal, L. and Belin-Depoux, M. (1991). La correlation entre les rythmes biologiques de l'association plante-fourmis: les cas des nectaries extra-floraux de *Malpighiaceae* americaines. *Comptes Rendus Hebdomadaires des Séances de l'Académie des Sciences, Paris series III*, **312**: 49–54.

Passera, L., Lachaud, J. P. and Gomel, L. (1994). Individual food source fidelity in the neotropical ponerine ant *Ectatomma ruidum* Roger (Hymenoptera–Formicidae). *Ethology Ecology and Evolution* **6**: 13–21.

Pemberton, P. W. (1998). The occurrence and abundance of plants with extrafloral nectaries, the basis for antiherbivore defensive mutualisms, along a latitudinal gradient in east Asia. *Journal of Biogeography* **25**: 661–668.

Pemberton, R. W. and Lee, J. H. (1996). The influence of extrafloral nectaries on parasitism of an insect herbivore. *American Journal of Botany* **83**: 1187–1194.
Peng, C. W. and Weiss, M. J. (1992). Evidence of an aggregation pheromone in the flea beetle, *Phyllotreta cruciferae* (Goeze) (Coleoptera, Chrysomelidae). *Journal of Chemical Ecology* **18**: 875–884.
Pettersson, E. M. (2001). Volatiles from potential hosts of *Rhopalicus tutela* a bark beetle parasitoid. *Journal of Chemical Ecology* **27**: 2219–2231.
Pettersson, E. M., Birgersson, G. and Witzgall, P. (2001). Synthetic attractants for the bark beetle parasitoid *Coeloides bostrichorum* Giraud (Hymenoptera: Braconidae). *Naturwissenschaften* **88**: 88–91.
Picard, F. and Rabaud, E. (1914). Sur le parasitisme externe des Braconidae. *Bulletin de la Société Entomologique de France* **83**: 266–269.
Piel, J., Atzorn, R., Gabler, R., Kuhnemann, F. and Boland, W. (1997). Cellulysin from the plant parasitic fungus *Trichoderma viride* elicits volatile biosynthesis in higher plants via the octadecanoid signalling cascade. *Febs Letters* **416**: 143–148.
Pilcher, C. D., Obrycki, J. J., Rice, M. E. and Lewis, L. C. (1997). Preimaginal development, survival, and field abundance of insect predators on transgenic *Bacillus thuringiensis* corn. *Environmental Entomology* **26**: 446–454.
Pohnert, G., Jung, V., Haukioja, E., Lempa, K. and Boland, W. (1999). New fatty acid amides from regurgitant of lepidopteran (Noctuidae, Geometridae) caterpillars. *Tetrahedron* **55**: 11275–11280.
Porter, S. D. (1989). Effects of diet on the growth of laboratory fire ant colonies (Hymenoptera, Formicidae). *Journal of the Kansas Entomological Society* **62**: 288–291.
Potting, R. P. J., Vet, L. E. M. and Dicke, M. (1995). Host microhabitat location by stem-borer parasitoid *Cotesia flavipes*: the role of herbivore volatiles and locally and systemically induced plant volatiles. *Journal of Chemical Ecology* **21**: 525–539.
Powell, W., Pennacchio, F., Poppy, G. M. and Tremblay, E. (1998). Strategies involved in the location of hosts by the parasitoid *Aphidius ervi* Haliday (Hymenoptera: Braconidae: Aphidiinae). *Biological Control* **11**: 104–112.
Price, P. W., Bouton, C. E., Gross, P., Mcpheron, B. A., Thompson, J. N. and Weis, A. E. (1980). Interactions among 3 trophic levels: influence of plants on interactions between insect herbivores and natural enemies. *Annual Review of Ecology and Systematics* **11**: 41–65.
Pyke, G. H. (1991). What does it cost a plant to produce floral nectar? *Nature* **350**: 58–59.
Ramachandran, R. and Norris, D. M. (1991). Volatiles mediating plant–herbivore–natural enemy interactions: electroantennogram responses of soybean looper, *Pseudoplusia includens*, and a parasitoid, *Microplitis demolitor*, to green leaf volatiles. *Journal of Chemical Ecology* **17**: 1665–1690.
Rapusas, H. R., Bottrell, D. G. and Coll, M. (1996). Intraspecific variation in chemical attraction of rice to insect predators. *Biological Control* **6**: 394–400.
Retana, J., Bosch, J., Alsina, A. and Cerdá, X. (1987). Foraging ecology of the nectarivorous ant *Camponotus foreli* (Hymenoptera, Formicidae) in a savanna-like grassland. *Misselània Zoològica* **11**: 187–193.
Reymond, P. and Farmer, E. E. (1998). Jasmonate and salicylate as global signals for defense gene expression. *Current Opinion in Plant Biology* **1**: 404–411.
Reymond, P., Weber, H., Damond, M. and Farmer, E. E. (2000). Differential gene expression in response to mechanical wounding and insect feeding in *Arabidopsis*. *Plant Cell* **12**: 707–719.
Rhoades, D. F. (1979). Evolution of plant chemical defense against herbivores. In *Herbivores: Their Interaction with Secondary Plant Metabolites*, eds. G. A. Rosenthal and D. H. Janzen, pp. 4–54. New York: Academic Press.

efficiency of *Microplitis croceipes* (Hymenoptera: Braconidae) in cotton. *Environmental Entomology* **26**: 617–623.

Steinberg, S., Dicke, M., Vet, L. E. M. and Wanningen, R. (1992). Response of the braconid parasitoid *Cotesia* (= *Apanteles*) *glomerata* to volatile infochemicals: effects of bioassay set-up, parasitoid age and experience and barometric flux. *Entomologia Experimentalis et Applicata* **63**: 163–175.

Steinberg, S., Dicke, M. and Vet, L. E. M. (1993). Relative importance of infochemicals from 1st and 2nd trophic level in long-range host location by the larval parasitoid *Cotesia glomerata*. *Journal of Chemical Ecology* **19**: 47–59.

Stephenson, A. G. (1982). The role of the extrafloral nectaries of *Catalpa speciosa* in limiting herbivory and increasing fruit production. *Ecology* **63**: 663–669.

Stintzi, A., Weber, H., Reymond, P., Browse, J. and Farmer, E. E. (2001). Plant defense in the absence of jasmonic acid: the role of cyclopentenones. *Proceedings of the National Academy of Sciences, USA* **98**: 12837–12842.

Stout, M. J. and Bostock, R. M. (1999). Specificity of induced responses to arthropods and pathogens. In *Induced Defenses Against Pathogens and Herbivores*, eds. A. A. Agrawal, S. Tuzun and E. Bent, pp. 183–211. St Paul, MO: APS Press.

Sudd, J. H. and Franks, N. R. (1987). *The Behavioural Ecology of Ants*. New York: Chapman & Hall.

Sullivan, B. T., Pettersson, E. M., Seltmann, K. C. and Berisford, C. W. (2000). Attraction of the bark beetle parasitoid *Roptrocerus xylophagorum* (Hymenoptera: Pteromalidae) to host-associated olfactory cues. *Environmental Entomology* **29**: 1138–1151.

Swift, S. and Lanza, J. (1993). How do *Passiflora* vines produce more extrafloral nectar after simulated herbivory? *Bulletin of the Ecological Society of America* **74**: 451.

Takabayashi, J., Dicke, M. and Posthumus, M. A. (1991a). Induction of indirect defense against spider-mites in uninfested lima-bean leaves. *Phytochemistry* **30**: 1459–1462.

(1991b). Variation in composition of predator-attracting allelochemicals emitted by herbivore-infested plants: relative influence of plant and herbivore. *Chemoecology* **2**: 1–6.

(1994). Volatile herbivore-induced terpenoids in plant mite interactions: variation caused by biotic and abiotic factors. *Journal of Chemical Ecology* **20**: 1329–1354.

Takabayashi, J., Takahashi, S., Dicke, M. and Posthumus, M. A. (1995). Developmental stage of herbivore *Pseudaletia separata* affects production of herbivore-induced synomone by corn plants. *Journal of Chemical Ecology* **21**: 273–287.

Takasu, K. and Lewis, W. J. (1995). Importance of adult food sources to host searching of the larval parasitoid *Microplitis croceipes*. *Biological Control* **5**: 25–30.

Tanowitz, B. D. and Koehler, D. L. (1986). Carbohydrate analysis of floral and extra-floral nectars in selected taxa of *Sansevieria* (Agavaceae). *Annals of Botany* **58**: 541–545.

Taylor, J. S. (1932). Report on cotton insect and disease infestation. II. Notes on the American boll worm (*Heliothis obselata* F.) on cotton and its parasite (*Microbracon brevicornis* Wesm.). *Scientific Bulletin of Research in Agriculture and Forestry of the Union of South Africa*, vol. 113.

Tennant, L. E. and Porter, S. D. (1991). Comparison of diets of 2 fire ant species (Hymenoptera, Formicidae): solid and liquid components. *Journal of Entomological Science* **26**: 450–465.

Thaler, J. S. (1999). Jasmonate-inducible plant defences cause increased parasitism of herbivores. *Nature* **399**: 686–688.

Thaler, J. S., Stout, M. J., Karban, R. and Duffey, S. S. (1996). Exogenous jasmonates simulate insect wounding in tomato plants (*Lycopersicon esculentum*) in the laboratory and field. *Journal of Chemical Ecology* **22**: 1767–1781.

Thaler, J. S., Fidantsef, A. L., Duffey, S. S. and Bostock, R. M. (1999). Trade-offs in plant defense against pathogens and herbivores: a field demonstration of chemical elicitors of induced resistance. *Journal of Chemical Ecology* **25**: 1597–1609.

Thomson, N. J. (1981). Reversed animal–plant interactions: the evolution of insectivorous and ant-fed plants. *Biological Journal of the Linnean Society* **16**: 147–155.

Tilman, D. (1978). Cherries, ants and tent caterpillars: timing of nectar production in relation to susceptibility of caterpillars to ant predation. *Ecology* **59**: 686–692.

Tobin, J. E. (1994). Ants as primary consumers: diet and abundance in the Formicidae. In *Nourishment and Evolution in Insect Societies*, vol. 9, eds. J. H. Hunt and C. A. Nalepa, pp. 279–307. Boulder, CO: Westview Press.

Treacy, M. F., Benedict, J. H., Segers, J. C., Morrison, R. K. and Lopez, J. D. (1986). Role of cotton trichome density in bollworm (Lepidoptera, Noctuidae) egg parasitism. *Environmental Entomology* **15**: 365–368.

Tscharntke, T., Thiessen, S., Dolch, R. and Boland, W. (2001). Herbivory, induced resistance, and interplant signal transfer in *Alnus glutinosa*. *Biochemical Systematics and Ecology* **29**: 1025–1047.

Tumlinson, J. H., Turlings, T. C. J. and Lewis, W. J. (1992). The semiochemical complexes that mediate insect parasitoid foraging. *Agricultural Zoological Review* **5**: 221–252.

Turlings, T. C. J. and Benrey, B. (1998). Effects of plant metabolites on the behavior and development of parasitic wasps. *Ecoscience* **5**: 321–333.

Turlings, T. C. J., Tumlinson, J. H. and Lewis, W. J. (1990). Exploitation of herbivore-induced plant odors by host-seeking parasitic wasps. *Science* **250**: 1251–1253.

Turlings, T. C. J., Tumlinson, J. H., Eller, F. J. and Lewis, W. J. (1991a). Larval-damaged plants: source of volatile synomones that guide the parasitoid *Cotesia marginiventris* to the microhabitat of its hosts. *Entomologia Experimentalis et Applicata* **58**: 75–82.

Turlings, T. C. J., Tumlinson, J. H., Heath, R. R., Proveaux, A. T. and Doolittle, R. E. (1991b). Isolation and identification of allelochemicals that attract the larval parasitoid, *Cotesia marginiventris* (Cresson), to the microhabitat of one of its hosts. *Journal of Chemical Ecology* **17**: 2235–2251.

Turlings, T. C. J., Mccall, P. J., Alborn, H. T. and Tumlinson, J. H. (1993a). An elicitor in caterpillar oral secretions that induces corn seedlings to emit chemical signals attractive to parasitic wasps. *Journal of Chemical Ecology* **19**: 411–425.

Turlings, T. C. J., Waeckers, F., Vet, L. E. M., Lewis, W. J. and Tumlinson, J. H. (1993b). Learning of host-finding cues by hymenopterous parasitoids. In *Insect Learning: Ecological and Evolutionary Perspectives*, eds. D. R. Papaj and A. Lewis, pp. 51–78. New York: Chapman & Hall.

Turlings, T. C. J., Loughrin, J. H., Mccall, P. J., Rose, U. S. R., Lewis, W. J. and Tumlinson, J. H. (1995). How caterpillar-damaged plants protect themselves by attracting parasitic wasps. *Proceedings of the National Academy of Sciences, USA* **92**: 4169–4174.

Turlings, T. C. J., Lengwiler, U. B., Bernasconi, M. L. and Wechsler, D. (1998). Timing of induced volatile emissions in maize seedlings. *Planta* **207**: 146–152.

Turlings, T. C. J., Alborn, H. T., Loughrin, J. H. and Tumlinson, J. H. (2000). Volicitin, an elicitor of maize volatiles in oral secretion of *Spodoptera exigua*: isolation and bioactivity. *Journal of Chemical Ecology* **26**: 189–202.

Turlings, T. C. J., Gouinguené, S., Degen, T. and Fritzsche Hoballah, M. E. (2002). The chemical ecology of plant–caterpillar–parasitoid interactions. In *Multitrophic Level Interactions*, eds. T. Tscharntke and B. Hawkins, pp. 148–173. Cambridge: Cambridge University Press.

Udayagiri, S. and Jones, R. L. (1992a). Flight behavior of *Macrocentrus grandii* Goidanich (Hymenoptera, Braconidae), a specialist parasitoid of European corn borer (Lepidoptera, Pyralidae): factors influencing response to corn volatiles. *Environmental Entomology* **21**: 1448–1456.

(1992b). Role of plant odor in parasitism of European corn borer by braconid specialist parasitoid *Macrocentrus grandii* Goidanich: isolation and characterization of plant synomones eliciting parasitoid flight response. *Journal of Chemical Ecology* **18**: 1841–1855.

van der Meijden, E. and Klinkhamer, P. G. L. (2000). Conflicting interests of plants and the natural enemies of herbivores. *Oikos* **89**: 202–208.

van Loon, J. J. A., de Vos, E. W. and Dicke, M. (2000a). Orientation behaviour of the predatory hemipteran *Perillus bioculatus* to plant and prey odours. *Entomologia Experimentalis et Applicata* **96**: 51–58.

van Loon, J. J. A., de Boer, J. G. and Dicke, M. (2000b). Parasitoid–plant mutualism: parasitoid attack of herbivore increases plant reproduction. *Entomologia Experimentalis et Applicata* **97**: 219–227.

van Tol, R. W. H. M., van der Sommen, A. T. C., Boff, M. I. C., van Bezooijen, J., Sabelis, M. W. and Smits, P. H. (2001). Plants protect their roots by alerting the enemies of grubs. *Ecology Letters* **4**: 292–294.

Venzon, M., Janssen, A. and Sabelis, M. W. (1999). Attraction of a generalist predator towards herbivore-infested plants. *Entomologia Experimentalis et Applicata* **93**: 305–314.

Vet, L. E. M. and Dicke, M. (1992). Ecology of infochemical use by natural enemies in a tritrophic context. *Annual Review of Entomology* **37**: 141–172.

Vet, L. E. M., Wäckers, F. L. and Dicke, M. (1991). How to hunt for hiding hosts: the reliability-detectability problem in foraging parasitoids. *Netherlands Journal of Zoology* **41**: 202–213.

Vet, L. E. M., Lewis, W. J. and Carde, R. T. (1995). Parasitoid foraging and learning. In *Chemical Ecology of Insects 2*, eds. R. T. Carde and W. J. Bell, pp. 65–101. New York: Chapman & Hall.

Vinson, S. B. (1968). Distribution of an oil carbohydrate and protein food source to members of imported fire ant colony. *Journal of Economic Entomology* **61**: 712–714.

(1981). Habitat location. In *Semiochemicals: Their Role in Pest Control*, eds. D. A. Nordlund, W. J. Lewis and R. L. Jones, pp. 51–77. New York: John Wiley & Sons.

Vinson, S. B., Elzen, G. W. and Williams, H. J. (1987). The influence of volatile plant allelochemicals on the third trophic level (parasitoids) and their herbivorous hosts. In *Insect–Plants*, eds. V. Labeyerie, G. Fabres and D. Lachaise, pp. 109–114. Dordrecht: Junk.

Wäckers, F. L. (1994). The effect of food-deprivation on the innate visual and olfactory preferences in the parasitoid *Cotesia rubecula*. *Journal of Insect Physiology* **40**: 641–649.

(2001). A comparison of nectar- and honeydew sugars with respect to their utilization by the hymenopteran parasitoid *Cotesia glomerata*. *Journal of Insect Physiology* **47**: 1077–1084.

(2003). The effect of food supplements on parasitoid–host dynamics. *Proceedings of the International Symposium on Biological Control of Arthropods* **1**: 226–231.

Wäckers, F. L. and Wunderlin, R. (1999). Induction of cotton extrafloral nectar production in response to herbivory does not require a herbivore-specific elicitor. *Entomologia Experimentalis et Applicata* **91**: 149–154.

Wäckers, F. L., Zuber, D., Wunderlin, R. and Keller, F. (2001). The effect of herbivory on temporal and spatial dynamics of foliar nectar production in cotton and castor. *Annals of Botany* **87**: 365–370.

Wadhams, L. J., Birkett, M. A., Powell, W. and Woodcock, C. M. (1999). Aphids, predators, and parasitoids. In *Insect–Plant Interactions and Induced Plant Defences*, eds. D. J. Chadwick and J. A. Goode, pp. 60–67. London: John Wiley & Sons.

Wagner, D. (1997). The influence of ant nests on *Acacia* seed production, herbivory and soil nutrients. *Journal of Ecology* **85**: 83–93.

Walling, L. L. (2000). The myriad plant responses to herbivores. *Journal of Plant Growth Regulation* **19**: 195–216.

Wegener, R., Schulz, S., Meiners, T., Hadwich, K. and Hilker, M. (2001). Analysis of volatiles induced by oviposition of elm leaf beetle *Xanthogaleruca luteola* on *Ulmus minor*. *Journal of Chemical Ecology* **27**: 499–515.

Weiler, E. W., Kutchan, T. M., Gorba, T., Brodschelm, W., Niesel, U. and Bublitz, F. (1994). The *Pseudomonas* phytotoxin coronatine mimics octadecanoid signaling molecules of higher-plants. *Febs Letters* **345**: 9–13.

Weissbecker, B., van Loon, J. J. A., Posthumus, M. A., Bouwmeester, H. J. and Dicke, M. (2000). Identification of volatile potato sesquiterpenoids and their olfactory detection by the two-spotted stinkbug *Perillus bioculatus*. *Journal of Chemical Ecology* **26**: 1433–1445.

Wertz, P. and Downing, D. (1989). Beta-glucosidase activity in porcine epidermis. *Biochimica et Biophysica Acta* **978**: 115.

Whitman, D. W. (1988). Allelochemical interactions among plants, herbivores, and their predators. In *Novel Aspects of Insect–Plant Interactions*, eds. P. Barbosa and D. Letourneau, pp. 207–248. New York: John Wiley & Sons.

(1994). Plant bodyguards: mutualistic interactions between plants and the third trophic level. In *Functional Dynamics of Phytophagous Insects*, ed. T. N. Ananthakrishan, pp. 207–248. New Dehli: Oxford and IBH Publishing.

Winz, R. A. and Baldwin, I. T. (2001). Molecular interactions between the specialist herbivore *Manduca sexta* (Lepidoptera, Sphingidae) and its natural host *Nicotiana attenuata*. IV. Insect-induced ethylene reduces jasmonate-induced nicotine accumulation by regulating putrescine *N*-methyltransferase transcripts. *Plant Physiology* **125**: 2189–2202.

Xu, N. and Chen, Z. (1999). Isolation and identification of tea plant volatiles attractive to tea geometrid parasitoids. *Acta Entomologica Sinica* **42**: 126–131.

Yu, S. (1989). Beta-glucosidase in four phytophagous Lepidoptera. *Insect Biochemistry* **19**: 103.

Zangerl, A. R. and Bazzaz, F. A. (1992). Theory and pattern in plant defense allocation. In *Plant Resistance to Herbivores and Pathogens*, eds. R. Fritz and E. L. Simms, pp. 363–392. Chicago, UL: University of Chicago Press.

Zangerl, A. R. and Rutledge, C. E. (1996). The probability of attack and patterns of constitutive and induced defense: a test of optimal defense theory. *American Naturalist* **147**: 599–608.

Zimmermann, M. (1932). Ueber die extra-floralen Nectarien der Angiospermen. *Botanisches Zentralblatt* **49**: 99–196.

Zwölfer, H. and Kraus, M. (1957). Biocoenotic studies on the parasites of two fir and two oak tortricids. *Entomophaga* **2**: 173–196.

3

Chemical ecology of astigmatid mites

Yasumasa Kuwahara
Division of Applied Life Sciences, Kyoto University, Japan

Introduction

Astigmatid mites form a suborder of the Acari, in the class Arachnida. Adult astigmatid mites are mostly oval or rod-like, less than 1 mm in length, and are opaque or transparent. Some species are economically important pests that attack a wide range of stored products and agricultural crops in fields and greenhouses. Others are common components of "house dust" and may contribute to health problems, causing atopic dermatitis and bronchial asthma. Most species, together with species in the suborder Oribatida, also function as scavengers of organic debris.

Pheromonal communication appears widespread among astigmatid mites. To date, the structures of 88 compounds, consisting of 26 monoterpenes, two sesquiterpenes, eight aromatic compounds, four aldehydes, a ketone, two novel fatty acids, a novel alkyl formate, and 14 fatty acid esters, have been conclusively identified from a total of 61 species of astigmatid mites belonging to 10 families, including 29 species that have not yet been formally described. Those unidentified species have been deduced to the genus level and are listed by the genus name with isolate names in parentheses, if necessary, such as *Histiostoma* sp. "shisetsu." Many of the compounds are found in a number of different species, in which they may have different behavioral roles. For example, compounds that function as alarm pheromones in 19 species also form part of the aggregation pheromone blend in four species, and the sex pheromone in 14 species. All of these semiochemicals, except lardolure (see below), appear to be emitted from the paired opisthonotal glands found in all astigmatid species. To date, six species have been shown to use two different types of pheromone, such as aggregation and alarm pheromones, aggregation and sex pheromones, and alarm and sex pheromones. In four of these species, different chemical components elicit each of the two different behavioral

Advances in Insect Chemical Ecology, ed. R. T. Cardé and J. G. Millar. Published by Cambridge University Press.

responses mediated by the different pheromone types. In the remaining two species, a single blend of two components triggers two different behavioral responses, with the response elicited dependent on dose and concentration.

Life history

The propagative forms of astigmatid mites (larvae, protonymph, tritonymph and adult) are weakly sclerotized and prefer humid conditions, whereas two types of deutonymph, phoretic and non-phoretic hypopi, are more robust because they must be able to survive less-favorable conditions. Phoretic hypopi are usually morphologically specialized and are a dispersive stage that attaches to a specific and predictable arthropod host to colonize patchily distributed new habitats. Their bodies are extensively sclerotized to withstand long periods of environmental stress while in transit. The change from the propagative forms to phoretic hypopi also is marked by a change in their hydrocarbon profiles from tridecane (**65**) to a mixture of (*Z*)-8-heptadecene (**77**) and (*Z*,*Z*)-6,9-heptadecadiene (**78**) in many species (Kuwahara *et al.*, 1994a), possibly to prevent desiccation (Fig. 3.1). The phoretic association to a particular host species is so specific that it is used as one of the criteria for mite identification. In contrast, non-phoretic hypopi represent a dormant form that often remains within the protonymphal cuticle, waiting passively for more favorable living conditions. Some non-phoretic hypopi are resistant to desiccation (Hughes, 1976).

Mite rearing methods

Most astigmatid mite species can be reared on moistened dry yeast at 20 °C under humid conditions in plastic petri dishes. Stored product mites and house dust mites prefer ~70% relative humidity (RH), whereas soil mites prefer 100% RH, and mites belonging to the Histiostomatidae and sections of the genus *Schwiebea* live submerged in water. To maintain humidity, rearing dishes are kept in zip-locked plastic bags. Other species, such as *Dermatophagoides farinae, Dermatophagoides pteronyssinus, Blomia tropicalis* and *Lardoglyphus konoi* (Sato *et al.*, 1993a), are maintained on diets of dried powdered fish and dry yeast (1:1), and *Carpoglyphus lactis* is reared on a mixture of sugar and yeast (1:1) (Kuwahara *et al.*, 1991a).

Chemistry of mite exudates

Opisthonotal glands

Astigmata and some Oribatida spp. are known to possess a pair of well-developed secretory glands, the opisthonotal glands, located in the opisthosoma

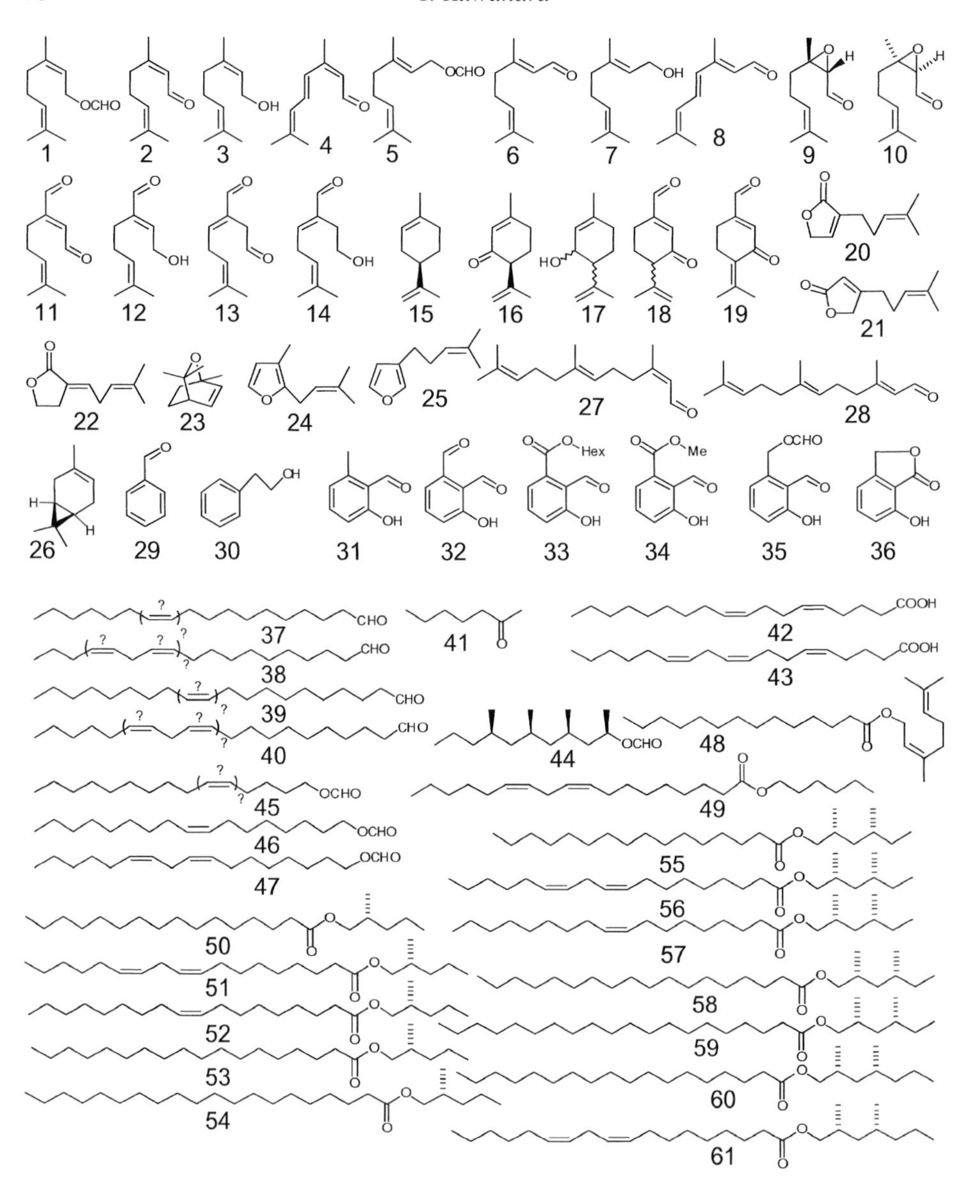

Fig. 3.1. Compounds found in astigmatid mites. **1**, Neryl formate (3,7-dimethyl-(*Z*)-2,6-octadienyl formate); **2**, neral (3,7-dimethyl-(*Z*)-2,6-octadienal); **3**, nerol (3,7-dimethyl-(*Z*)-2,6-octadienol); **4**, dehydroneral (3,7-dimethyl-(2*Z*,4*E*)-2,4,6-octatrienal); **5**, geranyl formate (3,7-dimethyl-(*E*)-2,6-octadienyl formate); **6**, geranial (3,7-dimethyl-(*E*)-2,6-octadienal); **7**, geraniol (3,7-dimethyl-(*E*)-2,6-octadienol); **8**, dehydrogeranial (3,7-dimethyl-(2*E*,4*E*)-2,4,6-octatrienal); **9**, (2*S*,3*S*)-epoxyneral (2*S*,3*S*)-epoxy-3,7-dimethyl-6-octenal); **10**, (2*R*,3*R*)-epoxyneral (2*R*,3*R*)-epoxy-3,7-dimethyl-6-octenal); **11**, α-acaridial (2(*E*)-(4-methyl-3-pentenyl)-butenedial); **12**, α-acariolal (2(*E*)-(4-methyl-3-pentenyl)-4-hydroxybutenal); **13**, β-acaridial (2(*E*)-(4-methyl-3-pentenylidene)-butanedial); **14**, (2(*E*)-(4-methyl-3-pentenylidene)-4-hydroxybutanal); **15**, limonene; **16**,

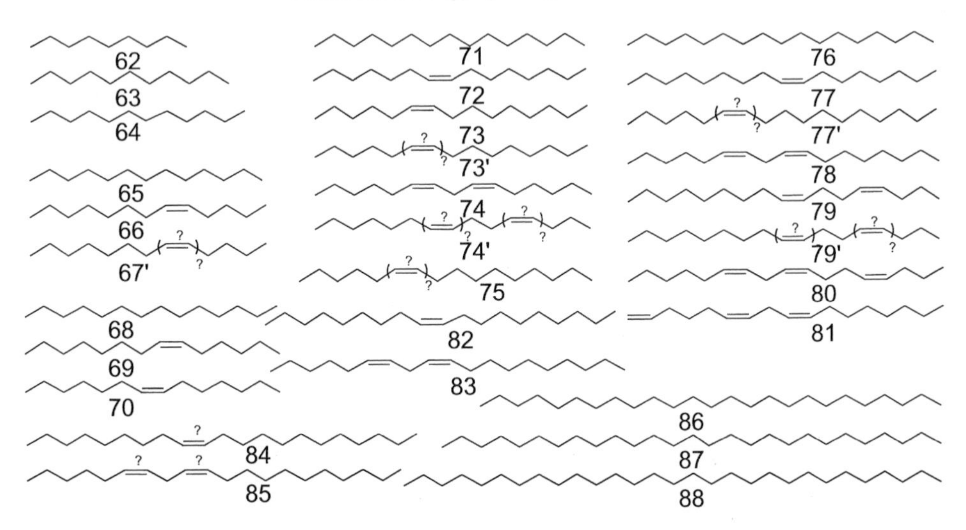

Fig. 3.1. (*cont.*) (*S*)(+)-isopiperitenone (3-methyl-6-(*S*)-isopropenyl-2-cyclohexen-1-one); **17**, isopiperitenol (3-methyl-6-isopropenyl-2-cyclohexen-1-ol); **18**, isorobinal (4-isopropenyl-3-oxo-1-cyclohexene-1-carbaldehyde); **19**, robinal (3-oxo-4-isopropylidene-1-cyclohexene-1-carbaldehyde); **20**, α,α-acariolide (3-(4′-methyl-3′-pentenyl)-2(5*H*)-furanone); **21**, α,β-acariolide (4-(4′-methyl-3′-pentenyl)-2(5*H*)-furanone); **22**, β-acariolide ((*E*)-2-(4′-methyl-3′-pentenylidene)-4-butanolide); **23**, dehydrocineole; **24**, rosefuran (3-methyl-2-(3-methyl-2-butenyl)furan); **25**, perillene (3-(4-methyl-3-pentenyl)furan); **26**, 3-carene; **27**, (*Z,E*)-farnesal; **28**, (*E,E*)-farnesal; **29**, benzaldehyde; **30**, β-phenylethanol; **31**, 2-hydroxy-6-methylbenzaldehyde; **32**, 3-hydroxybenzene-1,2-dicarbaldehyde; **33**, hexyl rhizoglyphinate (hexyl 2-formyl-3-hydroxybenzoate); **34**, methyl rhizoglyphinate (methyl 2-formyl-3-hydroxybenzoate); **35**, rhizoglyphinyl formate (2-formyl-3-hydroxybenzyl formate); **36**, 7-hydroxyphthalide (7-hydroxy isobenzofuranone); **37**, octadecenal; **38**, octadecadienal; **39**, eicosenal; **40**, eicosadienal; **41**, 2-heptanone; **42**, (*Z,Z*)-5,9-octadecadienoic acid; **43**, (*Z,Z,Z*)-5,9,12-octadecatrienoic acid; **44**, (*R,R,R,R*)-lardolure (1*R*,3*R*,5*R*,7*R*)-1,3,5,7-tetramethyldecyl formate); **45**, pentadecenyl formate; **46**, (*Z*)-8-heptadecenyl formate; **47**, (*Z,Z*)-8,11-heptadecadienyl formate; **48**, neryl myristate; **49**, hexyl linolate; **50**, (*S*)-2-methylpentyl palmitate; **51**, (*S*)-2-methylpentyl linolanate; **52**, (*S*)-2-methylpentyl linolate; **53**, (*S*)-2-methyl oleate; **54**, (*S*)-2-methyl stearate; **55**, (2*S*,4*S*)-2,4-dimethylhexyl palmitate; **56**, (2*S*,4*S*-2,4-dimethylhexyl linolanate; **57**, (2*S*,4*S*)-2,4-dimethylhexyl linolate; **58**, (2*S*,4*S*)-2,4-dimethylhexyl oleate; **59**, 2*S*,4*S*-2,4-dimethylhexyl stearate; **60**, (2*S*,4*S*)-2,4-dimethylheptyl stearate; **61**, (2*S*,4*S*)-2,4-dimethylheptyl linolate; **62**, nonane; **63**, undecane; **64**, dodecane, **65**, tridecane; **66**, (*Z*)-5-tridecene; **67′**, tridecene; **68**, tetradecane; **69**, (*Z*)-6-tetradecene; **70**, (*Z*)-7-tetradecene; **71**, pentadecane; **72**, (*Z*)-6-pentadecene; **73**, (*Z*)-7-pentadecene; **74**, (*Z,Z*)-6,9-pentadecadiene; **74′**, pentadecadiene; **75**, hexadecene; **76**, heptadecane; **77**, (*Z*)-8-heptadecene; **77′**, heptadecene; **78**, (*Z,Z*)-6,9-heptadecadiene; **79**, (*Z,Z*)-4,8-heptadecadiene; **79′**, heptadecadiene; **80**, (*Z,Z,Z*)-4,8,11-heptadecatriene; **81**, (*Z,Z*)-1,6,9-heptadecatriene; **82**, (*Z*)-9-nonadecene; **83**, (*Z,Z*)-6,9-nonadecadiene; **84**, heneicosene; **85**, heneicosadiene; **86**, pentacosane; **87**, heptacosane; **88**, nonacosane.

(Krantz, 1978). The gland consists of an orifice that opens onto the opisthosomal surface and a lumen connected to the orifice. In *D. farinae* and *D. pteronyssinus*, the gland is composed of a single cell, with an undulating cuticular lining approximately 0.1 μm thick on the luminal surface (Tongu *et al.*, 1986). Many mitochondria with well-developed cristae and a discontinuous membrane-like structure are distributed over the cytoplasm. The structure of the glands is probably similar in all mites in which they are found. The lumen of the gland usually is observable as a pair of highly refractive and colorless spots under a binocular microscope. In *Rhizoglyphus* spp., the oily liquid contents of the gland can be observed to ooze from the gland orifice, which is closed by a cuticular "hinged trapdoor" (Howard *et al.*, 1988), and to spread over the body surface. The gland exudate can be collected on a filter paper point. The gas chromatographic (GC) profile from the hexane rinse of the filter paper was the same as that from a hexane extract of mite bodies, prepared by dipping for 3 min. Curiously, many species of Astigmata are more impervious to organic solvents than insects. For example, it takes an astonishing 45 min to kill *Rhizoglyphus robini* by dipping in *n*-hexane, whereas insects die within seconds. The difference may lie in the relative scarcity of orifices on the mites, which hinders the penetration of solvent into the body. Based on the above observations and facts, it can be reasonably assumed that the GC profile of the hexane extract from a certain species is a relatively accurate reflection of the opisthonotal gland components. Furthermore, because the blend of compounds in the gland exudate is species specific, identification of each species from its chemical profile is feasible.

Compounds in the gland

Compounds present in the opisthonotal glands of Astigmata are listed in decreasing order of abundance by GC analyses in Table 3.1, and the structures are illustrated in Fig. 3.1. The table includes data on 61 species from 10 families, including 29 unidentified species, together with the semiochemical functions of the components, if known. Of the 88 compounds in Table 3.1, most have been conclusively identified by synthesis or comparison with authentic compounds. A minority of the compounds remain to be completely identified, as indicated in the table. These 88 compounds can be classified into the following nine groups: monoterpenes (**1–26**), sesquiterpenes (**27**, **28**), aromatics (**29–36**), aliphatic aldehydes (**37–40**), ketone (**41**), unusual fatty acids (**42**, **43**), aliphatic formates (**44–47**), fatty acid esters (**48–61**), and hydrocarbons (**62–88**). As an indication of the remarkable semiochemistry of these tiny animals, the following 28 compounds were first isolated and identified from Astigmata, with the structures unequivocally confirmed by synthesis: α-acaridial (**11**; Leal *et al.*, 1989c; Suzuki *et al.*, 1992), α-acariolal (**12**; Shimizu *et al.*, 2003), β-acaridial (**13**; Leal *et al.*, 1989d), β-acariolal (**14**; Shimizu

Table 3.1. *Taxonomy of Astigmata, and compounds identified in mite exudates or extracts*[a]

Taxonomy	Compounds identified[a]	Reference to biological function[b]	Reference to compounds identified and profile reported[b]
Pyroglyphoidea			
Pyroglyphidae			
Dermatophagoides farinae	**71**, **2**, **1**, **31** (sex, for male)*, **32**, **6**, **65**, **68**, **72**	Tatami *et al.*, 2001*	Kuwahara *et al.*, 1990; Kuwahara 1997
Dermatophagoides pteronyssinus	**6** (aggregation)*, **1**, **78**, **77**, **32**, **5**, **35**, **71**, **2**, **76**, **72**	Unpublished data*	Kuwahara *et al.*, 1990; Kuwahara, 1997; Sato *et al.*, 1993b†
Histiostomatoidea			
Histiostomatidae			
Histiostoma laboratorium	**6** (alarm)*, **77′**, **73′**, **79′**, **71**, **65**, **2**	Kuwahara *et al.*, 1991b*	
Histiostoma cyntandrae	**6**, **65**, **32**, **77′**, **71**		
Histiostoma piscium	**65**, **6**, **71**, **77′**, **32**		
Histiostoma sp. "shisetsu"	**6**, **32**, **65**, **77′**, **71**, **2**		
Histiostoma sp. "zyueki"	**6**, **26**, **77′**, **73′**, **79′**		
Sarasseniopus (*Anoelus*) *hughesi*	**6**		
Hemisarcoptoidea			
Carpoglyphidae			
Carpoglyphus lactis	**65**, **2** (alarm)*, **78**†, **32**, **6**, **36**, **71**, **73**, **88**, **87**, **86**	*Kuwahara *et al.*, 1980a*,b*, 1992a†	

(*cont.*)

Table 3.1. (*cont.*)

Taxonomy	Compounds identified[a]	Reference to biological function[b]	Reference to compounds identified and profile reported[b]
Chaetodactylidae			
Chaetodactylus nipponicus	**31**, **6**, **2**		
Sennertia sp.	**2**, **6**, **1**, **32**, **5**		
Winterschmidtiidae			
Oulenzia sp.	**65**, **2** (alarm)*, **64**, **71**, **6**, **73′**, **32**, **36**†, **3**	Unpublished data*	Shimizu and Kuwahara, 2001†
Glycyphagoidea			
Glycyphagidae			
Glycyphagus domesticus	**65**, **2** (alarm)*, **1**, **32**, **6**, **86**, **87**, **88**	Kuwahara *et al.*, 1991d*	
Echimyopodidae			
Blomia tropicalis	**2**, **65**, **6**		
Acaroidea			
Suidasiidae			
Suidasia medanensis	**83**, **2** (alarm)*, **82**, **32**, **27**†, **6**, **28**†, **1**, **85**, **84**, **35**	Leal *et al.*, 1989a*	Leal *et al.*, 1989a†
Suidasia sp.	**77′**, **79′**, **31**, **2**, **32**, **6**		
Tortonia sp.	**78** (alarm)*, **79**, **77**, **2**, **1**, **80**, **72**, **6**, **32**, **32**, **65**, **42**†, **43**†,	Kuwahara *et al.*, 1995a*	Kuwahara *et al.*, 1995b†

Tortonia sp. "aomori"	**79′**, **2**, **77′**, **1**, **32**, **6**		
Lardoglyphidae			
Lardoglyphus konoi	**2** (alarm)*, **65**, **6**, **44** (aggregation)†	Kuwahara *et al.*, 1980b*†, 1991a†; 1994b†; My-Yen *et al.*, 1980b†	Kuwahara *et al.*, 1980a*, 1982; Mori and Kuwahara, 1986a,b†
Acaridae, Acarinae			
Aleuroglyphus ovatus	**31** (sex, for male)*, **48**†, **65**, **13**, **2**, **32**, **9**, **6**	Kuwahara *et al.*, 1992b*; Shibata *et al.*, 1998*	Leal *et al.*, 1988a†
Acarus siro	**65**, **25**, **31**, **63**, **64**		Curtis *et al.*, 1982
Acarus immobilis	**65**, **31** (sex, for male)*, **24**†, **32** hydrocarbon mixture (sex, for female, **68**, **71**, **76–78**, **86–88**)*	Sato *et al.*, 1993c*; Shibata *et al.*, 1998*	Sato and Kuwahara, 1999†; Noguchi *et al.*, 1997*
Tyrophagus putrescentiae	**13** (sex, for male)†‡, **73**, **72**, **65**, **49**, **71**, **74**, **69**, **70**, **1** (alarm)*, **24**, **66**, **68**, **2** (alarm)*, **31**, **22**, **11**	Kuwahara *et al.*, 1975a*, 1979; My-Yen, 1980a*; unpublished data†	Kuwahara *et al.*, 1988a$; Leal *et al.*, 1989d‡; Morino *et al.*, 1997§
Tyrophagus neiswanderi	**73** (alarm)*, **72** (alarm)*, **13**, **65**, **24**†, **74**, **71**, **49**, **25**†, **11**, **1**, **66** (alarm)*, **32**, **68**, **69** (alarm)*, **70** (alarm)*	Kuwahara *et al.*, 1989b*	Leal *et al.*, 1989e†
Tyrophagus similis	**16** (alarm)*, **65**, **49**, **2**, **32**, **68**, **19**, **18**, **6**, **64**	Kuwahara *et al.*, 1987*	
Tyrophagus perniciosus	**65**, **31** (alarm)*, **9**†, **49**, **11**, **1**, **2**, **13**, **6**	Leal *et al.*, 1988b*	Leal *et al.*, 1989b†,c†; Suzuki *et al.*, 1992‡

(*cont.*)

Table 3.1. (*cont.*)

Taxonomy	Compounds identified[a]	Reference to biological function[b]	Reference to compounds identified and profile reported[b]
Tyrophagus longior	**65**, **13** (alarm)*, **49**, **11**, **73′**, **68**, **64**	Noguchi *et al.*, 1998*	
Tyrophagus sp.	**65**, **2**, **31**, **13**, **49**, **32**, **73**, **72**, **74′**, **71**, **6**, **1**, **67′**, **36**, **68**		
Tyroborus lini	**73′**, **16**, **49**, **65**, **1** (alarm)*, **2**, **71**, **18**, **74′**, **19**	Unpublished data	
Histiogaster sp.	**2** (sex, for male)*, **13**, **65**, **6**, **51**, **3**, **11**	Hiraoka *et al.*, 2002*	
Histiogaster sp. "A096"	**65**, **29**, **41**, **32**, **4**[†], **8** (alarm)*[†], **24**	Unpublished data*	Hiraoka *et al.*, 2001[†]
Histiogaster rotundus	**79′**, **65**, **32**, **1** (alarm)*, **11**, **13**, **2**, **24**, **77′**, **71**, **3**, **25**, **68**, **36**, **6***	Unpublished data*	
Tyreophagus sp. "shisetsu"	**77′**, **79′**, **2**, **65**, **31**, **6**, **32**, **71**		
Tyreophagus sp. "takeda"	**2**, **79′**, **71**, **77′**, **31**, **65**, **73′**, **6**, **32**		
Rhizoglyphus robini	**65**, **11** (sex, for male)[†], **1** (alarm)*, **13**, **19**[$], **33**[‡], **32**, **18**, **2**[§], **24**	Kuwahara *et al.*, 1988b*; unpublished data[†]; Baker and Krantz, 1984[§]	Leal *et al.*, 1990a[‡],b[§]
Rhizoglyphus setosus	**11**, **65**, **32**, **18** (sex, for male)[†], **13**, **19**, **1** (alarm)*	Akiyama *et al.*, 1997*; unpublished data[†]	
Rhizoglyphus sp. "oki"	**40**, **65**, **11**, **1** (alarm)*, **32**, **39**, **18**[†], **19**, **13**, **35**, **6**, **7**, **3**, **20**[‡], **21**[‡], **25**	Akiyama *et al.*, 1997*	Sakata *et al.*, 1996[†]; Kuwahara *et al.*, 1992c; Tarui *et al.*, 2002[‡]

Rhizoglyphus sp. "mori"	**11** (alarm)*, **65**, **1** (alarm)*, **32**, **18**, **13**, **33**†, **2** (alarm)*, **24**, **19**, **25**	Akiyama *et al.*, 1997*	Sakata and Kuwahara, 2001†
Rhizoglyphus sp. "risyoku"	**77′**, **13**, **79′**, **22**, **37**, **11**, **38**, **46**, **47**, **73′**, **76**		
Rhizoglyphus sp. "ryuudaikenaga"	**31**, **65**, **2**, **32**, **73′**		
Schwiebea elongata	**65**, **1**, **2** (aggregation†, alarm*), **32**, **6**, **31**	Kuwahara *et al.*, 2001*; unpublished data†	
Schwiebea similis	**65**, **11** (sex, for male)*, **1**, **32**, **13**, **2**, **36**	Unpublished data*	
Schwiebea sp. "okabe"	**11**, **65**, **1**, **32**, **2**, **13**, **36**, **6**, **31**, **25**		
Schwiebea sp. "chiba"	**77′**, **1** (sex, for female)*, **32**, **79′**, **15**, **6**, **76**	Unpublished data*	
Schwiebea araujoae (line, niwa)	**79′**, **20**†, **18**, **2**, **11**, **23**, **16**, **25**, **77′**, **13**, **19**	Tarui *et al.*, 2002†	
Schwiebea sp. "yoshida"	**2**, **77′**, **65**, **32**, **79′**, **1**, **3**, **71**, **6**		
Cosmoglyphus hughesi	**31** (sex, for male)*, **65**, **24**, **32**, **36**	Ryono *et al.*, 2001†	
Cosmoglyphus sp. "takeda"	**31**, **65**, **24**, **32**		
Caloglyphus spinitarsus	**31**, **65**, **24**		
Caloglyphus rodriguezi	**31**, **63** (sex, for male)*, **65**, **23**†, **32**	Mori *et al.*, 1995*, 1998a	Ayorinde *et al.*, 1984†
Sancassania shanghaiensis	**31**, **13**, **10**, **58**, **24**, **51**, **11**, **56**, **62**, **50**, **52**, **53**, **54**, **55**, **57**, **59**, **61**, **60**		Sakata *et al.*, 2001

(*cont.*)

Table 3.1. (*cont.*)

Taxonomy	Compounds identified[a]	Reference to biological function[b]	Reference to compounds identified and profile reported[b]
Caloglyphus sp. "sapporo"	**13**, **53**, **24**, **58**, **11**		
Caloglyphus sp. "kyotobyoinn"	**24**, **13**, **58**, **51**		
Caloglyphus sp. "amami"	**77′**, **13**, **79′**, **38**, **37**, **76**, **16**, **2**, **23**, **11**, **73′**, **75**		
Caloglyphus polyphyllae	**78**, **77**, **13** (sex, for male*, aggregation†), **81**$, **73′**, **11**, **12**‡, **24**, **65**, **71**, **76**, **14**§	Leal *et al.*, 1989f*; Shimizu *et al.*, 2001*	Shimizu *et al.*, 1999$; unpublished data‡§
Caloglyphus moniezi	**78**, **77**, **13**, **71**, **11**		
Caloglyphus sp. "MJ"	**78**, **77**, **10** (sex, for male)*, **13**, **76**, **11**, **73′**	Mori *et al.*, 1996*	Mori and Kuwahara, 1995*
Caloglyphus sp. "sasagawa"	**77**, **46**, **47**, **24** (sex, for male)*, **34**‡, **78**, **62**, **32**, **73′**, **45**, **30** (aggregation)†	Unpublished data*; Kuwahara, 1990†	Unpublished data‡
Caloglyphus sp. "kouchi-chiba"	**24**, **23**, **13**, **16**, **2**, **11**		
Caloglyphus sp. "HP"	**24** (sex, for male)*, **52**, **53**, **77′**, **32**, **23**	Mori *et al.*, 1998b*	
Caloglyphus sp. "sakata-new"	**24**, **52**, **23**		

[a]Compounds are listed in decreasing order as determined by gas chromatographic profiles (RIC traces), obtained with an HP 5989B mass spectrometer coupled to an HP 5890 series II+ gas chromatograph, using an HP-5 capillary column (0.32 mm × 30 m, 0.33 μm in film thickness), with a temperature program of 60 °C/2 min, 10 °C/min to 290 °C, hold for 5 min, using helium carrier gas at a flow rate of 1.23 ml/min. Unidentified species are listed by the genus name with "isolate names," if neccesary.

[b]Compound numbers correspond to structures shown in Fig. 3.1, and compounds detected in extracts are listed in decreasing order of abundance for each species, as determined by gas chromatographic analysis.

[c]Relevant citations are identified for those compounds where the biological activity (function) has been tested in bioassays and where a chemical structure (or profile) has been identified using the symbols *, †, ‡, $, § to link compound number and citation.

et al., 2003), isorobinal (**18**; structure elucidated by Sakata *et al.*, 1996), robinal (**19**; identification, Leal *et al.*, 1990b; synthesis, Kuwahara *et al.*, 1992c), α,α-acariolide (**20**; Tarui *et al.*, 2002), α,β-acariolide (**21**; Tarui *et al.*, 2002), β-acariolide (**22**; Morino *et al.*, 1997), dehydrocineole (**23**; Ayorinde *et al.*, 1984), 3-hydroxybenzene-1,2-dicarbaldehyde (**32**; Sakata and Kuwahara, 2001), hexyl and methyl rhizoglyphinate (**33** and **34**; Leal *et al.*, 1990a), rhizoglyphinyl formate (**35**; Sato *et al.*, 1993b), 7-hydroxyphthalide (**36**; Shimizu and Kuwahara, 2001), (*R*,*R*,*R*,*R*)-lardolure (**44**; identification and synthesis, Kuwahara *et al.*, 1982; stereochemistry, Mori and Kuwahara, 1986a,b), and fatty acid esters of (*S*)-2-methylpentanol (**50–54**), (2*S*,4*S*)-2,4-dimethylhexanol (**55–59**), and (2*S*,4*S*)-2,4-dimethylheptanol (**60**, **61**) (Sakata *et al.*, 2001).

Distribution of compounds among mite species

Exudates from histiostomatid mites are characterized by having geranial (**6**) as the major component. Likewise, hexyl linoleate (**49**) is the common component among species belonging to the genera *Tyrophagus* (Kuwahara *et al.*, 1988a) and *Tyroborus*. For these and other species, each species produces species-specific combinations of compounds, and GC profiles obtained by a standardized procedure provide a reliable method of identifying species. Furthermore, some species produce species-specific compounds. For example, *Aleuroglyphus ovatus* exudates are characterized by neryl myristate (**48**; Leal *et al.*, 1988a), *Caloglyphus polyphyllae* by (Z,Z)-1,6,9-heptadecatriene (**81**; Shimizu *et al.*, 1999), *Caloglyphus rodriguezi* by undecane (**63**; Mori *et al.*, 1995), and *Tyrophagus similis* by (*S*)(+)-isopiperitenone (**16**; Kuwahara *et al.*, 1987).

Functions and usage of compounds other than as semiochemicals

In addition to functioning as pheromones transmitting intraspecific signals, a number of mite-produced compounds appear to have other types of biological activity. For example, neral (**2**) and geranial (**6**) (Okamoto *et al.*, 1981), α- and β-acaridial (**11** and **13**; Kuwahara *et al.*, 1989a), and hexyl rhizoglyphinate (**33**; Leal *et al.*, 1990a) have antifungal properties. Alpha-acaridial (**11**) is also a strong sensitizer of atopic dermatitis, as demonstrated by skin patch tests (Sakurai *et al.*, 1997; Nakayama *et al.*, 1998).

Mite volatiles also are used as kairomones: neral (**2**), geranial (**6**), neryl formate (**1**) and tridecane (**65**) from the mold mite *Tyrophagus putrescentiae* function as an attractant for both sexes of the parasitoid *Lariophagus distinguendus*, which attacks the granary weevil *Sitophilus granarius* in wheat grains (Ruther and Steidle, 2000). The mold mite is attracted to a mixture of 2-alkanones (2-heptanone, 2-octanone,

2-nonanone, and 8-nonen-2-one) and 3-methylbutanol present in cheddar cheese (Yoshizawa *et al.*, 1970, 1971), and to *cis*- and *trans*-octa-1,5-dien-3-ol present in *Trichothecium roseum* and mushroom (Vanhaelen *et al.*, 1979, 1980). Several species of Astigmata are also attracted to common fatty acids and their esters, possibly as food attractants (Sato *et al.*, 1993a), and these compounds have been used to attract house dust mites (Ninomiya and Kawasaki, 1988).

The detection of a mite-produced chemical, tridecane (**65**), has also been used for practical purposes in the detection of infestations of mites such as *Acarus siro, Aeroglyphus robustus*, and *Lepidoglyphus destructor* in bin-stored wheat (Tuma *et al.*, 1990).

Biosynthesis of mite compounds

All of the monoterpenes (**1–26**) might reasonably be expected to be derived from the mevalonate pathway, as in other organisms, and there is some proof of this from several studies. For example, carbon-13 from 1-[^{13}C]-glucose was incorporated into the 2, 4, 6, and 8 positions in the chain, and the 3- and 7-methyl branches of neral (**2**) by *Carpoglyphus lactis* fed a mixture of 1-[^{13}C]-glucose and yeast (1:1) (unpublished data). (*R*,*R*,*R*,*R*)-Lardolure ((1*R*,3*R*,5*R*,7*R*)-1,3,5,7-tetramethyldecyl formate (**44**), see below) and the (*S*)-methyl-substituted alcohol moieties of esters (**50–61**) likely arise from polyketide pathways, from the condensation of methylmalonyl-coenzyme A units. Similar compounds with multiple methyl branches have been identified from other taxa. For example, 2,4,6,8-tetramethyldecanoic acid is a major component of the wax produced from the uropygial gland of geese (Simpson, 1980). Furthermore, insect pheromones such as serricornin (Chuman *et al.*, 1979), stegobinone (Kuwahara *et al.*, 1978), and the pheromones of nitidulid beetles (Bartelt and Weisleder, 1996) are also thought to be biosynthesized from polyketide precursors assembled from propionate instead of acetate units.

Summary of pheromonal functions

Although our knowledge of pheromones used by Astigmata is still limited, some patterns have begun to emerge. To date, all known pheromones of Astigmata fall into three classes: alarm pheromones, aggregation pheromones, and sex pheromones. Among the 61 species examined so far, sex pheromones are known for 14 species, alarm pheromones for 19 species, and aggregation pheromones in four species. As discussed below, two different types of pheromone have been found in six species, including two species that use a single compound for two different pheromonal functions, depending on concentration and the context in which it is produced.

Limited evidence from microsurgical ablation experiments suggests that semiochemicals are perceived by the external scapular setae (Leal *et al.*, 1989f). The morphology of the setae has been characterized by scanning electron microscopy (Leal and Mochizuki, 1990).

Alarm pheromones

Detection of alarm pheromone

An alarm pheromone, neryl formate (**1**), was demonstrated with the mold mite *T. putrescentiae* (Kuwahara *et al.*, 1975a, 1979) by observing the escape behavior of nearby mites in the culture after either disturbing or squashing mites with a paint brush or needle point; the squashed bodies were characterized by a grassy odor. Observations of escape behavior caused by exposing mites to higher concentrations or doses of compounds than they are likely to encounter in a natural context must be treated with caution. For example, *D. farinae* and *Aleuroglyphus ovatus* exhibited escape behavior upon exposure to citral (mixture of **2** and **6**; 1000 ppm soaked into a 3 mm × 3 mm filter paper, corresponding to approximately 1 μg) and neryl formate (**1**; 10 000 ppm, 10 μg). In contrast, neither species exhibited escape behavior in response to hemolymph from a squashed mite (Kuwahara *et al.*, 1980b).

Identification of the gland emitting the pheromone

Many species of astigmatid mite emit species-specific alarm pheromones when they are disturbed. This is easily demonstrated by the species-specific odor change produced upon shaking a petri dish of mites. This odor change can be documented more methodically by comparing the profile of volatiles produced by undisturbed and disturbed colonies. For example, disturbed colonies of the mold mite *T. putrescentiae* were found to produce an average of 102 times as much of the alarm pheromone neryl formate (**1**) as undisturbed colonies (Kuwahara *et al.*, 1979).

The lumen parts of the opisthonotal gland in mold mite specimens mounted in Faure's liquid are often found to develop yellow to brownish black colors after storage for more than 1 week, although the mechanism of this color development is obscure. Intensities of color developed in the lumen parts of the gland after 2 months of storage were evaluated and compared using 100 randomly sampled mite specimens from undisturbed and disturbed colonies (described as above). In the disturbed mite group, 80% of the specimens did not develop color, with the remainder showing light color development. In contrast, only 52% of the undisturbed mites showed no color, and the average color development in the remaining specimens was much darker than in the undisturbed mites. These results suggested that color development was related to the gland contents (Kuwahara *et al.*, 1979). Later, neral (**2**) and geranial (**6**) were identified as minor components of the alarm pheromone (My-Yen *et al.*, 1980a). Based on this fact, and without recognizing the presence of the other major aldehyde component β-acaridial (**13**; Leal *et al.*, 1989d), the gland was successfully stained purple by Purpald, a diagnostic reagent for detection of aldehydes, or black by treatment with silver nitrate, which produces a black residue (My-Yen *et al.*, 1980a). Similar staining experiments demonstrated the presence of aldehydes in the opisthonotal glands of *D. farinae, L. konoi, C. lactis* and *Al. ovatus* (Kuwahara *et al.*, 1980b), and neral (**2**) and geranial (**6**) were also identified in exudates from these species (Kuwahara *et al.*, 1980a). Neral (**2**) was also demonstrated to function as an alarm pheromone in *L. konoi* and *C. lactis* (Kuwahara *et al.*, 1980b).

These results suggest that the paired opisthonotal glands are the source of the alarm pheromone of the mold mite and related species of astigmatid and oribatid mites that possess opisthonotal glands. Scanning electron micrographs of the gland orifices of three species (*T. putrescentiae, Al. ovatus* and *C. lactis*) have revealed a cuticular, hinged "trapdoor" closure, which appears to regulate the release of gland contents (Howard *et al.*, 1988).

Bioassay methods for alarm pheromones

In early studies, the criterion for pheromonal activity was designated as the lowest concentration (ppm) of the candidate compound in a 10-fold dilution series that elicited a demonstrable response from test mites. Activity was expressed in terms of the dose of the test compound applied to the filter paper dispenser. More recently, more sophisticated bioassays using arenas have been developed, in which the movements of mites toward or away from test stimuli can be quantified (Nishimura *et al.*, 2002).

Chemistry of alarm pheromones

Since the first identification of an alarm pheromone for the mold mite *T. putrescentiae* as neryl formate (**1**) and determination of its threshold concentration to be in the 10–100 ppm range (Kuwahara *et al.*, 1975a), alarm pheromones have been identified and bioassayed in 20 out of the 61 species belonging to 7 out of 10 families of astigmatid mites that have been examined (Table 3.2). The alarm pheromone components represent a variety of chemical classes, including a monoterpene ester (neryl formate (**1**)), aldehydes (geranial (**6**), neral (**2**), α- and β-acaridials (**11** and **13**), and dehydrogeranial (**8**)), a ketone ((*S*)-(+)-isopiperitenone (**16**)), straight-chain alkenes ((*Z*)-6- and (*Z*)-7-tetradecenes and pentadecenes (**69**, **70**, **72**, and **73**),

Table 3.2. *Alarm pheromone components of Astigmatid mites*

Compound(s)	Dose	Species	Reference
Neryl formate (**1**)	10 ppm	*Tyrophagus putrescentiae*	Kuwahara *et al.*, 1975a
	0.05 ng	*Tyrobolus lini*	Unpublished data
	1 ng	*Histiogaster rotundus*	Unpublished data
	1 ppm	*Rhizoglyphus robini*	Kuwahara *et al.*, 1988b
	10 ng		Akiyama *et al.*, 1997
	100 ng	*Rhizoglyphus setosus*	Akiyama *et al.*, 1997
	10 ng	*Rhizoglyphus* sp. "oki"	Akiyama *et al.*, 1997
Neryl formate (**1**) + α-acaridial (**11**) + citral (**6** + **2**)	100 ng	*Rhizoglyphus* sp. "mori"	Akiyama *et al.*, 1997
Neral (**2**)	100–1000 ppm	*Carpoglyphus lactis*	Kuwahara *et al.*, 1980b
	10–100 ppm	*Glycyphagus domesticus*	Kuwahara *et al.*, 1991c
	10 ppm	*Suidasia medanensis*	Leal *et al.*, 1989a
	1 ppm	*Lardoglyphus konoi*	Kuwahara *et al.*, 1980b
	10–100 ng	*Schwiebea elongata*	Kuwahara *et al.*, 2001
		Oulenzia sp.	Unpublished data
Geranial (**6**)	1 ppm	*Histiostoma laboratorium*	Kuwahara *et al.*, 1991b
(*Z*)-6- and (*Z*)-7-Pentadecene (**72** + **73**), (*Z*)-6- and (*Z*)-7-tetradecene (**69** + **70**) + (*Z*)-5-tridecene (**66**)	100 ppm	*Tyrophagus neiswanderi*	Kuwahara *et al.*, 1989b
(*Z,Z*)-6,9-Heptadecadiene (**78**)	100 ppm	*Tortonia* sp.	Kuwahara *et al.*, 1995a
2-Hydroxy-6-methyl-benzaldehyde (**31**)	10 ppm	*Tyrophagus perniciosus*	Leal *et al.*, 1988d
Beta-acaradial (**13**)	50 ng	*Tyrophagus longior*	Noguchi *et al.*, 1998
(*S*)-(+)-Isopiperitenone (**16**)	100 ppm	*Tyrophagus similis*	Kuwahara *et al.*, 1987
(2*E*,4*E*)-Dehydrocitral (**8**)	1 ng	*Histiogaster* sp. "A096"	Unpublished data

Z,Z-6,9-heptadecadiene (**78**)), and an aromatic salicylaldehyde analog (2-hydroxy-6-methylbenzaldehyde (**31**)) (Table 3.1). All are volatile compounds with molecular masses ranging from 150 to 236 Da, and effective doses range from a few nanograms to about 1 μg. From a biological viewpoint, the chemical structures of the alarm pheromones may be conserved within a genus, as is the case in the genus *Rhizoglyphus*, in which all four species examined to date use the same compound (neryl formate (**1**)) as their alarm pheromone (Akiyama *et al.*, 1997). Conversely, alarm pheromones may be species specific, for example in the genus *Tyrophagus*, in which all five species examined produce quite different pheromones (Noguchi *et al.*, 1998). For a number of other species and genera, the situation is unclear; for example, geranial (**6**), which is the alarm pheromone of *Histiostoma laboratorium* (Kuwahara, *et al.*, 1991b), is commonly found as the major compound in exudates from Histiostomatidae, and it is likely an alarm pheromone component for congeneric species.

A pioneering study by Baker and Krantz (1984) demonstrated the use of an alarm pheromone for *R. robini* (identified at the time as neral (**2**) plus geranial (**6**)) in combination with acaricides for mite control. The results appeared promising in principle, but in practice, use of mite alarm pheromones for control of pests infesting stored products may not be feasible because of the strong and often offensive odors and flavors of the compounds.

Structure–activity relationships for alarm pheromones

Structure–activity studies have explored the effects of structural variations on the biological activity of alarm pheromone components for only one species, the mold mite *T. putrescentiae*. This mite uses neryl formate (**1**) as an alarm pheromone component, but the mite also responds to the analogs neral (**2**) and neryl acetate. In contrast, geranyl formate (**5**), geranial (**6**) and geranyl acetate, in which the conformation of the 2-(*Z*)-trisubstituted double bond is reversed to the 2-(*E*)-conformation, are totally inactive. Similarly, citronellal, citronellol, and citronellyl formate, in which the 2-(*Z*)-double bond has been reduced, and 3,7-dimethyloctyl formate, in which both double bonds have been reduced, were also inactive. Furthermore, the non-natural monoterpenoid 3,7-dimethyl-(*Z*)-2-octenyl formate, in which the double bond in the 6-position is now saturated, exhibited activity comparable to that of neryl formate (**1**) (Kuwahara, 1982). A further group of 15 model compounds modified at the methyl group on carbon 3 or with modified carbon chains, all with the basic (*Z*)-allylic formate structure, were also active, with the relative activities also being correlated with volatilities. Conversely, the corresponding (*E*)-isomers are inactive (Kuwahara and Sakuma, 1982a,b). These results indicate the vital importance of the 2-(*Z*)-trisubstituted double bond moiety in the molecule, while also indicating a surprising amount of plasticity in the terminal functional group. As might be expected for an airborne pheromone, volatility also affected the overall responses.

As discussed below, 1,3,7-trimethyl-(*Z*)-2,6-octadienyl formate and 1,3,7-trimethyl-(*Z*)-2-octenyl formate, prepared as mimics of the aggregation pheromone structure, had alarm pheromone-like activity, whereas 1,3,7-trimethyloctyl formate and simplified 1-methylalkyl formates showed aggregation pheromone-like activity with *C. lactis*, whose alarm pheromone is neral (**2**) (Honma *et al.*, 1995). The evidence suggests that for those species which use neral (**2**), geranial (**6**), or neryl formate (**1**) as alarm pheromones, the presence of (*Z*)- or (*E*)-allylic double bonds in the molecules are essential for biological activity.

In total, 41 out of 61 species examined possess neral (**2**) or geranial (**6**) as a major component of exudates. Determination of which isomer is the active principle is complicated by the fact that neral and geranial are easily isomerized to give an equilibrium mixture of the two aldehydes. Furthermore, the aldehydes can be enzymatically reduced to the corresponding alcohols during extraction of mites in solvents (e.g., hexane, benzene, and ether) for extended periods (>1 h) (Kuwahara *et al.*, 1983; Kizawa *et al.*, 1993). Characterization of the enzyme(s) is in progress. Given the dependence of biological activity on the specific conformation, it seems likely that mites produce either neral (**2**) or geranial (**6**) selectively and store the compound in the opisthonotal glands. To estimate the natural neral/geranial ratio, and minimize isomerization and reduction, extracts should be prepared by quick dips (3 min) of mites in hexane, followed by immediate GC analysis.

Aggregation pheromones

The aggregation pheromones of insects can be classified into two groups: those emitted by males or females to attract both sexes simultaneously for resource exploitation and/or mating (e.g., bark beetles: Byers, 1989) and those emitted to attract and arrest individuals of specific developmental stages, as found in the German cockroach *Blattella germanica* (Ishii and Kuwahara, 1967, 1968; Sakuma and Fukami, 1990, 1993).

Lardolure

Evidence for mite aggregation pheromones was first recognized during bioassays of alarm pheromone activity, using squashed mite bodies as the stimulus (Kuwahara *et al.*, 1980b). The hemolymph from *Lardoglyphus konoi* was found to induce aggregation after the alarm pheromone neral (**2**) present in the hemolymph had dissipated. Aggregation behavior was displayed not only by conspecifics but also by three related species: *C. lactis, Al. ovatus* and *T. putrescentiae* (Kuwahara *et al.*, 1980b). Hexane extracts of *L. konoi* bodies proved to be active with these species but not against *D. farinae* (My-Yen *et al.*, 1980b). Using *C. lactis* as test animals, the active compound (0.4 mg) was isolated from the hexane extract of separated mite bodies (280 ml) or from a mixture of mite bodies (66 ml) and culture medium (115 ml) (My-Yen *et al.*, 1980b), suggesting that the aggregation pheromone accumulated in the culture medium.

Following a lengthy purification procedure, a larger quantity (13.6 mg) of the pheromone was isolated from 25 g hexane-soluble materials extracted from 1.23 liters of mites and culture medium. The pheromone, active at 0.1 ppm against *C. lactic*, was identified as 1,3,5,7-tetramethyldecyl formate (**44**), with the structure confirmed by synthesis (Kuwahara *et al.*, 1982). The stereochemistry of lardolure was established as (1*R*,3*R*,5*R*,7*R*)-1,3,5,7-tetramethyldecyl formate (**44**, abbreviated to (*R*,*R*,*R*,*R*)-lardolure), by comparison of the natural pheromone with synthetic stereoisomers prepared in a partially stereo-controlled manner (Mori and Kuwahara, 1986b). The (*R*,*R*,*R*,*R*)-enantiomer displayed the same optical rotatory dispersion (ORD) spectrum and bioactivity as natural lardolure (Mori and Kuwahara, 1986a). (*R*,*R*,*R*,*R*)-Lardolure elicited aggregation of *L. konoi* (active at 10 ppm), and the biological activity was suppressed by the (*S*,*S*,*S*,*S*)-enantiomer. Surprisingly, (*R*,*R*,*R*,*R*)-lardolure elicted aggregation of *C. lactis* (as the kairomone) at much lower levels (active at 0.1 ppm), and *C. lactis* was insensitive to the unnatural enantiomer (Kuwahara *et al.*, 1991a). In further structure–activity studies, six compounds with the same alkyl residue as (*R*,*R*,*R*,*R*)-lardolure (**44**) and 13 with modified alkyl chains were bioassayed for pheromone activity against *L. konoi*, and for kairomonal activity against *C. lactis*. None of the analogs exhibited pheromonal activity, whereas a broad spectrum of kairomonal activity, up to one-tenth that of (*R*,*R*,*R*,*R*)-lardolure (**44**), was displayed by *C. lactis* (Kuwahara *et al.*, 1994b).

Because the structure of lardolure (**44**) bears some resemblance to neryl formate (**1**), further kairomonal structure–activity trials were conducted with *C. lactis* using analogs of neryl formate, including 1,3,7-trimethyl-(*Z*)-2,6-octadienyl formate, 1,3,7-trimethyl-(*Z*)-2-octenyl formate, 1,3,7-trimethyl-6-octenyl formate, 1,3,7-trimethyloctyl formate and 1-methylalkyl formates. All compounds without 2-(*Z*)-double bonds elicited aggregation responses, whereas compounds containing 2-(*Z*)-double bonds elicited responses typical of alarm pheromone activity. Conversely, all compounds possessing a 2-(*E*)-double bond moiety possessed neither alarm nor aggregation pheromone-like activity (Honma *et al.*, 1995).

Aggregation pheromones have been identified from four other astigmatid mite species, as mentioned below, including one unpublished case. Beta-phenylethanol (**30**), active at 100 ppm, was identified as a minor component of extracts of the unidentified *Caloglyphus* sp. "sasagawa" (Kuwahara, 1990). Curiously, this compound inhibits the attraction of another mite species, the bulb mite *R. robini*, to a mixture of alcohols produced by *Fusarium* cultures (Shinkaji *et al.*, 1988).

Levinson *et al.* (1991) reported that a nitrogenous waste, guanine (2-amino-6-hydroxypurine), functioned as an "aggregation pheromone" of the flour mite *Ac. siro*, together with ammonia as a kairomone. However, because guanine is non-volatile, it should probably be classified as an arrestant rather than an aggregation pheromone.

A number of other instances of mite aggregation phenomena are known, such as aggregation of hibernating *D. farinae* protonymphs (Reka *et al.*, 1992), aggregation of immobile deutonymphs (cyst-like hipopis) of *Chaetodactylus nipponicus* in the nest of the hornfaced bee *Osmia cornifrones*, and the aggregation of phoretic astigmatid deutonymphs on their specific carrier animal. However, the active principles mediating the formation of these aggregations remain to be identified.

Compounds with multiple functions

There is some evidence that a compound may serve different signaling functions depending on the context and concentration in which it is released. For example, neral (**2**) functions as an alarm pheromone for *Schwiebea elongata* at higher doses (Kuwahara *et al.*, 2001), whereas at lower doses, it has been shown to attract (and possibly arrest) females (Nishimura *et al.*, 2002).

In a second example, β-acaridial (**13**) was the first mite female sex pheromone identified for *C. polyphyllae* (Leal *et al.*, 1989f). At the same doses (1 ng) it also acts as an aggregation pheromone, causing mites to aggregate in groups when they are disturbed, such as when they are transferred to a fresh rearing dish (Shimizu *et al.*, 2001). Beta-acaridial (**13**) is present in the opisthonotal gland at 1–2 ng per mite, so a significant fraction of the contents of the reserve in the lumen must be discharged to stimulate the aggregation response.

Sites of pheromone production

Although neral (**2**), β-acaridial (**13**), and possibly β-phenylethanol (**30**) are components of the opisthonotal gland exudates, as mentioned above, the site of production of lardolure (**44**) is not yet known. In particular, lardolure is not detectable in hexane extracts prepared by dipping 1–30 mites (depending on body size) in 4 μl hexane for 3 min, indicating that it is not present on the cuticle or in the opisthonotal glands. Lardolure may arise from the digestive tract, as is the case with the putative pheromone of *Ac. siro*, guanine (Levinson *et al.*, 1991).

Mite sex pheromones

Chemistry

There is limited evidence of both male- and female-produced sex pheromones in astigmatid mites, with female sex pheromones probably mediating a greater proportion of reproductive activities than male-produced pheromones. At present, there are indications of male sex pheromones for two *Acarus* spp., which also have female-produced sex pheromones (Levinson *et al.*, 1989; Sato *et al.*, 1993c).

The male-produced sex pheromone that functions to attract females in *Acarus immobilis* is composed of a mixture (active at 100–1000 ng) of the male-specific hydrocarbons pentadecane (**71**), heptadecane (**76**), (*Z*)-8-heptadecene (**77**), and (*Z,Z*)-6,9-heptadecadiene (**78**), plus several other hydrocarbons (tridecane (**65**), pentacosane (**86**), heptacosane (**87**), and nonacosane (**88**)) present in extracts from both sexes. The female-produced sex pheromone that functions as a courtship-stimulating pheromone for males consists of 2-hydroxy-6-methylbenzaldehyde (**31**), which is active in 100–1000 ng doses (Sato *et al.*, 1993c). Although these doses are about 25 times greater than those naturally produced by the mites, fractions containing these compounds were the only ones that elicited behavioral responses from test animals. Consequently, it seems likely that the bioassay methods used in these experiments may not have been appropriate.

Reproductive behaviors of other astigmatid mites appear to be mediated by pheromonal components that are present in extracts from both sexes and from nymphs, but which only appear to elicit behavioral responses from males. Sex pheromones have been identified from nine species. For example, males, females, and nymphs of *A. immobilis* produce 2-hydroxy-6-methylbenzaldehyde (**31**) (2.86 ng/female, 3.78 ng/male and 0.22 ng/nymph; Sato *et al.*, 1993c). This compound elicited male mounting responses in bioassays at doses of 100 ng. The same compound was identified from *Al. ovatus* (12 ng/female, 16 ng/male, trace amounts in nymphs), and as for *A. immobilis*, it only elicited bioassay responses from males at unnaturally high doses of 0.5 μg (Kuwahara *et al.*, 1992b). Much larger amounts were identified from *Cosmoglyphus hughesi* (283 ng/female, 135 ng/male, 3.8 ng/protonymph), and in this case, biologically relevant doses of 100 ng were stimulatory to males (Ryono *et al.*, 2001). Recently, the same compound was found as the female sex pheromone of the house dust mite, *D. farinae* (6.5 ng/female, 0.8 ng/male, active within range of 5–50 ng dose) (Tatami *et al.*, 2001). A total of four species among nine shares the same compound as the pheromone.

In contrast to the cases above, the other five species each possess a different compound as their pheromone. Beta-acaridial (**13**) has been identified from a *Caloglyphus* species, *C. polyphyllae* (1.96 ng/female, 1.16 ng/male), and characteristic behavioral responses were elicited from males at 1 ng levels (Leal *et al.*,

1989f; Shimizu *et al.*, 2001). Undecane (**63**) from *C. rodriguezi* (14.2 ng/female, 2.9 ng/male, 0.066 ng/nymph) elicited responses from males at doses of 1 ng (Mori *et al.*, 1995). Responses to low, biologically relevant doses of mite-produced compounds also were obtained with two other *Caloglyphus* spp. The first, unidentified *Caloglyphus* sp. "MJ," produced (2*R*,3*R*)-epoxyneral (**10**) (25.7 ng/female, 18.0 ng/male, 3.4 ng/protonymph), which was active at doses of 0.1–1 ng (Mori and Kuwahara, 1995; Mori *et al.*, 1996). A second unidentified species, *Caloglyphus* sp. "HP," produced even larger amounts of its pheromonal component rosefuran (**24**) (87.6 ng/female, 10.4 ng/male), which was again active at 1 ng levels (Mori *et al.*, 1998b). Neral (**2**) (29 ng/female, 6 ng/male) has been identified as the female sex pheromone from an unidentified *Histiogaster* sp. and was active at 0.1–1 ng dose (Hiraoka *et al.*, 2002).

Two explanations for puzzling distributions between males and females

Although the above pheromones from the opisthonotal gland secretions trigger the male's mounting behavior, the compounds are present in extracts from both sexes and are also occasionally detectable in larval and nymphal stages. Among the *Caloglyphus* spp., females produce much larger amounts of the active components than males in three of the four species (ratios in unidentified *Caloglyphus* sp. "HP," 8.4:1; *C. rodriguezi*, 6.3:1; *C. polyphyllae*, 3.4:1), and the pheromones are adult specific in *Caloglyphus* sp. "HP" and *C. rodriguezi*. By comparison, the pheromone in unidentified *Caloglyphus* sp. "MJ" has a similar ratio (1.4:1) between the sexes (Mori and Kuwahara, 2000). Moreover, the sex pheromones are found even in protonymphs of unidentified *Caloglyphus* sp. "MJ" and in *C. polyphyllae*. Similar situations pertain with *C. hughesi* (ratio of pheromone in female versus male, 2.1:1; Ryono *et al.*, 2001), *Al. ovatus* (0.75:1; Kuwahara *et al.*, 1992b), and *A. immobilis* (0.76:1; Sato *et al.*, 1993c). To rationalize this unusual distribution of what appear to be sex pheromones, we have postulated an evolutionary trend among four *Caloglyphus* spp., from the initial state in which the sex pheromone is distributed in all stages (e.g., in *C. polyphyllae* and unidentified *Caloglyphus* sp. "MJ") to the derived state in which distribution has become biased toward females (e.g., in *C. rodriguezi* and unidentified *Caloglyphus* sp. "HP") (Kuwahara *et al.*, 1998; Mori and Kuwahara, 2000).

There is also another possible explanation for this puzzling distribution of sex pheromone components, based on recent observations of *C. polyphyllae*. When a group of these mites with all developmental stages present is transferred to a new culture medium and/or a place, the mites initially aggregate, as mentioned above. The active principle(s) mediating this temporary aggregation phenomenon has been identified as β-acaridial (**13**; active at 1 ng) (Shimizu *et al.*, 2001), which is also a sex pheromone component for this species. There are parallels with another species, *S. elongata*, in which neral (**2**) functions as an alarm pheromone at higher doses (Kuwahara *et al.*, 2001) whereas at lower doses it elicits aggregative responses (Nishimura *et al.*, 2002). These observations, coupled with knowledge of astigmatid mite biology, suggest that dose and context may be critically important in the interpretation of mite pheromonal signals. In the context of disturbance or colonization of new habitats, it would be beneficial for both sexes and nymphs to produce chemical signals associated with alarm or aggregation. In the context of reproduction in an undisturbed colonized habitat, the same chemical signal released by females may trigger mating behavior in males. Further clarification of these alternatives awaits data from other mite species.

Mites have high thresholds for sex pheromones

In contrast to insects, in which responses of males can be triggered by picogram to femtogram doses of female-produced pheromones (Kuwahara *et al.*, 1971, 1975b), mites have relatively high thresholds of response, with observable behavioral responses being triggered only by nanogram or higher doses. There are several possible explanations for this: (i) as a consequence of their small body size, mites have few pheromone receptors; (ii) mite pheromones need only act over relatively short distances because mites live gregariously in nature, depending on patchily and ephemerally distributed food sources; and (iii) astigmatid mites live in humid and complex habitats, in which pheromone may be readily absorbed by water droplets or by the substrate.

Furthermore, the reasons for the wide variation in the sensitivity of various mite species to their pheromones are unclear. At one end of the spectrum, males of species such as *C. rodriguezi* (Mori *et al.*, 1995), unidentified *Caloglyphus* sp. "MJ" (Mori *et al.*, 1996), unidentified *Caloglyphus* sp."HP" (Mori *et al.*, 1998b), *D. farinae* (Tatami *et al.*, 2001), and unidentified *Histiogaster* sp. (Hiraoka *et al.*, 2002) respond to a fraction of a female equivalent of pheromone. In contrast, responses have only been elicited from males of *Al. ovatus* (Kuwahara *et al.*, 1992b), or *A. immobilis* (Sato *et al.*, 1993c) in amounts equivalent to many tens of female equivalents. The answer to this puzzling dichotomy may lie with bioassay methods. Increasing experience with handling mites and the development of improved bioassay methodologies has improved sensitivities. For example, use of smaller bioassay chambers (7 mm internal diameter by 5 mm height, versus the 10 cm internal diameter by 15 mm high chambers used in earlier assays), and modified mite conditioning regimens have resulted in smaller doses being required to demonstrate biological activity. Nevertheless, even under ideal circumstances, the biological activities of mite pheromones are still relatively weak in comparison with insect pheromones, with nanogram or higher quantities being required for male mites in all species for which sex pheromones have been identified.

Dose–response relationships are convex curve

It is noteworthy that the dose–response curves for both male- and female-produced mite sex pheromones are convex (Sato *et al.*, 1993c; Mori *et al.*, 1995, 1996, 1998b; Tatami *et al.*, 2001; Hiraoka *et al.*, 2002), with an optimum range being flanked by regions of no response at low doses and by no response or unusual behavior of test mites, such as escape response or cessation of motion, at excessive doses. In contrast, in all known cases of alarm or aggregation pheromones, except the case of the aggregation pheromone of β-acaridial in *C. polyphyllae* (Shimizu *et al.*, 2001), the dose–response curves appear to be sigmoid, with no obvious upper limit at which response is curtailed (Kuwahara *et al.*, 1980b; My-Yen *et al.*, 1980b). This suggests that the different types of pheromone may be perceived or processed in different ways.

Recent advances in pheromone research in astigmatid mites

For many years, conventional wisdom held that each mite species could have only one kind of pheromone, such as an alarm, sex, or aggregation pheromone, in part because the only identified sources of pheromones were the paired opisthonotal glands. *L. konoi* provided an exception, having an aggregation pheromone (lardolure (**44**)) and an alarm pheromone (neral (**2**)), with the alarm pheromone being produced from the opisthonotal glands and possibly the lardolure produced from another internal source. However, it is now known that astigmatid mite species are capable of producing complex pheromones with multiple functions. In the sections above, I have mentioned three categories of double pheromone systems

1. A single compound that is responsible for two pheromonal functions depending upon the context and habitat
2. A single compound that has different pheromonal functions at different doses
3. Two compounds present in the opisthonotal glands, with each having separate pheromonal functions.

In such a system, the two pheromone functions would be expected to have differing dose–response relationships, with one being sigmoid and the other convex.

C. polyphyllae is an example of the first category. This species produces β-acaridial (**13**) as the major component of its exudates. This compound functions as the female sex pheromone (Leal *et al.*, 1989f), and males can discriminate females from males (Mori and Kuwahara, 2000). In another context, this compound functions as an aggregation pheromone with a convex dose–response (Shimizu *et al.*, 2001). From an evolutionary viewpoint, the sex pheromone system of this species may be at a relatively primitive stage, with the pheromones being found in both sexes, but a higher amount in females (3.4:1 ratio, females to males). The compound still has a major role as an aggregation pheromone, which is shared by both sexes and by all developmental stages. To date, in four of nine species where the female sex pheromones have been identified, the quantitative ratios of the pheromone produced by females versus males is still low (less than 2:1, see above), and the compound(s) still retain significant activity as aggregation pheromones.

S. elongata is an example of the second category. At high doses, neral (**2**) acts as an alarm pheromone (Kuwahara *et al.*, 2001), whereas at lower doses the compound attracts females (Nishimura *et al.*, 2002). Because females of this species can reproduce parthenogenetically, the reason for this function as an attractant is not clear. Possible function may be as an aggregation pheromone. In addition, there are many other mite species in which the major component of the opisthonotal gland secretions functions as the alarm pheromone, and it is entirely possible that these compounds may be found to have additional pheromonal functions at lower doses.

The bulb mite *R. robini* is an example of category 3. The species has males and females, and neryl formate (**1**) is present as a minor component in the glands of both sexes and functions as an alarm pheromone (Kuwahara *et al.*, 1988b; Akiyama *et al.*, 1997). With extracts containing the alarm pheromone, the activity of the sex pheromone is suppressed, and only the alarm activity is manifested. However, when the alarm pheromone is removed by fractionation of the extract, sex pheromone activity is observed in the fraction containing α-acaridial (**11**), which is one of the major components of the crude extract. In this species, α-acaridial (**11**; 388 ng/female) displays a convex dose–response curve, with doses of 5–10 ng eliciting sexual activity from males, whereas males presented with doses of 50 ng or more show no response. Neryl formate (**1**), (10 ng/female) is active as the alarm pheromone at 10 ng. Based on these results, we hypothesize that gradual emission of α-acaridial (**11**) results in sex pheromonal activity, whereas total emission at one time as a defensive response releases a sufficient dose of neryl formate (**1**) to trigger alarm responses (Fig. 3.2) (Mizoguchi *et al.*, 2003). Elucidation of analogous systems of alarm and sex pheromones is now in progress with related species. For example, with the mold mite *T. putrescentiae*, we have identified β-acaridial (**13**, the major component, with a convex dose–response curve) as a possible sex pheromone (unpublished data), along with a minor component, neryl formate (**1**), which acts as an alarm pheromone with a sigmoidal dose–response curve. Isorobinal (**18**) has also been identified from *Rhizoglyphus setosus* as a possible female sex pheromone,

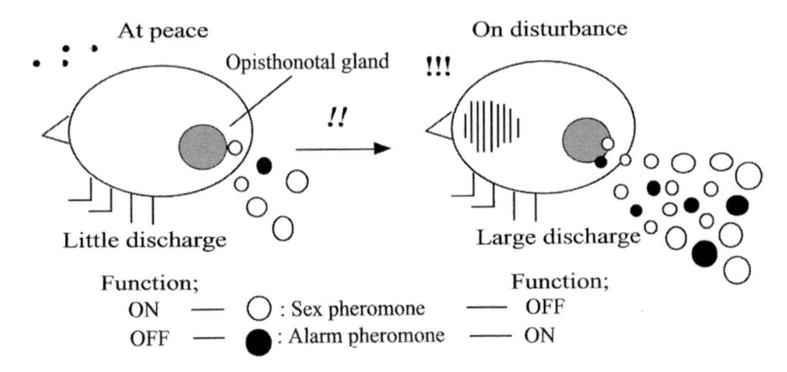

Fig. 3.2. Proposed double pheromone system using on excretory gland. A combination of the pheromones showing a convex and a sigmoidal dose–response relationships.

and its activity is manifested after removal of the alarm pheromone neryl formate (**1**) from the extract (unpublished data).

A bifunctional sex–aggregation pheromone system may also exist with unidentified *Caloglyphus* sp. "sasagawa," in which the major component, rosefuran (**24**), acts as a female sex pheromone with a sigmoidal dose–response curve (unpublished data), and β-phenylethanol (**30**), a minor component, functions as an aggregation pheromone.

An alternative combination of an aggregation–alarm pheromone system is found in *L. konoi.* This species produces the aggregation pheromone lardolure (**44**) as a minor component, and the alarm pheromone neral (**2**) as a major component. If both pheromones are presented simultaneously, aggregation pheromone activity manifests only after the alarm pheromone has dissipated. Furthermore, it seems likely that the mites can regulate the production of each component independently, because the evidence suggests that each is produced at a different site. Therefore, lardolure may be produced essentially continuously, and its function as an attractant is temporarily overridden by the rapid release of the alarm pheromone under adverse circumstances.

Consequently, there are actually three different scenarios that can occur in category 3, and it seems likely that these or similar scenarios will be found to be common phenomena among Astigmata as careful, quantitative studies covering wide ranges of dosage are carried out. The combination of two active principles, or a single active compound with two active ranges, provides mechanisms by which these tiny animals with bodies less than 1 mm in length can utilize sophisticated semiochemical signaling systems. With these intriguing examples as indicators, we are now reinvestigating opisthonotal gland exudates in search of further examples among astigmatid mites of pheromones with multiple roles.

Acknowledgements

I am grateful to professors Ring T. Cardé and Jocelyn G. Millar for critically reading the manuscript.

References

Akiyama, M., Sakata, T., Mori, N., Kato, T., Amano, H. and Kuwahara, Y. (1997). Chemical ecology of astigmatid mites. XLVI. Neryl formate, the alarm pheromone of *Rhizoglyphus setosus* Manson (Acarina: Acaridae) and the common pheromone component among four *Rhizoglyphus* mites. *Applied Entomology and Zoology* **32**: 75–79.

Ayorinde, O., Wheeler, J. W. and Duffield, R. M. (1984). Synthesis of dehydrocineole, a new monoterpene from the acarid mite *Caloglyphus rodriguezi* (Arachnida: Acari). *Tetrahedron Letters* **25**: 3525–3528.

Baker, G. T. and Krantz, W. (1984). Alarm pheromone production of the bulb mite, *Rhizoglyphus robini* Claparede, and its possible use as a control adjuvant in lily bulb. In *Acarology VI*, vol. 2, eds. D. A. Griffiths and C. E. Bowman, pp. 686–692. Chichester, UK: Ellis Horwood.

Bartelt, R. J. and Weisleder, D. (1996). Polyketide origin of pheromones of *Carpophilus davidsoni* and *C. mutilatus* (Coleoptera: Nitidulidae). *Bioorganic and Medicinal Chemistry* **4**: 429–438.

Byers, J. A. (1989). Chemical ecology of bark beetles. *Experientia* **45**: 271–283.

Chuman, T., Kohno, M., Kato, K. and Noguchi, M. (1979). 4,6-Dimethyl-7-hydroxynonan-3-one, a sex pheromone of the cigarette beetle (*Lasioderma serricorne* F.). *Tetrahedron Letters* **20**: 2361–2364.

Curtis, R. F., Hobson-Frohock, A., Fenwick, G. R. and Berreen, J. M. (1982). Volatile compounds from the mite *Acarus siro* in food. *Journal of Stored Products Research* **17**: 197–203.

Hiraoka, H., Mori, N., Nishida, R. and Kuwahara, Y. (2001). (4*E*)-Dehydrocitrals [(2*E*,4*E*)- and (2*Z*,4*E*)-3,7-dimethyl-2,4,6-octatrienals] from acarid mite *Histiogaster* sp. A096 (Acari: Acaridae). *Bioscience, Biotechnology and Biochemistry* **65**: 2749–2754.

Hiraoka, H., Mori, N., Okabe, K., Nishida, R. and Kuwahara, Y. (2002). Chemical ecology of astigmatid mites LXVII. Neral [*Z*-3,7-dimethyl-2,6-octadienal]: the female sex pheromone of an acarid mite, *Histiogaster* sp. (Acari: Acaridae). *Journal of Acarological Society of Japan* **11**: 17–26.

Honma, L. Y., Kuwahara, Y., Sato, M., Matsuyama, S. and Suzuki, T. (1995). Structurally hybridized compounds between the aggregation and alarm pheromones of mite triggering conflicted behavior of *Carpoglyphus lactis*. *Journal of Pesticide Science* **20**: 265–271.

Howard, R. W., Kuwahara, Y., Suzuki, H. and Suzuki, T. (1988). Pheromone study on acarid mites. XII. Characterization of the hydrocarbons and external gland morphology of the opisthonotal glands of six species of mites (Acari: Astigmata). *Applied Entomology and Zoology* **23**: 58–66.

Hughes, A. M. (1976). *The Mites of Stored Food and Houses*. (*Technical Bulletin 9*.) London: HMSO.

Ishii, S. and Kuwahara, Y. (1967). An aggregation pheromone of the German cockroach *Blattella germanica* L. (Orthoptera: Blattellidae). *Applied Entomology and Zoology* **2**: 203–217.

Ishii, S. and Kuwahara, Y. (1968). Aggregation of German cockroach (*Blattella germanica*) nymph. *Experientia* **24**: 88–89.
Kizawa, Y., Kuwahara, Y., Matsuyama, S. and Suzuki, T. (1993). Mite body catalyzes isomerization and reduction of neral (alarm pheromone component). Common phenomenon? *Journal of Acarological Society Japan* **2**: 67–74.
Krantz, G. W. (1978). *A Manual of Acarology*, 2nd edn. Corvallis, OR: OSU Book Stores.
Kuwahara, Y. (1982). Pheromone study on acarid mites VII. Structural requisites in monoterpenoids for inducing the alarm pheromone activity against the mold mite, *Tyrophagus putrescentiae* (Schrank) (Acarina: Acaridae). *Applied Entomology and Zoology* **17**: 127–132.
(1990). Pheromone studies on astigmatid mites: alarm, aggregation and sex. In *Modern Acarology*, eds. F. Dusbabek and V. Bukva, pp. 43–52. The Hague: Academia, Prague and SPB Academic.
(1997). Volatile compounds produced by two species of *Dermatophagoides* mites. *Skin Research* **39**(suppl. 19): 52–55. (In Japanese with English summary.)
Kuwahara, Y. and Sakuma, L. (1982a). Pheromone study on acarid mites VIII. Primary (*Z*)-2-alkenyl formate responsible for the alarm pheromone activity against the mold mite, *Tyrophagus putrescentiae* (Schrank) (Acarina: Acaridae). *Applied Entomology and Zoology* **17**: 263–268.
(1982b). Synthesis of alarm pheromone analogues of the mold mite, *Tyrophagus putrescentiae*, and their biological activities. *Agricultural and Biological Chemistry* **46**: 1855–1860.
Kuwahara, Y., Kitamura, C., Takahashi, S., Hara, H., Ishii, S. and Fukami, H. (1971). Sex pheromone of the almond moth and the Indian meal moth: *cis*-9, *trans*-12-tetradecadienyl acetate. *Science* **171**: 801–802.
Kuwahara, Y., Ishii, S. and Fukami, H. (1975a). Neryl formate: alarm pheromone of the cheese mite, *Tyrophagus putrescentiae* (Schrank) (Acarina, Acaridae). *Experientia* **31**: 1115–1116.
Kuwahara, Y., Fukami, H., Ishii, S., Matsumura, F. and Burkholder, W. E. (1975b). Studies on the isolation and bioassay of the sex pheromone of the drugstore beetle, *Stegobium paniceum* (Coleoptera; Anobiidae). *Journal of Chemical Ecology* **1**: 413–422.
Kuwahara, Y., Fukami, H., Howard, R., Ishii, S., Matsumura, F. and Burkholder, W. E. (1978). Chemical studies on the Anobiidae: sex pheromone of the drugstore beetle, *Stegobium paniceum* (L) (Coleoptera). *Tetrahedron* **34**: 1769–1774.
Kuwahara, Y., Fukami, H., Ishii, S., Matsumoto, K. and Wada, Y. (1979). Pheromone study on acarid mites II. Presence of the alarm pheromone in the mold mite, *Tyrophagus putrescentiae* (Schrank) (Acarina: Acaridae) and the site of its production. *Japanese Journal of Sanitary Zoology* **30**: 309–314.
(1980a). Pheromone study on acarid mite III. Citral: isolation and identification from four species of acarid mite, and its possible role. *Japanese Journal of Sanitary Zoology* **31**: 49–52.
Kuwahara, Y., Matsumoto, K. and Wada, Y. (1980b). Pheromone study on acarid mite IV. Citral: composition and function as an alarm pheromone and its secretory gland in four species of acarid mites. *Japanese Journal of Sanitary Zoology* **31**: 73–80.
Kuwahara, Y., My-Yen, L. T., Tominaga, Y., Matsumoto, K. and Wada, Y. (1982). 1,3,5,7-Tetramethyldecyl formate, lardolure: aggregation pheromone of the acarid mite, *Lardoglyphus konoi* (Sasa et Asanuma) (Acarina: Acaridae). *Agricultural and Biological Chemistry* **46**: 2283–2291.

Kuwahara, Y., Suzuki, H., Matsumoto, K. and Wada, Y. (1983). Pheromone study on acarid mites XI. Function of mite body as geometrical isomerization and reduction of citral (the alarm pheromone). *Applied Entomology and Zoology* **18**: 30–39.

Kuwahara, Y., Akimoto, K., Leal, W. S., Nakao, H. and Suzuki, T. (1987). Isopiperitenone: a new alarm pheromone of the acarid mite, *Tyrophagus similis* (Acarina, Acaridae). *Agricultural and Biological Chemistry* **51**: 3441–3442.

Kuwahara, Y., Leal, W. S., Akimoto, K., Nakano, Y. and Suzuki, T. (1988a). Pheromone study on acarid mites XVI. Identification of hexyl linolate in acarid mites and its distribution among the genus *Tyrophagus*. *Applied Entomology and Zoology* **23**: 338–344.

Kuwahara, Y., Shibata, C., Akimoto, K., Kuwahara, M. and Suzuki, T. (1988b). Pheromone study on acarid mites. XIII. Identification of neryl formate as an alarm pheromone from the bulb mite, *Rhizoglyphus robini* (Acarina: Acaridae). *Applied Entomology and Zoology* **23**: 76–80.

Kuwahara, Y., Leal, W. S., Suzuki, T., Maeda, M. and Masutani, T. (1989a). Antifungal activity of *Caloglyphus polyphyllae* sex pheromone and other mite exudates. Pheromone study on astigmatid mites, XXIV. *Naturwissenschaften* **76**: 578–579.

Kuwahara, Y., Leal, W. S., Nakono, Y., Kaneko, Y., Nakao, H. and Suzuki, T. (1989b). Pheromone study on astigmatid mites XXIII. Identification of the alarm pheromone on the acarid mite, *Tyrophagus neiswanderi* and species specificities of alarm pheromones among four species of the same genus. *Applied Entomology and Zoology* **24**: 424–429.

Kuwahara, Y., Leal, W. S. and Suzuki, T. (1990). Pheromone study on astigmatid mites XXVI. Comparison of volatile components between *Dermatophagoides farinae* and *D. pteronyssinus* (Astigmata, Pyroglyphidae). *Japanese Journal of Sanitary Zoology* **41**: 23–28.

Kuwahara, Y., Matsumoto, K., Wada, Y. and Suzuki, T. (1991a). Chemical ecology on astigmatid mites. XXIX. Aggregation pheromone and kairomone activity of synthetic lardolure (1*R*,3*R*,5*R*,7*R*)-1,3,5,7-tetramethyldecyl formate and its optical isomers to *Lardoglyphus konoi* and *Carpoglyphus lactis* (Acari: Astigmata). *Applied Entomology and Zoology* **26**: 85–89.

Kuwahara, Y., Sato, T. and Suzuki, T. (1991b). Chemical ecology on astigmatid mites. XXXI. Geranial as the alarm pheromone of *Histiostoma laboratorium* Hughes (Astigmata: Histiostomidae). *Applied Entomology and Zoology* **26**: 501–504.

Kuwahara, Y., Koshii, T., Okamoto, M., Matsumoto, K. and Suzuki, T. (1991c). Chemical ecology on astigmatid mites. XXX. Neral as the alarm pheromone of *Glycyphagus domesticus* (De Geer) (Acarina: Glyciphagidae). *Japanese Journal of Sanitary Zoology* **42**: 29–32.

Kuwahara, Y., Leal, W. S., Kurosa, K., Sato, M., Matsuyama, S. and Suzuki, T. (1992a). Chemical ecology on astigmatid mites XXXIII. Identification of (*Z,Z*)-6,9-heptadecadiene in the secretion of *Carpoglyphus lactis* (Acarina, Carpoglyphidae) and its distribution among astigmatid mites. *Journal of the Acarological Society of Japan* **1**: 95–104.

Kuwahara, Y., Sato, M., Koshii, T. and Suzuki, T. (1992b). Chemical ecology of astigmatid mites XXXII. 2-Hydroxy-6-methyl-benzaldehyde, the sex pheromone of the brown-legged grain mite, *Aleuroglyphus ovatus* (Troupeau) (Acarina: Acaridae). *Applied Entomology and Zoology* **27**: 253–260.

Kuwahara, M., Suzuki, K. and Hiramatsu, A. (1992c). Synthesis of robinal, a highly conjugated monoterpenoid from the mite *Rhizoglyphus robini*. *Bioscience, Biotechnology and Biochemistry* **56**: 1510–1511.

Kuwahara, Y., Matsuyama, S., Suzuki, T., and Okabe, K. (1994a). Chemical ecology of astigmatid mites XXXIX. Chemical dimorphism between hypopus and propagative forms in three species. *Journal of Acarological Society Japan* **3**: 13–20.

Kuwahara, Y., Asami, N., Morr, M., Matsuyama, S. and Suzuki, T. (1994b). Chemical ecology of astigmatid mites XXXVIII. Aggregation pheromone and kairomone activity of lardolure and its analogues against *Lardoglyphus konoi* and *Carpoglyphus lactis*. *Applied Entomology and Zoology* **29**: 253–257.

Kuwahara, Y., Ohshima, M., Sato, M., Kurosa, K., Matsuyama, S. and Suzuki, T. (1995a). Chemical ecology of astigmatid mites XL. Identification of the alarm pheromone and new C17 hydrocarbons from a *Tortonia* sp., a pest attacking the nest of *Osmia cornifrones*. *Applied Entomology and Zoology* **30**: 177–184.

Kuwahara, Y., Samejima, M., Sakata, T. *et al.* (1995b). Chemical ecology of astigmatid mites XLIV. Identification of (*Z,Z,Z*)-5,9,12-octadecatrienoic acid and (*Z,Z*)-5,9-octadecadienoic acid as possible biosynthetic precursors of new hydrocarbons (Z,Z,Z)-4,8,11-heptadecatriene and (Z,Z)-4,8-heptadecadiene found in the astigmatid mite, *Tortonia* sp. *Applied Entomology and Zoology* **30**: 433–441.

Kuwahara, Y., Mori, N., Shimizu, K., Tanaka, C. and Tsuda, M. (1998). Pheromone studies on astigmatid mites: recent progress – a comparison of molecular phylogeny, distribution and function of female sex pheromone in *Caloglyphus* spp. (Acarina: Acaridae). *Journal of Asia-Pacific Entomology* **1**: 9–15.

Kuwahara, Y., Ibi, T., Nakatani, Y. *et al.* (2001). Chemical ecology of astigmatid mites LXI. Neral, the alarm pheromone of *Schwiebea elongata* (Acari: Acaridae). *Journal of Acarological Society Japan* **10**: 19–25.

Leal, W. S. and Mochizuki, F. (1990). Chemoreception in astigmatid mites. *Naturwissenschaften* **77**: 593–594.

Leal, W. S., Kuwahara, Y. and Suzuki, T. (1988a). Neryl myristate from the acarid mite, *Aleuroglyphus ovatus* (Acarina, Acaridae). *Agricultural and Biological Chemistry* **52**: 1299–1300.

Leal, W. S., Nakano, Y., Kuwahara, Y., Nakao, H. and Suzuki, T. (1988b). Pheromone study on acarid mites XVII. Identification of 2-hydroxy-6-methyl-benzaldehyde as the alarm pheromone of the acarid mite, *Tyrophagus perniciosus* (Acarina: Acaridae), and its distribution among related mites. *Applied Entomology and Zoology* **23**: 422–427.

Leal, W. S., Kuwahara, Y., Suzuki, T. and Kurosa, K. (1989a). The alarm pheromone of the mite *Suidasia medanensis* Oudemans, 1924 (Acariformes, Suidasiidae). *Agricultural and Biological Chemistry* **53**: 2703–2709.

Leal, W. S., Kuwahara, Y., Suzuki, T., Nakano, Y. and Nakao, H. (1989b). Identification and synthesis of 2,3-epoxyneral, a novel monoterpene from the acarid mite, *Tyrophagus perniciosus* (Acarina, Acaridae). *Agricultural and Biological Chemistry* **53**: 295–298.

Leal, W. S., Kuwahara, Y., Nakano, Y., Nakao, H. and Suzuki, T. (1989c). 2(*E*)-(4-Methyl-3-pentenyl)-butanedial, α-acaridial. A novel monoterpene from the acarid mite *Tyrophagus perniciosus* (Acarina, Acaridae). *Agricultural and Biological Chemistry* **53**: 1193–1196.

Leal, W. S., Kuwahara, Y. and Suzuki, T. (1989d). 2(E)-(4-Methyl-3-pentenylidene)-butanedial, β-acaridial: a new type of monoterpene from the mold mite, *Tyrophagus putrescentiae* (Acarina, Acaridae). *Agricultural and Biological Chemistry* **53**: 875–878.

Leal, W. S., Kuwahara, Y., Suzuki, T. and Nakao, H. (1989e). Chemical taxonomy of economically important *Tyrophagus* mites (Acariformes, Acaridae). *Agricultural and Biological Chemistry* **53**: 3279–3284.

Leal, W. S., Kuwahara, Y., Suzuki, T. and Kurosa, K. (1989f). β-Acaridial, the sex pheromone of the acarid mite *Caloglyphus polyphyllae*. Pheromone study of acarid mites, XXI. *Naturwissenschaften* **76**: 332–333.

Leal, W. S., Kuwahara Y. and Suzuki, T. (1990a). Hexyl 2-formyl-3-hydroxybenzoate, a fungitoxic cuticular constituent of the bulb mite *Rhizoglyphus robini*. *Agricultural and Biological Chemistry* **54**: 2593–2597.

(1990b). Robinal, a highly conjugated monoterpenoid from the mite, *Rhizoglyphus robine*. Chemical ecology of astigmatid mites, XXVII. *Naturwissenschaften* **77**: 387–388.

Levinson, A. R., Levinson, H. Z. and Oelker, U. (1989). Two sex pheromones mediate courtship and mating in the flour mite. *Naturwissenschaften* **76**: 176–177.

Levinson, H. Z., Levinson, A. R. and Mueller, K. (1991). Functional adaptation of two nitrogenous waste products in evoking attraction and aggregation of flour mites (*Acarus siro* L.). *Anzeiger für Schaedlingskunde Planzenschutz Umweltschutz* **64**: 55–60.

Mizoguchi, A., Mori, N., Nishida, R., and Kuwahara, Y. (2003). α-Acaridial, a female sex pheromone from an alarm pheromone emitting mite *Rhizoglyphus robini*. *Journal of Chemical Ecology* **29**: 1681–1690.

Mori, K. and Kuwahara, S. (1986a). Synthesis of both the enantiomers of lardolure, the aggregation pheromone of the acarid mite, *Lardoglyphus konoi*. *Tetrahedron* **42**: 5539–5544.

(1986b). Stereochemistry of lardolure, the aggregation pheromone of the acarid mite, *Lardoglyphus konoi*. *Tetrahedron* **42**: 5545–5550.

(1995). Synthesis of (2*R*,3*R*)-epoxyneral, a sex pheromone of the acarid mite, *Caloglyphus* sp. (Astigmata: Acaridae). *Tetrahedron Letters* **36**: 1477–1478.

(2000). Comparative studies of the ability of males to discriminate between sexes in *Caloglyphus* spp. *Journal of Chemical Ecology* **26**: 1299–1309.

Mori, N., Kuwahara, Y., Kurosa, K., Nishida, R. and Fukushima, T. (1995). Chemical ecology of astigmatid mites XLI. *n*-Undecane: the sex pheromone of the acarid mite *Caloglyphus rodriguezi* Samsinak (Acarina: Acaridae). *Applied Entomology and Zoology* **30**: 415–423.

Mori, N., Kuwahara, Y. and Kurosa, K. (1996). Chemical ecology of astigmatid mites XLV. (2*R*,3*R*)-Epoxyneral: sex pheromone of the acarid mite *Caloglyphus* sp. (Acarina: Acaridae). *Bioorganic and Medicinal Chemistry* **4**: 289–295.

(1998a). Rosefuran: the sex pheromone of the acarid mite *Caloglyphus* sp. *Journal of Chemical Ecology* **24**, 1771–1779.

Mori, N., Fukui, M. and Kuwahara, Y. (1998b). Mating behavior of the astigmatid mite, *Caloglyphus rodriguezi* Samsinak (Acarina: Acaridae). *Applied Entomology and Zoology* **33**: 385–390.

Morino, A., Kuwahara, Y., Matsuyama, S. and Suzuki, T. (1997). (*E*)-2-(4′-Methyl-3′-pentenylidene)-4-butanolide, named β-acariolide: a new monoterpene lactone from the mold mite, *Tyrophagus putrescentiae* (Acarina: Acaridae). *Bioscience, Biotechnology and Biochemistry* **61**: 1906–1908.

My-Yen, L. T., Matsumoto, K., Wada Y. and Kuwahara, Y. (1980a). Pheromone study on acarid mites V. Presence of citral as a minor component of the alarm pheromone in the mold mite, *Tyrophagus putrescentiae* (Schrank, 1781) (Acarina: Acaridae). *Applied Entomology and Zoology* **15**: 474–477.

My-Yen, L. T., Wada, Y., Matsumoto, K. and Kuwahara, Y. (1980b). Pheromone study on acarid mites VI. Demonstration and isolation of an aggregation pheromone in *Lardoglyphus konoi* Sasa et Asanuma. *Japanese Journal of Sanitary Zoology* **31**: 249–254.

Nakayama, H., Kumei, A. and Sakurai, M. (1998). Treatment of atopic dermatitis by mite elimination (environmental improvement). *Japan Pediatric Dermatology* **17**: 97–102. (In Japanese.)

Ninomiya, Y. and Kawasaki, T. (1988). Acarid attractant. *Japan Kokai Tokkyo Koho*, Syo 63-230605 (27 September 1988).

Nishimura, K., Shimizu, N., Naoki Mori, N. and Kuwahara, Y. (2002). Chemical ecology of astigmatid mites LXIV. The alarm pheromone neral functions as an attractant in *Schwiebea elongata* (Banks) (Acari: Acaridae). *Applied Entomology and Zoology* **37**: 13–18.

Noguchi, S., Mori, N., Kuwahara, Y. and Sato, M. (1997). Facile synthesis of 2-hydroxy-6-methylbenzaldehyde, an alarm and sex pheromone component of astigmatid mites. *Bioscience, Biotechnology and Biochemistry* **61**: 1546–1547.

Noguchi, S., Mori, N., Kurosa, K. and Kuwahara, Y. (1998). Chemical ecology of astigmatid mites XLIX. β-Acaridial (2(*E*)-(4-methyl-3-pentenylidene)-butanedial), the alarm pheromone of *Tyrophagus longior* Gervais (Acarina: Acaridae). *Applied Entomology and Zoology* **33**: 53–57.

Okamoto, M., Matsumoto, K., Wada, Y. and Kuwahara, Y. (1981). Studies on antifungal effect of mite alarm pheromone citral. 2. Antifungal effect of the hexane extracts of the grain mites and some analogues of citral. *Japanese Journal of Sanitary Zoology* **32**: 265–270. (In Japanese with English summary.)

Reka, S. A., Suto, C. and Yamaguchi, M. (1992). Evidence of aggregation pheromone in the feces of house dust mite, *Dermatophagoides farinae*. *Japanese Journal of Sanitary Zoology* **43**: 339–341.

Ruther, J. and Steidle, J. L. M. (2000). Mites as matchmakers: semiochemicals from host-associated mites attract both sexes of the parasitoid *Lariophagus distinguendus*. *Journal of Chemical Ecology* **26**: 1205–1217.

Ryono, A., Mori, N., Okabe, K. and Kuwahara, Y. (2001). Chemical ecology of astigmatid mites LVIII. 2-Hydroxy-6-methylbenzaldehyde: sex pheromone of *Cosmoglyphus hughesi* (Acari: Acaridae). *Applied Entomology and Zoology* **36**: 77–81.

Sakata, T. and Kuwahara, Y. (2001). Structure elucidation and synthesis of 3-hydroxybenzene-1,2-dicarbaldehyde from astigmatid mites. *Bioscience, Biotechnology and Biochemistry* **65**: 2315–2317.

Sakata, T., Kuwahara, Y. and Kurosa, K. (1996). 4-Isopropenyl-3-oxo-1-carboxyaldehyde, isorobinal: a novel monoterpene from the mite *Rhizoglyphus* sp. (Astigmata: Rhizoglyphinae). *Naturwissenschaften* **83**: 427.

Sakata, T., Okabe, K. and Kuwahara, Y. (2001). Structure elucidation of twelve novel esters between five fatty acids and three branched new alcohols along with four monoterpenoids from *Sancassania shanghaiensis* (Acari: Acaridae). *Bioscience, Biotechnology and Biochemistry* **65**: 919–927.

Sakuma, M. and Fukami, H. (1990). The attraction of the German cockroach, *Blattella germanica* (L.) (Dictyoptera: Blattellidae) to their aggregation pheromone. *Applied Entomology and Zoology* **25**: 355–368.

(1993). Aggregation arrestant pheromone of the German cockroach, *Blattella germanica* (L.) (Dictyoptera: Blattellidae). Isolation and structural elucidation of blattellastanoside-A and -B. *Journal of Chemical Ecology* **19**: 2521–2541.

Sakurai, M., Nakayama, H. and Kumei, A. (1997). Results of patch tests and scratch patch tests with crushed live mites and α-acaridial. *Skin Research* **39**(suppl. 19): 56–60. (In Japanese with English summary.)

Sato, M. and Kuwahara, Y. (1999). Identification of rosefuran from flour mite *Acarus immobilis*. *Kagawa-Daigaku-Nougakubu-Gakuzyutsu-Hokoku* **51**: 31–35.

Sato, M., Kuwahara, Y., Matsuyama, S. and Suzuki, T. (1993a). Chemical ecology of astigmatid mites XXXVII. Fatty acid as food attractant of astigmatid mites, its scope and limitation. *Applied Entomology and Zoology* **28**: 565–569.

(1993b). 2-Formyl-3-hydroxybenzyl formate (rhizoglyphinyl formate), a novel salicylaldehyde analog from the house dust mite *Dermatophagoides pteronyssinus* [Astigmata, Pyroglyphidae]. *Bioscience, Biotechnology and Biochemistry* **57**: 1299–1301.

Sato, M., Kuwahara, Y., Matsuyama, S., Suzuki, T., Okamoto, M. and Matsumoto, K. (1993c). Male and female sex pheromones produced by *Acarus immobilis* Griffiths (Acaridae: Acarina). Chemical ecology of astigmatid mites XXXIV. *Naturwissenschaften* **80**: 34–36.

Shibata, S., Kuwahara, Y., Sato, M., Matsuyama, S. and Suzuki, T. (1998). Sex pheromone activity of 2-hydroxy-6-methylbenzaldehyde analogs against males of two astigmatid mites, *Aleuroglyphus ovatus* and *Acarus immobilis*. *Journal of Pesticide Science* **23**: 34–39.

Shimizu, N. and Kuwahara, Y. (2001). 7-Hydroxyphthalide (7-hydroxy isobenzofuranone): a new salicyl lactone from *Oulenzia* spp. (Astigmata: Winterschmitiidae). *Bioscience, Biotechnology and Biochemistry* **65**: 990–992.

Shimizu, N., Mori, N. and Kuwahara, Y. (1999). Identification of the new hydrocarbon (*Z,Z*)-1,6,9-heptadecatriene as the secretory component of *Caloglyphus polyphyllae* (Astigmata:Acaridae). *Bioscience, Biotechnology and Biochemistry* **63**: 1478–1480.

(2001). Aggregation pheromone activity of the female sex pheromone, β-acaridial, in *Caloglyphus polyphyllae* (Acari: Acaridae). *Bioscience, Biotechnology and Biochemistry* **65**: 1724–1728.

Shimizu, N., Tarui, H., Mori, N. and Kuwahara, Y. (2003). (*E*)-2-(2-Hydroxyethylidene)-6-methyl-5-heptenal (α-acariolal) and (*E*)-2-(2-hydroxyethyl)-6-methyl-2, 5-heptadienal (β-acariolal), two new types of isomeric monoterpenes from *Caloglyphus polyphyllae* (Acari: Acaridae). *Bioscience and Biotechnical Biochemistry* **67**: 308–313.

Shinkaji, N., Okabe, K., Amano, H. and Kuwahara, Y. (1988). Attractant isolated from culture filtrates of *Fusarium oxysporum* Schl., f. sp. *alli* for the robine bulb mite, *Rhizoglyphus robini* Claparade (Acarina: Acaridae). *Japanese Journal of Applied Entomology and Zoology* **32**: 55–59.

Simpson, T. J. (1980). Biosynthesis of polyketides. In *Biosynthesis*, vol. 6, ed. J. D. Bullock, London: The Royal Society of Chemistry.

Suzuki, T., Matsuyama, S. and Kuwahara, Y. (1992). A simple synthesis of α-acaridial. *Bioscience, Biotechnology and Biochemistry* **56**: 1888–1889.

Tatami, K., Mori, N., Nishida, R. and Kuwahara, Y. (2001). 2-Hydroxy-6-methylbenzaldehyde: the female sex pheromone of the house dust mite *Dermatophagoides farinae* (Astigmata: Pyroglyphidae). *Japanese Journal of Sanitary Zoology* **52**: 269–277.

Tarui, H., Ryono, A., Mori, N., Okabe, K. and Kuwahara, Y. (2002). 3-(4-Methyl-3-pentenyl)-2(5*H*)-furanone, α,α-acariolide and 4-(4-methyl-3-pentenyl)-2(5*H*)-furanone, α,β-acariolide: new monoterpene lactones from the astigmatid mites, *Schwiebea araujoae* and *Rhizoglyphus* sp. (Astigmata: Acaridae). *Bioscience, Biotechnology and Biochemistry* **66**: 135–140.

Tongu, Y., Ishii, A. and Oh, H. (1986). Ultrastructure of house-dust mites, *Dermatophagoides farinae* and *D. pteronyssinus*. *Japanese Journal of Sanitary Zoology* **37**: 237–244.

Tuma, D., Sinha, R. N., Muir, W. E. and Abramson, D. (1990). Odor volatiles associated with mite-infested bin-stored wheat. *Journal of Chemical Ecology* **16**:713–724.
Vanhaelen, M., Vanhaelen, F. R., Geeraerts, J. and Wirtthlin, T. (1979). *cis*- and *trans*-Octa-1,5-dien-3-ol, new attractants to the cheese mite *Tyrophagus putrescentiae* (Schrank) (Acarina, Acaridae) identified in *Trichothecium roseum* (Fungi imperfecti). *Microbios* **23**: 199–212.
Vanhaelen, M., Vanhaelen, F. R. and Geeraerts, J. (1980). Occurrence in mushroom (Homobasidiomycetes) of *cis*-octa-1,5-dien-3-ol and *trans*-octa-1,5-dien-3-ol, attractant to the cheese mite *Tyrophagus putrescentiae* (Schrank) (Acarina, Acaridae). *Experientia* **36**: 406–407.
Yoshizawa, T., Yamamoto, I. and Yamamoto, R. (1970). Attractancy of some methyl ketones isolated from Cheddar cheese for cheese mites. *Botyuu-Kagaku* **35**: 43–45.
(1971). Synergistic attractancy of cheese components for cheese mites, *Tyrophagus putrescentiae*. *Botyuu-Kagaku* **36**: 1–7.

4

Semiochemistry of spiders

Stefan Schulz

Institute of Organic Chemistry, the Technical University of Braunschweig, Germany

Introduction

Spiders are an important order of carnivorous arachnids having a great impact on many ecosystems. Because most of their prey consists of insects, they can play an important role in controlling pest insects in agricultural crops. There are currently about 36 000 described species, out of an estimated overall number of 60 000–80 000 species (Platnick, 1999). Unlike the situation with insects (Francke and Schulz, 1999), pheromones and other semiochemicals of arachnids, and especially spiders, have received little attention from researchers. What information is available on the use of semiochemicals by spiders will be reviewed and discussed in this chapter.

Spider pheromones

Spiders are not well known for their ability to communicate with pheromones, partly because of their relatively restricted movement compared with flying insects. Such limited motility makes observations of such phenomena less straightforward. Furthermore, few spider species are serious pests, and far more research has been focussed on their elaborate use of silk, which has fascinated humans for millenia. However, the problem of finding mates is essentially the same for both insects and spiders. That is, a relatively small animal living in a spacious environment needs efficient mechanisms to find a mate for reproduction. Flying insects are better equipped to respond to olfactory cues for mate location, being free to move in three dimensions, whereas spiders are often restricted to walking on vegetation, which makes it more difficult to follow odor plumes through complex plant architecture. The positive identification of a potential partner also becomes more important,

Advances in Insect Chemical Ecology, ed. R. T. Cardé and J. G. Millar. Published by Cambridge University

because encounters with unsuitable objects are more likely than in odor-directed flight toward a target. There is substantial and widespread evidence for the use of close-range and contact pheromones by the Araneae, as summarized in Table 4.1. Examples of pheromone use are now known for 18 out of the 108 known spider families (Platnick, 2001), and from more than 90 species. There are further anecdotal observations suggesting use of pheromones in spiders, but this review will focus on cases for which there is experimental evidence.

One of the earliest records of pheromones in spiders was made by Lendl (1887–88), who described an odor emanating from female *Geolycosa vultuosa* (as *Trochosa infernalis*) (Lycosidae), which attracted males. Subsequent work by Bristowe and Locket (1926) described male lycosid courtship displays (in the absence of the females) in response to contact chemical stimuli. Kaston (1936) demonstrated that female appendages released courtship in males, and that the activity was lost after extraction with ether. These observations were confirmed for several other species by various authors (reviews: Leborgne, 1981; Tietjen and Rovner, 1982; Pollard *et al.*, 1987; Stewart, 1988).

Silk production is common to all spiders and, therefore, plays an important role in the behavior of many species. It can be regarded as an extension of a spider's sensory system (Witt, 1975), and many of the studies listed in Table 4.1 refer to contact pheromones present on the web, with Lycosidae and Salticidae as the most well-studied families. However, hard evidence for a contact pheromone is lacking in many reports. There are several factors that have to be considered. Silk is a relatively rare material in nature, being produced mainly by spiders and some insects. Therefore, even contact with silk may serve as a signal indicating the presence of a conspecific, with signal specificity being conferred by factors such as the shape and physical behavior of the silk. For example, male *Araneus sclopetarius* (Araneidae) are arrested by contact with con- or heterospecific silk from females. Furthermore, they are able to discriminate between these silk types when given a choice (Roland, 1984). The crab spider *Misumena vatia* follows any silk lines encountered, regardless of the origin of the silk (Anderson and Morse, 2001). However, the social spider *Stegodyphus sarasinorom* (Eresidae) prefers conspecific silk to an *Araneus* thread, but not to *Amaurobius* or *Eresus* silk (Roland, 1984). In the Ctenidae, about 50% of the male *Cupiennius salei* reacted with courtship behavior to conspecific silk carrying no pheromone (Tichy *et al.*, 2001). The architecture of an orb web can also be used for size assessment of its owner, as has been shown for *Nephilengys cruentata* (Tetragnathidae) (Schuck-Paim, 2000). In the salticid *Portia labiata*, individuals can assess the fighting ability of conspecifics by their draglines (Clark and Jackson, 1995a; Clark *et al.*, 1999), whereas closely related species cannot. However, the described effects are not necessarily evoked by pheromones.

Table 4.1. *Studies on spider pheromones listing reports from a variety of sources, ranging from anecdotal observations to detailed behavioral studies including chemical tests*

Species[a]	Type of study[b]	Comments	Reference
2 **Atypidae**			
Sphodros abboti	O	Probable female contact sex pheromone; silk bound	Coyle and Shear, 1981
5 **Hexathelidae**			
Atrax infensus	V	Males attracted to hidden females	Hickman, 1964
34 **Dysderidae**			
Dysdera crocata	W	Contact sex pheromone	Pollard *et al.*, 1987
47 **Erisidae**			
Stegodyphus sarasinorum	O,S	Social species; individuals recognize conspecifics by chemical cues	Kullmann, 1972; Roland, 1984
55 **Theridiidae**			
Latrodectus hesperus	S	Silk of both sexes change behavior in the opposite sex	Ross and Smith, 1979
Latrodectus mactans	S	Female silk of *L. hesperus* induces courtship in males	Ross and Smith, 1979
Latrodectus revivensis	S	Silk of both sexes changes behavior in the opposite sex	Anava and Lubin, 1993
Steatoda (as *Teutana*) *grossa*	W	Female web releases courtship behavior in males	Gwinner-Hanke, 1970
61 **Linyphiidae**			
Frontinella communis (as *pyramitela*)	W	Males respond behaviorally to female webs	Suter and Renkes, 1982; Suter and Hirscheimer, 1986
	S,W	Male determines sex of conspecifics by cuticular chemical cues	Suter *et al.*, 1987
Linyphia tenuipalpis	I: **4, 5**	Female sex pheromone induces web reduction behavior of males	Schulz and Toft, 1993b
Linyphia triangularis	I: **4, 5**	Female sex pheromone induces web reduction behavior of males	Schulz and Toft, 1993b

Table 4.1. (*cont.*)

Species[a]	Type of study[b]	Comments	Reference
Neriene emphana	P	**4**, **5** present on female web	S. Schulz and S. Toft, unpublished data
Neriene (as *Linyphia*) *litigosa*	V	Volatile female sex pheromone	Watson, 1986
	P	**4**, **5** present on female web	S. Schulz and P. J. Watson, unpublished data
Neriene montana	P	**4**, **5** present on female web	S. Schulz and S. Toft, unpublished data
Neriene peltata	P	**4**, **5** present on female web	S. Schulz and S. Toft, unpublished data
Microlinyphia impigra	P	**4**, **5** present on female web	S. Schulz and S. Toft, unpublished data
62 **Tetragnathidae**			
Metellina segmentata	O	Male mate guarding released by pheromone on the silk of females	Prenter *et al.*, 1994
Nephila clavata	E,V	Males respond to pheromone from newly moulted females	Miyashita and Hayashi, 1996
63 **Araneidae**			
Araneus trifolium	V	Females emit volatile sex pheromone	Olive, 1982
Argiope aurantia	V	Females attracted to webs of *A. trifolium*	Enders, 1975; Olive, 1982
Argiope trifolium	O	Females emit volatile sex pheromone	Olive, 1982
Cyrtophora cicatrosa	V	Female and web emit sex pheromone	Blanke, 1973, 1975
64 **Lycosidae**			
Alopecosa (as *Tarentula*) *barbipes*	O	Male stimulated by female contact pheromone	Bristowe and Locket, 1926
Geolycosa vultuosa (as *Trochosa infernalis*)	O	Both sexes attracted to female odor	Lendl, 1887–88
Hogna (as *Lycosa*) *carolinensis*	O	Female silk releases courtship behavior in males	Farley and Shear, 1973

(*cont.*)

Table 4.1. (*cont.*)

Species[a]	Type of study[b]	Comments	Reference
Hogna (as *Lycosa*) *helluo*	O	Female cuticle releases courtship behavior in males	Kaston, 1936
Hogna (as *Lycosa*) *longitarsis*	W	Following of female draglines and courtship behavior by males	Lizotte and Rovner, 1989
Lycosa (as *Trochosa*) *singoriensis*	O	Detectable female odor; excitatory for males	Kolosvary, 1932
Pardosa amentata	S	Female draglines release courtship behavior in males	Richter *et al.*, 1971; Dijkstra, 1976
Pardosa hortensis	S	Female contact pheromone releases courtship behavior in males	Robert and Krafft, 1981
Pardosa lapidicina	W	Female draglines release courtship behavior in males	Dondale and Hegdekar, 1973
Pardosa milvina	E	Contact stimulus releases change in behavior in males	Kaston, 1936
	V	Volatile female sex pheromone	Searcy *et al.*, 1999
Pardosa modica	W	Female cuticle releases courtship behavior in males	Kaston, 1936
Pardosa moesta	W	Female draglines release courtship behavior in males	Hegdekar and Dondale, 1969
Pardosa ramulosa	O	Female draglines release searching behavior in males	Sarinana *et al.*, 1971
Pardosa saxatilis	W	Female draglines release courtship behavior in males	Kaston, 1936; Hegdekar and Dondale, 1969
Pardosa xerampelina	S	Female silk releases courtship behavior in males	Dumais *et al.*, 1973
Rabidosa (as *Lycosa*) *puntulata*	O	Males follow female draglines	Tietjen, 1977; Tietjen and Rovner, 1980

Table 4.1. (*cont.*)

Species[a]	Type of study[b]	Comments	Reference
Rabidosa (as *Lycosa*) *rabida*	O,S	Males follow female draglines, courtship behavior by males	Kaston, 1936; Rovner, 1968; Tietjen, 1977, 1979a; Tietjen and Rovner, 1980
Schizocosa avida	W	Female draglines release courtship behavior in males	Hegdekar and Dondale, 1969
Schizocosa crassipalpata (as *crassipalpis*)	W	Female draglines release courtship behavior in males	Hegdekar and Dondale, 1969
Schizocosa crassipes	W	Female draglines and cuticle release courtship behavior in males	Kaston, 1936
Schizocosa ocreata	V	Males respond to hidden females	Tietjen, 1979b
	E	Males respond to extracts of male silk	Ayyagari and Tietjen, 1986
	O	Males respond with courtship to female silk	Ayyagari and Tietjen, 1986
Schizocosa saltatrix	V	Males respond to hidden females	Tietjen, 1979b
Schizocosa (as *Lycosa*) *tristani*	W	Males follow female draglines; courtship behavior by males	Lizotte and Rovner, 1989
66 **Pisauridae**			
Dolomedes scriptus	E	Courtship by males elicited by ether soluble compounds from females	Kaston, 1936
Dolomedes triton	S,E	Male dragline following; behavior change after contact with water that housed females or their draglines	Roland and Rovner, 1983
74 **Ctenidae**			
Cupiennius coccineus	P	Compound **3** on female silk	M. D. Papke, S. Schulz and H. Tichy, unpublished data
	S	Conspecific silk evokes courtship	Barth and Schmitt, 1991

(*cont.*)

Table 4.1. *(cont.)*

Species[a]	Type of study[b]	Comments	Reference
Cupiennius getazi	P	Compound **3** on female silk	M. D. Papke, S. Schulz and H. Tichy, unpublished data
	S	Conspecific silk evokes courtship	Barth and Schmitt, 1991
Cupiennius salei	I: **3**	Female silk-bound sex pheromone elicits courtship behavior of males	Rovner and Barth, 1981; Papke *et al.*, 2000; Tichy *et al.*, 2001
75 **Agenelidae**			
Agenelopsis aperta	I: **1**	Female volatile sex pheromone attracts males and releases courtship	Riechert and Singer, 1995; Papke *et al.*, 2001
Agelena consociata	S	Social spider; silk attractive to conspecifics	Krafft, 1970, 1971
	E	Cuticular pheromone for recognition of conspecifics	Krafft, 1974
Tegenaria atrica	E	Female silk and cuticle elicit courtship behavior in males	Prouvost *et al.*, 1999
	O	Hydrocarbon patterns possibly mediate intraspecific agonistic interactions	Trabalon *et al.*, 1996; Pourié and Trabalon, 2001
Tegenaria domestica	S	Male orientation towards female silk	Roland, 1984
	E	Male courtship induced by female cuticular extract	Trabalon *et al.*, 1997
Tegenaria pagana	S	Male orientation towards female silk	Roland, 1984
	E	Male courtship induced by female cuticular extract	Trabalon *et al.*, 1997
Tegenaria parietina	S	Males exhibit courtship on empty female webs	Krafft, 1978
82 **Dictynidae**			
Dictynia calcarata	O	Female silk releases courtship behavior in males	Jackson, 1978

Table 4.1. (*cont.*)

Species[a]	Type of study[b]	Comments	Reference
83 **Amaurobiidae**			
Amaurobius fenestralis	S	Males exhibit courtship on empty female webs	Krafft, 1978
Amaurobius similis	S	Males exhibit courtship on empty female webs	Krafft, 1978
Coelotes terrestris	S	Males exhibit courtship on empty female webs	Krafft, 1978; Roland, 1984
Mallos trivittata (as *trivittatus*)	S	Female silk release courtship behavior in males	Jackson, 1978
Mallos gregalis	W	Silk attracts conspecifics	Jackson, 1982
91 **Clubionidae**			
Clubiona cambridgei	W	Contact sex pheromone	Pollard *et al.*, 1987
106 Philodromidae			
Tibellus oblongus	S	Female legs attractive to males; ether-washed legs are not attractive	Kaston, 1936
107 **Thomisidae**			
Diaea socialis	E	Social spider; volatile silk bound pheromone attracts conspecifics	Evans and Main, 1993
108 **Salticidae**			
Anasaitis (as *Corythalia*) *canosa*	S	Female silk changes behavior of males, volatiles prolong courtship	Jackson, 1987; Pollard *et al.*, 1987
Bavia aericeps	S	Female silk changes behavior of males	Jackson, 1987
Brettus cingulatus	V	Males locate females by volatile pheromones	Pollard *et al.*, 1987
Carrholtus xanthogrammus	S	Female silk used for mate location by males	Yoshida and Suzuki, 1981
Corythalia xanthopa	O	Males court recently deceased females without contact	Crane, 1949
Cosmophasis micarioides	S	Female silk changes behavior of males	Jackson, 1987
Cyrba algerina	O	Female volatile pheromone releases courtship in males	Pollard *et al.*, 1987

(*cont.*)

Table 4.1. (*cont.*)

Species[a]	Type of study[b]	Comments	Reference
Euophrys parvula	S	Female silk changes behavior of males	Jackson, 1987
Euryattus sp.	S	Female silk changes behavior of males	Jackson, 1987
Helpis minitabunda	S	Female silk changes behavior of males	Jackson, 1987
Jacksonoides (as *Lagnus*) *kochicanosa*	S	Female silk changes behavior of males	Jackson, 1987
Lyssomanes viridis	O	Female volatile pheromone releases courtship in males	Pollard *et al.*, 1987
Marpissa marina	S	Female silk changes behavior of males	Jackson, 1987
Menemerus sp.	S	Female silk changes behavior of males	Jackson, 1987
Mopsus mormon	S	Female silk changes behavior of males	Jackson, 1987
Myrmarachne lupata	S	Female silk changes behavior of males	Jackson, 1987
Natta horizontalis (as *Cyllobelus rufopictus*)	S	Female silk changes behavior of males	Jackson, 1987
Phidippus audax	W	Female silk changes behaviour of males	Pollard *et al.*, 1987
Phidippus johnsoni	S	Female silk induces courtship by males	Jackson, 1986, 1987
	W	Female silk induces courtship by males	Pollard *et al.*, 1987
Phidippus octopunctatus (as *opifex*)	S	Female silk changes behavior of males	Jackson, 1987
Phidippus pulcher (as *otiosus*)	S	Female silk changes behavior of males	Jackson, 1987
Phidippus regius	S	Female silk changes behavior of males	Jackson, 1987
Phidippus whitmani	S	Female silk changes behavior of males	Jackson, 1987
Plexippus paykulli	S	Female silk changes behavior of males	Jackson, 1987
Portia africana	V	Female odor changes behavior of conspecifics	Willey and Jackson, 1993

Table 4.1. (*cont.*)

Species[a]	Type of study[b]	Comments	Reference
Portia fimbriata	S	Female silk induces courtship by males	Jackson, 1987; Clark and Jackson, 1995b
	V	Female odor changes behavior of conspecifics	Willey and Jackson, 1993
Portia labiata	S	Female silk induces courtship by males	Jackson, 1987; Clark and Jackson, 1995b
	V	Female odor changes behavior of conspecifics	Willey and Jackson, 1993
Portia schultzi	S	Female silk changes behavior of males	Jackson, 1987
	V	Female odor changes behavior of conspecifics	Willey and Jackson, 1993
Pseudicius sp.	S	Female silk changes behavior of males	Jackson, 1987
Simaetha paetula	S	Female silk changes behavior of males	Jackson, 1987
Simaetha thoracica	S	Female silk changes behavior of males	Jackson, 1987
Thiodina sylvana	S	Female silk changes behavior of males	Jackson, 1987
Trite auricoma	S	Female silk changes behavior of males	Jackson, 1987
Trite planiceps	S	Males respond selectively to female silk, probably a bound pheromone	Jackson, 1987; Taylor, 1998

[a] Species names were used according to Platnick (2001), as are the numbers before the family names.
[b] Study types: E, behavioral test with extracts; S, behavioral test with silk, substratum, or bodies; O, behavioral observation; I, identification of pheromone with behavioral test; W, behavioral tests with extracted silk or bodies; A, presence of pheromone components of related species confirmed; V, behavioral test for volatile pheromone; P, unpublished communication.

To exclude the possibility of response toward tactile cues delivered from silk, silk has been extracted with various solvents and then retested in behavioral experiments (marked W in Table 4.1). This procedure implies pheromone usage but is not conclusive. The *experimentum crucis*, a behavioral test of the extract for pheromonal activity, often was not performed, particularly in older studies. A complicated issue is that solvent extraction of silk can alter its physical characteristics, as shown in detailed experiments by Shao and Vollrath (1999). They immersed dragline silk from various species into solvents of different polarity, ranging from chaotropic solvents through water to 1-butanol. Their results showed that the silk structure may be altered, especially when it is strained in the solvent, causing changes in its physical properties. Polar solvents such as water had the greatest effect, whereas the effect of less-polar 1-butanol was negligible. It has been suggested that the effects result from breakage of hydrogen bonds between the protein strands of the silk by solvent molecules (Shao and Vollrath, 1999). Therefore, loss of activity upon solvent extraction suggests but does not conclusively prove the existence of soluble pheromones. The silks of orb weavers contain large amounts of small water-soluble compounds, which are essential for web function (e.g., Vollrath *et al.*, 1990; Higgins *et al.*, 2001).

Extraction with apolar solvents such as diethyl ether removes lipids present on the silk of many species (Schulz, 1999) and, therefore, alters the "feel" of the silk. This also may eliminate response, even when no semiochemicals are used. Also, spiders probably can detect physical properties such as tensile strength of the fibers, as discussed by Tietjen (1977) and Tietjen and Rovner (1980) for *Rabidosa rabida* (Lycosidae). For example, rain or dew may change the silk's properties, providing information on the age of the silk. Consequently, solvent extraction might indeed inactivate the silk, even if a pheromone does not exist. This may explain why extraction of "active" silk with solvents in some cases yields both "inactive" silk and inactive extracts. It is also possible that pheromones may be degraded by solvent extraction, although this is rare in pheromone research. Overall, the intent of this discussion is to emphasize the importance of separation of pheromones from their original matrices, followed by retesting of the extracts for evidence of active principles. Only then can the existence of a pheromone be proven conclusively.

Proof of the existence of a volatile pheromone is more straightforward, but experimental separation of olfactory from visual, vibrational, and acoustic cues remains essential. Mate location may be mediated by pheromones in many spider species, because spiders, like insects, often face the task of locating a mate at some distance.

Attempts to characterize spider pheromones chemically began in the mid 1980s. Ayyagari and Tietjen (1986) partially isolated and fractionated a pheromone that mediated agonistic interactions between *Schizocosa ocreata* males. Only simple chemical tests were used to characterize the active fraction, rather than modern spectroscopic techniques. Since then, three pheromones have been identified (see Linyphiidae, Ctenidae, and Agenelidae; Table 4.1) and several other attempts have been made (see Thomisidae and Agenelidae; Table 4.1). In the following sections, studies performed in the different spider families, arranged according to the phylogeny of Coddington and Levi (1991), will be discussed.

Mygalomorphae

Purse-web spiders (Atypidae) capture their prey through the walls of a silk tube, thus hiding from the environment. Males stopped wandering around on sites from which female silk tubes had been removed (Coyle and Shear, 1981), suggesting a contact sex pheromone. However, the response could have been evoked by remnants of the silk tube. In the Hexathelidae, male *Atrax infensus* were attracted by volatiles from females, and this behavior was exploited to trap males, using traps with hidden females (Hickman, 1964).

Dysderidae

Courtship behavior of *Dysdera crocata*, a specialized feeder on woodlice (Psocoptera), is released by contact with nest silk or the cuticle of a live or even dead female. Washing with ether rendered these substrates inactive, suggesting the presence of a lipophilic contact pheromone (Pollard *et al.*, 1987).

Eresidae

Only about 20 spider species are social (Foelix, 1996), and some of them belong to the genus *Stegodyphus*. It is tempting to speculate that chemical stimuli may play the same important role for these spiders as in their social insect counterparts. Nevertheless, the social spiders never reach the level of eusociality, and examples of chemical communication in these spiders are rare. *S. sarasinorum* accepts all kinds of prey, even paraffin drops, when vibrational cues are offered, except parts of or whole conspecifics (Kullmann, 1972). This may indicate the presence of a communal pheromone, which is probably not species specific, because populations of *Stegodyphus mimosarum* and *Stegodyphus dumicola* can be mixed without antagonism (Seibt and Wickler, 1988). The pheromone is probably not volatile, because odor had no effect on *S. sarasinorum* (Roland, 1984), and the silk is probably not involved, because this species cannot discriminate well between conspecific and heterospecific silk (Roland, 1984). Other social species are discussed below.

Dictynidae

Males react with courtship response when placed on empty female webs of *Dictyna calcarata*. In some cases, this spider also reacted to heterospecific female silk, but less often than to conspecific female silk. No response was observed towards conspecific or heterospecific male silk (Jackson, 1978).

Clubionidae

Only one species in this large family has been investigated. Female *Clubiona cambridgei* silk induces courtship by males. The silk is active even in the absence of the female, but loses its activity upon extraction with ether (Pollard *et al.*, 1987), suggesting the presence of a lipid-soluble pheromone.

Salticidae

These jumping spiders are known for their excellent vision, but there is evidence from several species that they also use pheromones for communication, as suggested by Crane (1949). More than 30 species now have been investigated, making this the most thoroughly studied family. The earlier work has been reviewed and will not be discussed in detail here (see Table 4.1 for list of species; Pollard *et al.*, 1987; Jackson, 1987). In addition, the variability of response to pheromones by individual males has been addressed (Jackson and Cooper, 1990).

Trite planiceps females hide in rolled-up leaves and are thus concealed from searching males. Males associate with female but not male silk (Taylor, 1998). Removal of silk from leaves resulted in prolonged, but still successful, searching for the entrance hole (Taylor, 1998), but this effect could also be the result of signals other than pheromonal cues. Similar results were obtained for *P. labiata* and *Portia fimbriata* (Clark and Jackson, 1995b). Males of both species spent more time on filter paper treated with silk of conspecific females than on untreated filter paper, whereas females did not discriminate. Seven-day-old silk was no longer active, suggesting degradation or evaporation of a possible pheromone, but the lack of activity could also be because of changes in the physical and chemical properties of silk. Interestingly, the cannibalistic *P. labiata* females spent more time on their own silk than on silk of other conspecific females (Clark and Jackson, 1994). Furthermore, individuals can assess the fighting ability of conspecifics just from dragline silk (Clark *et al.*, 1999).

Portia spp. also respond to volatile cues. These spiders are web-invading araneophagic spiders and use aggressive mimicry by sending vibrational cues through the prey spider's web. This aggressive behavior was reduced when a conspecific female was present in the web, but not by the presence of conspecific males or juveniles. The web itself did not play a role, because the effect was also observed by hidden stimulus spiders, not present on the web itself. The four species *Portia africana*, *P. fimbriata*, *P. labiata*, and *Portia schultzi* respond only to female conspecific odors (Willey and Jackson, 1993). Therefore, the evidence suggests that most of the salticids that have been investigated deposit pheromonal cues on the silk. Olfactory pheromones seem to be less abundant: Pollard *et al.* (1987) report that from 41 investigated species only seven used volatile pheromones.

Lycosidae

The wolf spiders (Lycosidae) have also been the subject of a number of studies. Kaston (1936) showed that autotomized legs of females induced courtship in several species. Legs of males and ether-extracted legs of females were not active. Several studies have observed that male lycosids can follow trails left by conspecific females, probably guided by a pheromone. As early as 1926, Bristowe and Locket showed that male *Alopecosa* (as *Tarentula*) *barbipes* were excited by female trails, even when no silk was present because the spinnerets had been sealed (Bristowe and Locket, 1926). Later studies proved that dragline silk from females induced courtship by males. Male *Hogna* (as *Lycosa*) *longitarsis* and *Schizocosa* (as *Lycosa*) *tristani* both follow fresh dragline silk of conspecific females and intiate courtship. Male silk, older-female silk, and silk sprayed with hexane were inactive, whereas silk sprayed with ethanol or water remained active (Lizotte and Rovner, 1989), suggesting a non-polar silk-bound pheromone. In contrast, the active silk of female *Pardosa lapidicina* became inactive after washing with water (Dondale and Hegdekar, 1973), and in several other species extraction with diethyl ether or benzene led to inactive silk samples (Hegdekar and Dondale, 1969). However, the solvent extracts proved inactive, and so it cannot be determined whether the solvent changed the silk characteristics, degraded the active component, or the concentration of the extract that was tested was too low. In contrast, Kaston (1936) showed that ether extract of *Pardosa milvina* legs changed the behavior of conspecific males. *Pardosa hortensis* males also reacted to female silk (Robert and Krafft, 1981), as has been shown for several other lycosids (Kaston, 1936; Richter *et al.*, 1971; Dumais *et al.*, 1973; Farley and Shear, 1973; Dijkstra, 1976; Tietjen, 1977, 1979a; Tietjen and Rovner, 1980; Ayyagari and Tietjen, 1986).

The only evidence to date for a male pheromone in spiders was presented by Ayyagari and Tietjen (1986), who obtained an extract from male *S. ocreata* silk that mediated agonistic encounters between males. The pheromone was not identified, but the authors concluded that two different components must be present, one apolar and the other of medium polarity.

Evidence for volatile pheromones in lycosids is limited, despite the fact that Lendl (1887–88) noted attraction of males toward females in *G. vultuosa* (as *T. infernalis*) and Kolosvary (1932) reported an odor emanating from female *Lycosa* (as *Trochosa*) *singoriensis* that induced courtship by males. Tietjen (1979b) found that hidden females altered the behavior of conspecific males, but no real attraction toward the odor source was observed. The most conclusive evidence was presented by Searcy *et al.* (1999), who proved attraction of *P. milvina* males toward females using Y-olfactometer tests and pitfall traps containing adult females. Conspecific subadults or males were not attractive.

Thomisidae and Philodromidae

In the crab spiders only *Diaea socialis* has been investigated for the presence of pheromones (Evans and Main, 1993). This is an unusual social species, because it does not build snare webs and, therefore, cannot use vibrational cues for group communication. The spiders live together after hatching until the adult stage, a period of approximately 18 months. Different colonies can be mixed without any agonistic behavior. Bioassays using silk samples, washed silk, and silk extracts provided conclusive evidence for a pheromone attracting both juvenile and female spiders to the silk. By choice tests, it was shown that the pheromone is silk-bound, ether-extractable and probably volatile, but identification attempts were unsuccessful. Juveniles and gravid females responded differently to different silk types and, therefore, probably had different chemical cues, pointing to a complex system. Females are not attracted to juvenile silk, while juveniles are. Gravid females are repelled by the pheromone. Other crab spiders probably also use pheromones (see below for a discussion). For example, Kaston (1936) showed that male *Tibellus oblongus* (Philodromidae) reacted with vibration of legs and palpi after contact with autotomized legs of conspecific females, but not after contact with male or ether-extracted female legs.

Agenelidae

Females of the desert spider *Agenelopsis aperta* emit a volatile pheromone that attracts conspecific males (Riechert and Singer, 1995). This pheromone was identified as 8-methyl-2-nonanone (**1**; Fig. 4.1), a previously unknown arthropod semiochemical. It was found by headspace analysis and abdominal washings of females 2 weeks after their final molt, when they become sexually receptive; it was absent in females of other age classes. The pheromone attracted males in a three-choice arena system at doses as low as 500 ng (Papke *et al.*, 2001). Another female-specific ketone, 6-methyl-3-heptanone (**2**), was not attractive. Very low doses of **1** (10^{-9} mg/ml applied to a filter paper placed in empty juvenile female webs) also induced courtship behavior in males (Papke *et al.*, 2001). The normal behavioral sequence was followed, except for phases which required input from the female. The ED_{50} value (mean effective dose) of **1** was 5.5×10^{-4} mg/ml hexane. In contrast, ketone **2** only induced a response in some males at unnaturally high concentrations

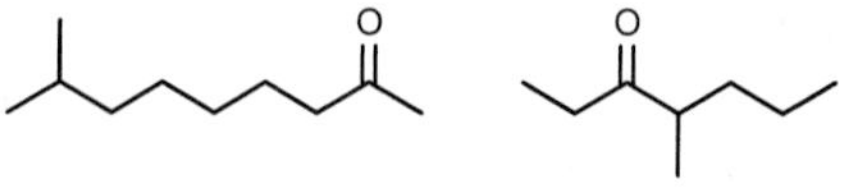

Fig. 4.1. Volatile ketones released by females of *Agenelopsis aperta* 2 weeks after their final molt. Compound **1** showed high pheromonal activity whereas **2** was inactive.

(100 mg/ml solvent). Experiments with a mixture of both compounds did not show any synergistic effect of **2** (Papke *et al.*, 2001).

Other studies have demonstrated contact pheromones on the silk and cuticle of several agenelid species. Males of the solitary spider *Tegenaria atrica* recognize webs of conspecific females, and males are significantly more mobile on webs of receptive females, show prolonged contact with their first pair of legs, and display prolonged courtship behavior (Prouvost *et al.*, 1999). Solvent-washed webs or females did not evoke any lasting responses from males, and responses to receptive females were significantly different from the responses to unreceptive females. Pentane extracts of both silk and females were partitioned between pentane and methanol, and both fractions induced behavioral changes in males, but the more polar methanol fraction was clearly more active. Analysis by gas chromatography and mass spectrometry (GC–MS) showed differences in the cuticular profiles between receptive and unreceptive females, with some acids and their esters in the methanol fraction and some hydrocarbons in the pentane fraction showing significant quantitative differences. Compounds identified on both silk and the cuticle included palmitic, linoleic, and oleic acids, their methyl esters, and tricosane. Several novel spider compounds were found in the methanol fraction, including long-chain methyl ketones and 1-alkanols, and the vinyl ketone 1-docosen-3-one. In total, about 90 compounds were identified as constituents of the surface lipids (Prouvost *et al.*, 1999). The methanol fraction also contained some more polar components, including diols, ketols, and glycerides that are not easily detected by ordinary GC–MS methods (M. D. Papke, M. Trabalon and S. Schulz, unpublished results). Further behavioral tests with subfractions will be necessary to elucidate the full suite of chemical cues involved in the courtship behavior of *T. atrica*.

Similar investigations with the related species *Tegenaria pagana* and *Tegenaria domestica* (Trabalon *et al.*, 1997) determined that males of both species reacted to attractive females with longer contact times compared with that for unattractive or solvent-washed females. The response was species specific, and when pentane extracts were partitioned between pentane and methanol, most of the activity was found in the methanol fraction. The methanol fractions contained fatty acids and their methyl esters and showed quantitative differences between the two physiological states of the females with respect to myristic acid, methyl palmitate, palmitic acid, methyl stearate, and methyl oleate in *T. domestica*, and octadecadienoic acid in *T. pagana*. The species showed different hydrocarbon profiles, but it is not yet certain whether all components present in the extracts were detected with the analytical methods used. The results suggest involvement of chemical stimuli in courtship release, but other factors such as conspecific female silk or substratum-bound cues may play important roles (Roland, 1984).

Male *Tegenaria parietina* have been observed to perform courtship on empty female webs (Krafft, 1978). In *T. atrica*, cuticular lipid profiles were linked to changes in behavior in young spiderlings, which initially live together but become increasingly aggressive as they age and their solitary period begins. This behavioral change has been linked to changes in hydrocarbon profiles, which were analyzed with principal component analysis. It has been hypothesized that female tolerance is dependent on the hydrocarbon profile, based on results obtained from behavioral experiments with spiders of different age groups (Trabalon *et al.*, 1996). Rich food sources allow normally solitary adult spiders to be reared in groups. The normal aggressive behavior against conspecifics is reduced. The analyses of the cuticular lipids showed marked difference between isolated "agressive" and grouped "tolerant" spiders, manifested in 6 out of 75 components (Pourié and Trabalon, 2001).

In the social species *Agelena consociata*, silk galleries and cocoons are attractive to conspecifics. Trail following on silk was also observed (Krafft, 1970, 1971). Furthermore, conspecifics washed with ethanol and diethyl ether were attacked as prey, whereas intact spiders or vibrating lures covered with extracts were not (Krafft, 1974).

Amaurobidae

The social spider *Mallos gregalis* is arrested by ether-soluble components of silk of conspecifics (Jackson, 1982). However, courtship behavior is not released by female silk (Jackson, 1978), in contrast to the behavior seen with silk of female *Mallos trivittata* (as *trivittatus*) silk. Male silk is inactive (Jackson, 1978). *Coelotes terrestris* males recognized female silk or a substratum on which females had been running (Roland, 1984) and males also exhibited courtship on empty female webs (Krafft, 1978). Courtship behavior, mainly vibratory signals in empty female webs, has also been observed in *Amaurobius similis* and *Amaurobius fenestralis*. This behavior is only performed on conspecific webs. A third species, *Amaurobius ferox*, exhibited no responses to silk cues (Krafft, 1978).

Ctenidae

The wandering spider *Cupiennius salei* (Ctenidae) deposits a pheromone on its silk. This cue triggers vibrational signaling by males, to which a receptive female

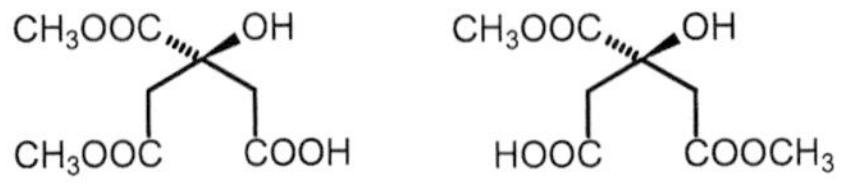

CH_3OOC OH CH_3OOC OH

CH_3OOC COOH HOOC $COOCH_3$

(*S*)-**3** (*R*)-**3**

Fig. 4.2. Pheromone of *Cupiennius salei*, cupilure (**3**). Only the (*S*)-enantiomer is active.

responds (Papke *et al.*, 2000; Tichy *et al.*, 2001). The vibrational cues are used by the male to guide him to the female (Rovner and Barth, 1981; Barth and Schmitt, 1991). Thus, the pheromone does not have a direct role in orientation of the male to the female. Silk from males or juvenile females do not elicit the response.

The behavior of the males was used in a bioassay-guided identification of the pheromone, which was extractable by polar solvents such as water or methanol. The bioassay apparatus consisted of a cardboard channel with silk samples treated with different extracts. A positive response consisted of a male producing a typical drumming behavior upon contact with a stimulus. Because males responded to some extent to silk without pheromone, individual spiders were tested first with inactive silk, followed by extract-treated silk (Tichy *et al.*, 2001). GC–MS analyses of the active silk extracts were not successful, even using several derivatization procedures. However, the pheromone structure was deduced by ^{1}H NMR spectroscopy of d_4-methanol extracts of 10 mg silk samples obtained by drawing silk from the silk glands with the help of an electrically driven reeling machine (Tichy *et al.*, 2001). The extracts contained almost pure pheromone, with only minor amounts of other components. By comparison with synthetic samples, derivatization, and gas chromatography on a chiral cyclodextrin phase, the pheromone was identified as a 95:5 mixture of (*S*)- and (*R*)-enantiomers of dimethyl citrate (cupilure (**3**), Fig. 4.2). Only the (*S*)-enantiomer was active (Tichy *et al.*, 2001). Cupilure is probably present in ionized form on the silk ($pK_a \sim 3.5$), because solvent extracts were neutral. The ionized carboxyl group may be conjugated to basic amino acids of the silk proteins. This would also explain why the pheromone is not volatile, despite its relatively low molecular weight, and is easily washed from the silk by water, including rain. In the tropical habitat of *C. salei* this would ensure the presence of cupilure only on freshly laid silk.

Cupilure was also detected on silk from the related and sympatric species *Cupiennius getazi* and *Cupiennius coccineus* (M. D. Papke, S. Schulz and H. Tichy, unpublished data). Barth and Schmitt (1991) have shown that the three *Cupiennius* spp. can discriminate between hetero- and conspecific silk, but vibrational cues are more important in discrimination than silk composition. The ability to discriminate conspecific silk may originate from its mechanical or physical characteristics, rather than from pheromone components. Separation of the species is also enhanced by different, but overlapping, activity periods.

Pisauridae

Dolomedes triton males live mainly on water surfaces and can follow female draglines, even on water. Males also respond to water that has been in contact with female integument with announcement displays, which are enhanced in the presence of silk (Roland and Rovner, 1983). Earlier experiments had suggested the presence of an ether-soluble pheromone on the integument of the congener *Dolomedes scriptus* that released courtship in males, as did the silk of females (Kaston, 1936).

Araneidae

Orb-weavers are one of the most successful spider families with more than 2600 species. To date, volatile pheromones have been proven in three species. In very convincing experiments Blanke (1973, 1975) showed that *Cyrtophora cicatrosa* female emit a volatile pheromone that attracts males over distances of up to 50 cm, and higher concentrations of the pheromone induced courtship behavior on the web. As is the case with *A. aperta* (see p. 124), only older females (10 to 20 days after the final molt) emit the pheromone, and pheromone production ceases after copulation. Olive (1982) used caged mature females to attract males of *Araneus trifolium* and *Argiope trifasciata*. Interestingly, *Argiope aurantia* males were also attracted to caged *A. trifasciata*, suggesting a common or related pheromone. Under natural conditions the species are probably isolated by different diurnal activity cycles. That *A. aurantia* emits a pheromone had been previously suspected (Enders, 1975).

Linyphiidae

The web-reduction pheromone released by females of the European linyphiid *Linyphia triangularis* was the first pheromone identified from a spider (Schulz and Toft, 1993b). As first observed by van Helsdingen (1965) in the linyphiid *Lepthyphantes leprosus*, the male of the related American species *L. litigosa* modified the web of an unmated female by severing threads and rolling up large parts of it after arrival. Watson (1986) showed that this behavior reduces the release of the male-attracting pheromone, which is emitted from the silk surface. This behavior is only shown in less-dense populations where males search for females. In high populations, mate-guarding of immature females occurs and copulation takes places after the final molt, as with European linyphiids (Toft, 1989).

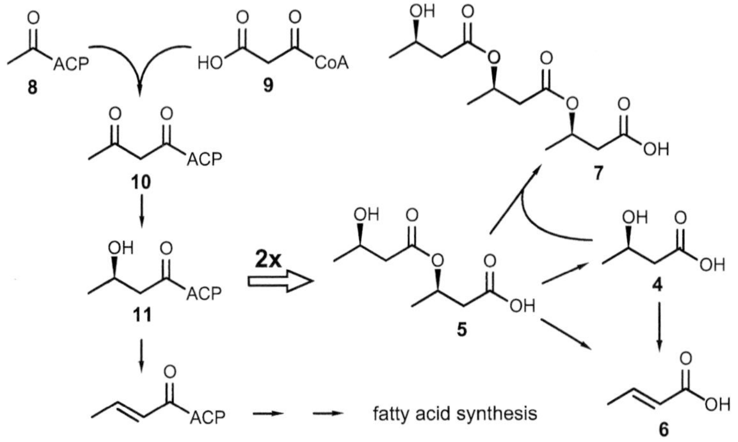

Fig. 4.3. Biosynthesis, structure, and degradation of the pheromones **4** and **5** of *Linyphia triangularis* and *Linyphia tenuipalpis*. ACP, acyl carrier protein; CoA, coenzyme A.

The web-reduction behavior of *L. triangularis* can be induced by empty webs of unmated females. The activity can be transferred to inactive webs (e.g., those of juveniles or mated females) by spraying them with dichloromethane extracts of active webs. The web-reduction pheromone was identified using GC–MS methods, together with chemical derivatization of extracts, gas chromatography on chiral phases, and, finally, synthesis of the active enantiomer (Schulz and Toft, 1993b). The pheromone is the condensation product of two molecules of (*R*)-3-hydroxybutyric acid (**4**), that is, (3*R*,3*R*′)-3-(3-hydroxybutyryloxy)butyric acid (**5**; Fig. 4.3). In laboratory assays, web-reduction behavior was consistently elicited by **5**, and also by the monomer **4**. The pheromone **5** occurred in amounts as high as 5 μg/web. Minor amounts of **4** (between 0.03 and 0.05 μg/web) and trace quantities of crotonic acid (**6**) also were present, possibly from degradation of **5**. This compound is unstable, degrading to **6** and **4**. The presence of the trimeric product **7** in older silk samples (Fig. 4.3) provides a further indication that this degradation actually takes place.

Both **4** and **6** attracted male spiders (S. Toft and S. Schulz, unpublished data), and they are probably the volatile pheromones responsible for male attraction that were suggested by Watson (1986). The system may be remarkably ingenious, because females probably produce the estolide **5** during web construction, then, on the silk, **5** slowly degrades to form the volatile male attractants **4** and **6**. Thus, **5** serves the dual function of a web-reduction pheromone, and as a "propheromone" or slow-release precursor of the attractants. In this way, having spun her web, the female spider is then free for other tasks, such as subduing prey, while still being assured of a steady release of pheromone to attract males. However, this scenario still needs to be verified with bioassays, because the first field trapping experiments with both **4** and **6** were not successful (S. Toft, unpublished results). However, there are other possible reasons for the failure of the field trials.

The pheromones discussed so far are also active in the sympatric species, *Linyphia tenuipalpis* (Schulz and Toft, 1993a), which inhabits the same microhabitat; interspecific web takeover is known to occur in nature (Toft, 1987). Experiments with webs of unmated females showed that male *L. triangularis* start to cut threads in 59% of conspecific webs, whereas the thread-cutting response falls to 33% for heterospecific webs. *L. tenuipalpis* males were attracted to 57% of conspecific female webs, whereas the response to heterospecific webs was only 7%. A third species from another genus, *Neriene emphana*, responded in 55% of the cases to *L. triangularis* webs, and in 16% to *L. tenuipalpis* webs (Schulz and Toft, 1993a). These results, together with the occurrence of **4** and **5** in several other linyphiids (see Table 4.1), indicate that the attractant pheromone is not species specific. Consequently, species discrimination must be mediated by other signals, such as different lipid patterns of the silk. This aspect is discussed further below.

The sex pheromone is interesting from a biosynthetic perspective (see Fig. 4.3) because it is closely connected with primary metabolism. That is, the monomer **4** is an intermediate in fatty acid biosynthesis. Condensation of acetyl-ACP (**8**; ACP, acyl carrier protein) with malonyl-CoA (**9**; CoA, coenzyme A) yields acetoacyl-ACP (**10**). Enantioselective reduction with NADPH leads to (*R*)-3-hydroxybutyryl-ACP (**11**). Two units of this precursor could then be condensed to form the pheromone **5**, which then degrades to **4** and **6** as described above. Alternatively, **4** can also be formed by direct hydrolysis of intermediate **11**.

Suter and co-workers have investigated in detail the chemical cues involved in the behavior of *Frontinella communis* (as *pyramitela*). The silk of adult female releases courtship in males (see also p. 135), whereas methanol-washed silk, silk from immature females, or heterospecific silk are inactive. The pheromone also changes the behavior of males from negatively to positively geotactic (Suter and Renkes, 1982). Web geometry plays no role in eliciting courtship, because single threads or whole webs are equally capable of inducing courtship. A third group of webs, those of adult males, do not induce courtship but do induce negative geotaxis. Therefore, different pheromones seem to be involved in setting the direction of geotaxis and in the induction of courtship (Suter and Hirscheimer, 1986). It was suggested that cuticular and web lipid patterns may play a role in mediating this behavior. Males respond with many abdomen flexions to tarsal contact with a male carcass. In contrast, contact with a female carcass results in foreleg waving. These differences disappeared when the bodies were washed with methanol and hexane prior to testing (Suter *et al.*, 1987).

Theridiidae

Theridiid spiders build webs that are related to the sheet webs of linyphiids. Webs of females released courtship stridulation by male *Steatoda* (as *Teutana*) *grossa*, and the activity was removed by extraction with ether (Gwinner-Hanke, 1970). The well-known black widow spiders also use pheromones. Males of *hesperus* initiate courtship on webs of conspecific females, but not on male webs; conversely, females displayed enhanced activity on the webs of males. Volatile cues were not involved. Male *L. hesperus* also showed reduced responses to webs of the related species *Latrodectus mactans*, and vice versa (Ross and Smith, 1979), suggesting similar pheromones, possibly as mixtures of lipids. Similar behaviors were observed with the desert species *Latrodectus revivensis* (Anava and Lubin, 1993), but in contrast to the studies with the other two species, males did not respond to the silk of juvenile females. Females did not respond to male silk but acted aggressively when encountering the silk of other females while being inactive toward their own silk, suggesting that females may produce unique chemical signatures. In social insects, variable patterns of cuticular hydrocarbon are used for colony recognition (Vander Meer *et al.*, 1998), and analogous chemical pattern recognition mechanisms may also play a role in *L. revivensis*. Preliminary experiments with 1-methoxyalkanes identified in the silk lipids of female *L. revivensis* showed that these compounds evoked a searching behavior in the males, as did γ-aminobutyric acid, a common component of orb webs, whereas the more volatile compound pyrrolidone was repellent at higher concentrations (M. D. Papke, S. Schulz, and Y. Lubin, unpublished results). In contrast, several different colonies of the social spider *Anelosimus eximus* accepted individuals from other colonies, unlike many eusocial insects. Nevertheless, there were quantitative differences between colonies in the composition of the cuticular lipids (Pasquet *et al.*, 1997), which consisted mainly of unusual propyl esters of fatty acids (Bagnères *et al.*, 1997).

Tetragnathidae

Two species have been investigated in the Tetragnathidae family, which builds orb webs like the Araneidae. Male *Metellina segmentata* guard females for some time before they mate, with a distinct preference for the largest females. The decision to guard was made by testing the periphery of the web, indicating that cues in the silk signal an attractive female; the presence or absence of the female did not play a role (Prenter *et al.*, 1994). The authors argued that physical cues related to web architecture are unlikely to be perceived, because only the border of the web was tested, and natural variation in web construction is high. The only other explanation would be the presence of a female pheromone on the silk (Prenter *et al.*, 1994).

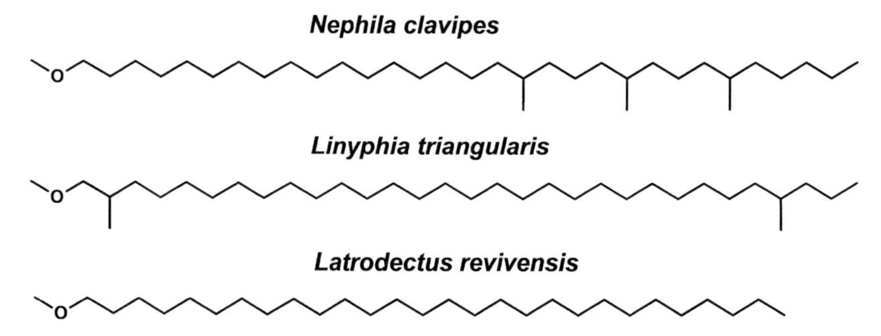

Fig. 4.4. Structures of typical 1-methoxyalkanes present on the silk and cuticle of several spider species.

Feeding remnants may also play a role, because larger females probably catch more prey.

The small males of *Nephila clavata* cohabit with the much larger females in their web, and copulation takes place immediately after the final molt. Olfactometer tests showed that freshly molted females are much more attractive than older ones. Acetone extracts of the body surface of freshly molted females contained the active principle. Miyashita and Hayashi (1996) suggested that the volatile signal may originate from the molting fluid.

Lipids

As can be seen from the examples discussed so far, pheromones present on spider cuticle or silk frequently play an important role in spider communication, but limited information is available about their composition. Lipids, whose primary function is regulation of water content, also may have important roles in communication.

The composition of lipids from the silk and cuticule has been reviewed by Schulz (1997a, 1999). These lipids consist primarily of alkanes, as found in other arthropods, with 2-methylalkanes with an even number of carbon atoms in the chain being most abundant, with lesser amounts of alcohols, acids, aldehydes, and wax esters. Recently, a thorough analysis of the silk lipids of *N. clavipes* (Schulz, 2001) revealed a unique class of lipids from spider silk and cuticle, consisting of straight-chain and branched methyl ethers (1-methoxyalkanes, Fig. 4.4) with chain lengths between 25 and 45 carbon atoms.

Methods for their analysis have been published (Schulz, 1997b, 2001). So far, representatives of this class of compounds have been found in the Linyphiidae (Schulz and Toft, 1993a), the Tetragnathidae (Schulz, 2001), and the Theridiidae (M. D. Papke, S. Schulz, and Y. Lubin, unpublished results). Although the families share some alkanes, each uses different groups of these ethers. For example, specific

Table 4.2. *Occurrence of alkanes and 1-methoxyalkanes on the silk lipids of linyphiid spiders*

Compounds[a]	Species content[b]				
	LTR	LTE	MIM	NMO	NEM
Methyl ethers					
2,22-Dimethyl-C_{25}-OMe	•	–	○	○	○
26-Methyl-C_{27}-OMe	–	–	–	–	•
26-Methyl-C_{28}-OMe	•	○	○	•	–
2,26-Dimethyl-C_{28}-OMe	○	○	○	•	–
26-Methyl-C_{29}-OMe	○	○	○	○	○
2,14,26-Trimethyl-C_{29}-OMe	–	–	–	–	•
28-Methyl-C_{30}-OMe	○	○	○	•	–
2,28-Dimethyl-C_{30}-OMe	•	○	–	•	–
2,28-Dimethyl-C_{31}-OMe	•	•	–	–	–
2,30-Dimethyl-C_{32}-OMe	○	○	○	•	–
2,30-Dimethyl-C_{33}-OMe	–	•	•	○	–
18,22,26-Trimethyl-C_{36}-OMe	–	–	–	•	–
Alkanes					
2-Methylhexacosane	•	•	•	•	•
2-Methyloctacosane	•	•	•	•	•

[a]Carbon chain lengths of the ethers are indicated by subscript numbers (e.g., C_{25} for pentacosane). LTR, *Linyphia triangularis*; LTE, *Linyphia tenuipalpis*; MIM, *Microlinyphia impigra*; NMO, *Neriene montana*; NEM, *Neriene emphana*; Me, methyl.
[b]Presence of individual compounds: •, as major components in silk lipid extract; ○, as minor or trace components; –, absence.

mixtures of the methyl ethers have been found on the webs of several linyphiid species, whereas the patterns of alkanes were relatively similar (see Table 4.2).

It has been suggested that the ethers, compounds unique to spiders, may provide reliable signals for pattern recognition and species determination. In contrast, a pattern of hydrocarbons, as used in several insect species, might be susceptible to contamination from cuticular hydrocarbons from insect prey remnants, which might alter the blends produced by the spiders and deposited on the webs (Schulz, 1997a, 1999).

Another group of compounds unique to spiders, namely propyl esters of long-chain, multiply branched fatty acids, have been found in the cuticular lipids of the spider *A. eximus* (Bagnères *et al.*, 1997).

Overview of spider pheromones

From our limited knowledge of spider pheromones, at least three different categories of releaser pheromones can be identified: those associated with triggering courtship behaviors, species recognition, and intersexual attraction. It is likely that other types of pheromone will be identified as more species are studied, particularly social spiders requiring the behaviors of the colony to be organized and integrated.

Many studies have provided evidence that pheromones on the silk and/or the cuticle of females stimulate courtship or related behaviors in males, as do the few pheromones that have been fully identified to date. Contact with the pheromone and with silk or cuticle is usually necessary to evoke the proper courtship responses. Olfactory cues, such as those in *A. aperta*, may release courtship when the male is in contact with silk. There are insufficient data to draw conclusions about which classes and types of chemical might typically be used for the various kinds of pheromone, particularly as the few identified pheromone structures vary widely in polarity, volatility, and other chemical characteristics.

A particularly thorny problem that remains to be resolved is the species-recognition process that is mediated by physical or tactile cues associated with silk or the cuticle, as well as pheromones on these two substrates. To date, the lipid mixtures associated with silk or cuticle seem to display the most variable structures and blends, making them good candidates for species recognition. In contrast, the more polar components appear to be less species specific and so are less likely to be the key factors in intraspecific recognition.

Finally, volatile attractants that act over a distance, analogous to many insect sex attractant pheromones, have been identified from several spider species. To date, this class of spider pheromones has received very limited study, and there is no doubt that many more volatile attractants remain to be discovered.

Two of the three attractant pheromones identified to date are very close structurally to those used in primary metabolism. The biosynthesis of the estolide **5** probably starts from 3-hydroxybutyric acid (**4**), an intermediate in fatty acid biosynthesis (Fig. 4.3). Condensation of two units furnishes the pheromone **5**. The formation of cupilure (**3**; Fig. 4.2) can be easily explained by two methylations from ubiquitous citric acid. Both compounds are unlike any known insect pheromones, whereas the third known attractant pheromone (ketone **1**; Fig. 4.1), bears some resemblance to some insect pheromones. A proper comparison of the differences and similarities between insect and arachnid pheromones will require the identification of representative compounds from most of the families of both groups of organisms.

Spider attractants discovered serendipitously

During a field study of the attraction of scavenging flies to the defensive compounds of true bugs, Aldrich and Barros (1995) found considerable numbers of crab spiders (Thomisidae) in their traps. More detailed study showed that both (*E*)-2-octenal (**12**) and (*E*)-2-decenal (**13**) were attractive to four American crab spider species:

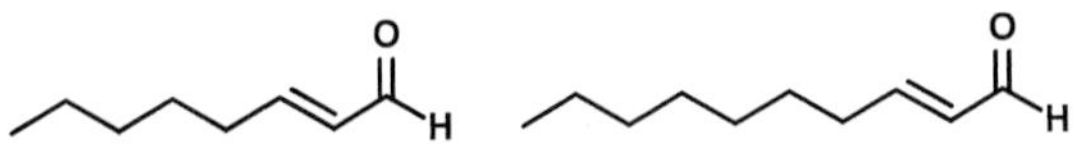

12 13

Fig. 4.5. Aldehyde attractants of *Xysticus* spp., (*E*)-2-octenal (**12**) and (*E*)-2-decenal (**13**).

Xysticus ferox, *Xysticus discursans*, *Xysticus triguttatus*, and *Xysticus auctificus* (Aldrich and Barros, 1995) (Fig. 4.5). Only males were caught, suggesting that these aldehydes might be related to the sex pheromones of the spiders. Furthermore, several related compounds were not attractive, indicating a considerable degree of specificity to the attractants.

Males of a European species, *Xysticus kochii*, also were attracted to **12** and **13** (S. Toft and S. Schulz, unpublished data). The aldehydes **12** and **13** were not found in odors collected from females, although the saturated analogs octanal and decanal were present. However, neither they nor any of the other identified compounds attracted males (S. Schulz and S. Toft, unpublished data), and the reason for the attraction of male crab spiders to aldehydes **12** and **13** remains unknown.

Kairomones used by spiders

There are several documented cases of spiders exploiting kairomones, and interestingly, exploitation can go either way. That is, some predatory species use kairomones to locate their prey, whereas in other cases, prey species are capable of detecting and fleeing from kairomones produced by the predators that are hunting them. For example, the wolf spider *Hogna helluo* locates its prey, female *P. milvina* spiders, using volatile cues (probably the *P. milvina* sex pheromone, see p. 123). Curiously, *H. helluo* does not respond to volatiles from another prey species, the domestic cricket *Acheta domesticus* (Persons and Rypstra, 2000), but non-volatile chemical cues assist in location of both prey species. Conversely, to hinder detection, *P. milvina* responds to chemical cues from its predator by reducing locomotion (Persons and Rypstra, 2001; Barnes *et al.*, 2002). Even more interesting, the diet of the predator plays a role, because chemical cues from *H. helluo* fed exclusively on *P. milvina* induced greater inhibition of locomotion than cues from the predatory spider when it was fed exclusively on crickets (Persons *et al.*, 2001).

The cursorial spider *Habronestes bradleyi* (Zoodariidae) is a specialist predator of the meat ant *Iridomyrmex purpureus*. These ants are highly aggressive toward intruders and recruit conspecifics with an alarm pheromone, 6-methyl-5-hepten-2-one (sulcatone, **14**; Fig. 4.6). This alarm pheromone, which is released by a few workers injured during ritualized agonistic interactions, is used as a prey-location kairomone by the spider (Allan *et al.*, 1996). The spider's predation behavior is

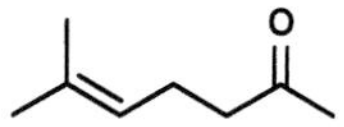

Fig. 4.6. Sulcatone (**14**), the kairomone used by the zoodariid spider *Habronestes bradleyi*.

closely correlated with the agonistic behavior of the ants. Other ant species are not attacked. The kairomone is detected by receptors on the first pair of legs, as determined by ablation experiments.

Other zoodariid spiders that eat ants or termites possess a well-developed femoral organ, which is used to immobilize or kill prey. Even superficial contact with the organ results in paralysis or death, suggesting that the femoral organ contains a powerful toxin (Jocqué and Billen, 1987; Jocqué and Dippenaar-Schoeman, 1992). However, *H. bradleyi* does not have the femoral organs (Jocqué and Billen, 1987).

Kairomones are also used for prey location by other spider families. The myrmecophagic jumping spider *Habrocestum pulex* (Salticidae) is attracted by the alarm pheromone **14** (Clark *et al.*, 2000). This species also recognizes soil on which ants had been running, as shown by increased activity, such as agitated walking, undirected leaping, and posturing with body raised.

Another kairomonal interaction, the detection of chemical cues from predators, has been found in the interaction between *Argyrodes trigonum* (Theridiidae), a kleptoparasitic species that lives in the web of and occasionally consumes its linyphiid host *F. communis* (as *F. pyramitela*). Upon contact with *A. trigonum* cuticular chemicals, *F. communis* responds by retreating, in contrast to the attack behavior elicited by cuticular compounds from other prey species (Suter *et al.*, 1989).

Chemical cues from spiders are also used by non-arachnid predators to locate their prey. For example, the ichneumonid wasp *Gelis festinans* parasitizes the spider *Erigone atra*, which lives in wheat fields. Contact with the silk of its host elicits increased searching behavior from the parasitoid, whereas contact with silk from other spider species does not, indicating a high degree of specificity (Baarlen *et al.*, 1996). It has been suggested that silk lipids may play a role in these interactions, but as yet there is no hard evidence.

Spider allomones

One of the most fascinating uses of semiochemicals by spiders is mimicry of lepidopteran pheromones by bolas spiders (Araneidae), first observed by Hutchinson (1903). The biology of these spiders has been reviewed in detail by Yeargan (1994). Briefly, bolas spiders (Mastophoreae) do not build webs like other Araneidae but instead produce a short hanging line with a drop of glue at the end. When a male

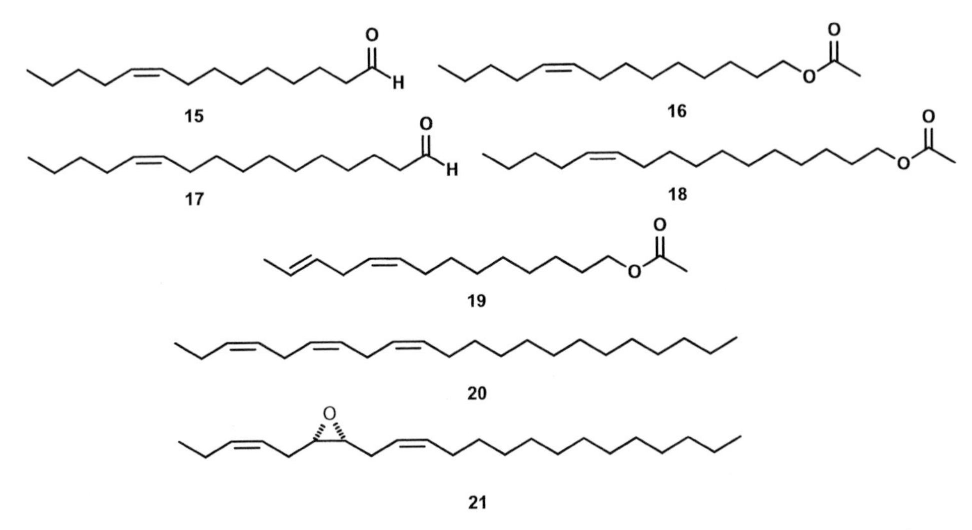

Fig. 4.7. Volatiles identified from the bolas spider *Mastophora cornigera* and *Mastophora hutchinsoni* (**15**–**19**) and the pheromone components of *Tetanolita myenesalis* (**20**, **21**). **15**, (*Z*)-9-Tetradecenal; **16**, (*Z*)-9-tetradecenyl acetate; **17**, (*Z*)-11-hexadecenal; **18**, (*Z*)-11-hexadecenyl acetate; **19**, (9*Z*,12*E*)-9,12-tetradecadienyl acetate; **20**, (3*Z*,6*Z*,9*Z*)-3,6,9-henicosatriene; **21**, (3*Z*,9*Z*)-(6*S*, 7*R*)-6,7-epoxy-3,9-henicosatriene.

moth approaches, the spider rapidly flicks its sticky bolas at the prey. If the moth is hit, which occurs only rarely, the glue holds the struggling victim so that it can be bitten by the spider. The male moths are attracted by allomones emanating from the spiders. The fact that only male Lepidoptera from a subset of the available species are caught prompted the hypothesis that the allomones might be mimicing the sex pheromones of the prey species (Eberhard, 1977). In a groundbreaking study Stowe *et al.* (1987) corroborated this hypothesis, identifying the lepidopteran sex pheromone components (*Z*)-9-tetradecenal (**15**), (*Z*)-9-tetradecenyl acetate (**16**), (*Z*)-11-hexadecenal (**17**), and possibly (*Z*)-11-hexadecenyl acetate (**18**) from odors collected from hunting *Mastophora cornigera* (Fig. 4.7).

Even more extraordinary, the authors presented evidence that the attractant mixture produced may vary with time both within and between individuals, suggesting that the spiders have some degree of control over the composition of the blend. This would explain why at least 19 different moth species are captured by *M. cornigera*, even when compounds that attract one species are known to be inhibitory to a second prey species. For example, the noctuid moth *Peridroma saucia* is attracted to blends of **16** and **18**, but attraction is inhibited by **17**. This type of situation is well documented with lepidopteran sex pheromones and is one of the mechanisms that allows related species to share some pheromone components while still maintaining species-specific attractant blends (Roelofs, 1995). Thus, *M. cornigera* must either be able to tune its blend specifically to match closely the required blend, or it must rely on the appearance of an occasional or accidental visitor attracted by an off-blend. Even in the latter case, the capture rate might be sufficient to supply the spider's nutritional needs.

The situation may be less complicated with other bolas spiders that attract fewer prey species. For example, *Mastophora hutchinsoni* catches four different moth species, of which *Lacinipolia renigera* and *Tetanolita myenesalis* are the major prey (Yeargan, 1994). The sex pheromone of *L. renigera* is made up of a 100:4 mixture of **16** and (9*Z*,12*E*)-9,12-tetradecadienyl acetate (**19**; Fig. 4.7), and both components are required to attract male moths. Analysis of the volatiles released by individual spiders in the field confirmed the presence of **16** and presented circumstantial but conclusive evidence for the presence of **19**, in a ratio similar to that produced by female moths (Gemeno *et al.*, 2000), confirming aggressive mimicry of the moth's sex pheromone by the spider. The minor prey species *Nephelodes minimans* and *Parapediasia teterrella* are attracted by the aldehyde **17** and the corresponding acetate **18**, and by **17** and its (*Z*)-9-isomer, respectively. These compounds have not been detected in the headspace around hunting spiders but may be produced in amounts below the detection limit of the analytical methods used.

Compounds **15–19** belong to the major class of moth pheromones that are biosynthesized by the action of a desaturase on a fatty acyl precursor followed by modification of the polar head group of the alkyl chain (Francke and Schulz, 1999). Another large group of moths use a second class of pheromones, long-chain polyenes and their epoxides (Francke and Schulz, 1999; Millar, 2000), which are formed by a somewhat different biosynthetic route, the details of which are still under study. The second major prey species of *M. hutchinsoni*, *T. myenesalis*, uses these types of compound in its pheromone, specifically, a 1:1 mixture of (3*Z*,6*Z*,9*Z*)-3,6,9-henicosatriene (**20**) and (3*Z*,9*Z*)-(6*S*,7*R*)-6,7-epoxy-3,9-henicosatriene (**21**) (Haynes *et al.*, 1996) (Fig. 4.7). The available evidence suggests that the spider is able to synthesize these types of lepidopteran pheromone as well, but they have not yet been conclusively identified from the spider. Furthermore, the spider must be able to change its blend, because the pheromone blend of *L. renigera* interferes with the attraction of *T. myenesalis* (Gemeno *et al.*, 2000) that is, it should not be producing both types of pheromone simultaneously.

The use of allomones to attract prey is not restricted to adult or last instar stages of *M. hutchinsoni*. Because juvenile spiderlings are too small to catch moths, they attract moth flies (Diptera: Psychodidae). They do not use a bolas but instead seize their prey with their forelegs (Yeargan and Quate, 1996), and each spider species specializes on a specific prey species. Allomones that might mediate the attraction of prey have not yet been identified, and, indeed, the possible pheromones of psychodid moth flies are still largely unknown. To date, homosesquiterpene pheromones have been identified only from psychodid sandflies in the genus *Lutzomyia* (Hamilton *et al.*, 1999a–c). However, *Lutzomyia* belongs to the subfamily Phlebotominae, in which it is known that males attract females with pheromones, whereas the bolas spiderlings preferentially catch *Psychoda*, in the subfamily Psychodinae. Furthermore, the spiderlings preferentially catch male *Psychoda*, again suggesting a link between the prey's pheromone and the spider-produced allomone (Yeargan and Quate, 1996). Stowe (1986, 1988) and Yeargan (1994) have provided a thorough discussion of the prey capture strategies of bolas spiders.

It must be pointed out that the analysis of bolas spider allomones is extremely difficult because the spiders produce them in minute amounts (1–2 ng/h; Stowe *et al.*, 1987). In addition, the spiders must be sampled individually because they are solitary hunters.

Attraction of prey by chemical means is not restricted to the six genera (Gertsch, 1947) that make up the Mastophoreae, namely *Ordgarius*, *Dichrostichus*, *Mastophora*, *Agatostichus*, *Cladomelea*, and *Acantharanea*. The tribe Cyrtarachneae is related to bolas spiders and produces horizontal webs with spanning threads covered with viscid droplets and a preformed breaking joint at one end. When a moth flies past these threads, the joints break, entangling the moth in the web. It has been suggested that species with reduced webs would be most likely to use pheromones for efficient attraction of prey (Robinson, 1982). The Celaenieae (Araneidae) with the genera *Celaenia* and *Taczanowskia* also attract moths, catching their prey with their outstretched legs (Stowe, 1986). A similar behavior has been observed from *Kaira alba* (Araneinae), a species not related to the other ones, which also catches male moths (Stowe, 1986). Attraction of male moths has also been observed occasionally to ordinary orb webs of *Araneus aurantia*, *A. trifasciata*, and *A. trifolium* (Horton, 1979), suggesting that the web might release a mimic of the sex pheromone of the attracted moth, *Hemileuca lucina*, but attraction by other cues could not be excluded. Another possible example of the attraction of prey by mimicry of their sex pheromones was encountered during the analysis of spider silk lipids of *L. triangularis* (Schulz and Toft, 1993a), which were found to contain small amounts of (*Z*)-9-tricosene and (*Z*)-11-octadecenyl acetate, especially from subadult spiders. Both are known pheromone components of flies (Francke and Schulz, 1999), often attracting both sexes.

Another interesting use of semiochemicals for prey attraction was observed in the social spider *Mallos gregalis* (Dyctinidae). These spiders prey on carrion-feeding flies. Yeasts grow well in the remnants of previously trapped flies, which the spiders leave in their webs unlike other social spiders. The odors produced by fermentation of the fly carcasses attract more flies. Interestingly, the yeasts do not grow in normal dead flies or flies eaten by other spiders, indicating a mutualistic relationship between the spiders and the yeast (Tietjen *et al.*, 1987). Furthermore, mimicry of food odors may be an excellent strategy for preying on several species simultaneously, whereas pheromone mimicry, although more reliable, may also be more selective. Several examples of such strategies are known in the animal kingdom (Stowe, 1988; Dettner and Liepert, 1994).

Chemical mimicry of spiders

The bolas spider allomones described above are examples of aggressive chemical mimicry. Another form of mimicry seems to operate in some myrmecophilic spiders that live all or part of their lives in ant nests. Cushing (1997) lists 31 such species from 13 families. The mechanism by which the spiders become integrated into the ant nest is largely unknown. Recently, the behavior of the spider *Gamasomorpha maschwitzi* (Oonopidae), which lives in colonies of the southeast Asian army ant *Leptogenys distinguenda* (Ponerinae), was investigated (Witte *et al.*, 1999). These spiders are not myrmecomorphic and are considerably smaller than the ants, which never attack them. *G. maschwitzi* either rides on the ants or closely follows them; the spider is also able to follow fresh pheromone trails for short distances on its own. Integration into ant colonies is often accompanied by similar profiles of epicuticular compounds of guest and host (Dettner and Liepert, 1994). Often it is not clear whether this similarity is formed by transfer of chemicals from the ants to the guests during contact, or actively mimiced by the guest via its own biosynthesis, as has been shown in some cases. The cuticular lipids of *G. maschwitzi* and *L. distinguenda* were found to be virtually identical, with unbranched alkenes being the major components, accompanied by the respective alkanes. Even the specific mixtures of double-bond isomers characteristic for this ant species (1-, 5-, 7-, and 9-pentacosene together with 6,9-pentacosadiene) were present in the same proportions on the spider cuticle (S. Schulz, V. Witte, and U. Maschwitz, unpublished data). It seems likely that the spiders acquire these during their close contact with the host, but *de novo* synthesis cannot be ruled out. Although it has not been explicitly tested, in all likelihood the close match with the host's hydrocarbon pattern plays a role in the acceptance of the spider into the colony. Recently, it has been shown that the salticid spider *Cosmophasis bitaeniata*, which preys on the larvae of the green tree ant *Oecophylla smaragdina*, mimics the cuticular hydrocarbon pattern of its host, which consists of mono- and dimethylalkanes. Recognition bioassays with hydrocarbon extracts showed that O. *smaragdina* workers did not react aggressively to nestmate or spider extracts but did in response to non-nestmate extracts (Allan *et al.*, 2002).

Spider toxins

Spiders do not usually have defensive secretions, although some *Mastophora* and *Cyrtarachne* spp. emit a distinctive odor when handled, possibly emanating from regurgitant (Hutchinson, 1903; Eberhard, 1980). Instead, many spiders rely mainly

Fig. 4.8. Pheromone sensillum of *Cupiennius salei*.

on their poisonous chelicerae, which are normally used for prey capture, for defense. For example, the toxin of the black widow spider *Latrodectus mactans* contains two similar proteins, α-latrotoxin and α-latroinsectotoxin, both of which bind to presynaptic nerve membranes and induce extensive neurotransmitter release. These toxins have identical action, but they differ in their target organisms: α-latrotoxin is only toxic to vertebrates, whereas α-latroinsectotoxin is only toxic to insects (Grishin, 1996). Spider toxins are usually proteins or conjugates between amino acids, aromatic acids, and polyamines (called acylpolyamines). They have been extensively studied, and the reader is directed to the numerous recent reviews of their chemistry and effects (Usherwood and Blagbrough, 1991; Blagbrough *et al.*, 1992; McCormick and Meinwald, 1993; Schäfer *et al.*, 1994; Mueller *et al.*, 1995; Magazanik, 1996; Schulz, 1997a; Guggisberg and Hesse, 1998; Escoubas *et al.*, 2000).

Perception and production of spider semiochemicals

The mechanisms of perception, biosynthesis, regulation, and genetics of the spider pheromones are largely unknown, as are their sites of production. Recently, the sensillum that detects the pheromone of *C. salei* has been identified with the help of synthetic cupilure (**3**) (Tichy *et al.*, 2001). It is a cuticular tip-pore sensillum (Fig. 4.8) that is located on the pedipalps and is similar in design to other contact chemoreceptors. The sensillum hairs contain dendrites that enervate 19 chemosensitive cells (Foelix and Chu-Wang, 1973), which are used mainly for food assessment. Whether they are also used for olfaction is not known. The tarsal organs originally were thought to be involved in olfaction, based on electrophysiological studies (Dumpert, 1978), but later it was shown that they are in fact hygroreceptors (Tichy and Ehn, 1994). The sensitivity of the pheromone sensillum has been estimated.

The contact area is about 0.03 μm^2, on which about 180 000 pheromone molecules are present on the silk under natural conditions.

Even less is known about the possible sites of production of spider pheromones. Silk-associated pheromones are probably produced in the silk gland system. For example, pheromone was found on silk drawn from immobile, anaesthetized *C. salei*. In some cases, pores in the cuticle have been suggested to be pheromone-emitting structures, as in *Alopecosa cuneata* (Kronestedt, 1986) (see also the femoral organ of zoodariid spiders (Jocqué and Billen, 1987; Jocqué and Dippenaar-Schoeman, 1992)). Epidermal glands of the genital area of female spiders have also been suggested to be involved in pheromone production (Kovoor, 1981).

Analytical methods for the analysis of spider pheromones

The methods used to analyze spider semiochemicals are similar to those used with insects. However, a complicating factor is that spiders normally have to be held individually, so that large-scale rearing and collection of silk can be laborious. Collection of silk from natural sources is not useful because the silk absorbs many compounds present in the environment, as well as dust particles and debris, particularly when glue is present on the fibers. In fact, analysis of natural samples can even show whether the webs originate from places near odorous flowers or spruce trees (S. Schulz and S. Toft, unpublished results). For the identification of the *Linyphia* pheromone **5** (Schulz and Toft, 1993b), a large number of frames were built in which the spiders spun their webs in the laboratory. The webs were then collected and combined to obtain sufficient material for analysis (about 20 webs per sample). Silk from spiders that do not build webs can be obtained by winding the silk onto a reel. For example, *C. salei* spiders were anaesthetized with carbon dioxide and the silk was pulled from the silk glands and wound onto a rotating glass rod (Tichy *et al.*, 2001). Anaesthetizing may alter the physiological state of the spider, and therefore, silk collected in this fashion may differ from silk produced under more natural conditions. For example, rapid reeling of silk may lead to a depletion of any active compound normally present in the silk, even though this did not appear to be the case with *C. salei*.

Volatile pheromones from spiders are more difficult to analyze because the site of the pheromone-producing organs is not known. Good results have been obtained by collection of volatiles released into the airspace around a spider. For example, single individuals of *A. aperta* confined in 50 ml glass chambers emitted enough material for analysis (Papke *et al.*, 2001). Probable pheromone components were deduced by comparison of samples collected from individuals that had been shown to be attractive or unattractive to males. Abdominal wipes afforded similar quantities of the pheromone, but together with large amounts of cuticular components, which complicated the analysis. It is important to confine the spiders in vials that are large enough to allow the spiders to exhibit normal behaviors. Similar techniques were used for the identification of the bolas spiders' allomones.

Because of the small amounts of sample that are usually obtained, coupled GC–MS is the method of choice for analysis of volatile pheromones. The analysis of the less-volatile lipids and polar pheromone components may require derivatization and microchemical tests, both to improve chromatographic characteristics and to provide information about the structures. It is likely that chromatographic techniques with high separation power and high sensitivity for polar compounds, such as coupled capillary electrophoresis–mass spectrometry, will prove useful for analysis of spider extracts in future studies.

Conclusion

The available information unequivocally demonstrates the widespread use of semiochemicals by spiders, including both pheromones and other types of semiochemical such as kairomones and allomones. Many fundamental questions have yet to be addressed, such as the site of pheromone biosynthesis or storage, the mechanisms of signal perception, and the biosynthetic pathways used for spider semiochemicals. Much more research, involving both biologists and chemists, is needed to get a more comprehensive picture of the use of semiochemicals by spiders. With the limited information available to us, it is premature to speculate on the possible practical applications of spider semiochemicals, such as the attraction of beneficial spiders into agricultural crops, trapping of potentially dangerous spiders in houses, or improvement in breeding procedures for spiders.

Acknowledgements

I thank Susan Riechert and Sören Toft for many useful comments and corrections. G. B. Edwards supplied me with an obscure reprint.

References

Aldrich, J. R. and Barros, T. M. (1995). Chemical attraction of male crab spiders (Araneae, Thomisidae) and kleptoparasitic flies (Diptera, Milichiidae and Chloropidae). *Journal of Arachnology* **23**: 212–214.

Allan, R. A., Elgar, M. A. and Capon, R. J. (1996). Exploitation of an ant chemical alarm signal by the zodariid spider *Habronestes bradleyi* Walckenaer. *Proceedings of the Royal Society of London, Series B* **263**: 69–73.

Allan, R. A., Capon, R. J., Brown, W. V. and Elgar, M. A. (2002). Mimicry of host cuticular hydrocarbons by salticid spider *Cosmophasis bitaeniata* that preys on larvae of tree ants *Oecophylla smaragdina*. *Journal of Chemical Ecology* **28**: 835–848.

Anava, A. and Lubin, Y. (1993). Presence of gender cues in the web of a widow spider, *Latrodectus revivensis*, and a description of courtship behavior. *Bulletin of the British Arachnological Society* **9**: 119–122.

Anderson, J. T. and Morse, D. H. (2001). Pick-up lines: cues used by male crab spiders to find reproductive females. *Behavioral Ecology* **12**: 360–366.

Ayyagari, L. R. and Tietjen, W. J. (1986). Preliminary isolation of male inhibitory pheromone of the spider *Schizocosa ocreata* (Araneae, Lycosidae). *Journal of Chemical Ecology* **13**: 237–244.

Baarlen, P. V., Topping, C. J. and Sunderland, K. D. (1996). Host location by *Gelis festinans*, an eggsac parasitoid of the linyphiid spider *Erigone atra*. *Entomologica Experimentalis et Applicata* **81**: 155–163.

Bagnères, A. G., Trabalon, M., Blomquist, G. J. and Schulz, S. (1997). Waxes of the social spider *Anelosimus eximus*: abundance of novel *n*-propyl esters of long-chain methyl-branched fatty acids. *Archives of Insect Biochemistry and Physiology* **36**: 295–314.

Barnes, M. C., Persons, M. H. and Rypstra, A. L. (2002). The effect of predator chemical cue age on antipredator behavior in the wolf spider *Pardosa milvina* (Araneae: Lycosidae). *Journal of Insect Behavior* **15**: 269–281.

Barth, F. G. and Schmitt, A. (1991). Species recognition and species isolation in wandering spiders (*Cupiennius* spp.; Ctenidae). *Behavioral Ecology and Sociobiology* **29**: 333–339.

Blagbrough, I. S., Brackley, P. T. H., Bruce, M. *et al.* (1992). Arthropod toxins as leads for novel insecticides: an assessment of polyamine amides as glutamate antagonists. *Toxicon* **30**: 303–322.

Blanke, R. (1973). Nachweis von Pheromonen bei Netzspinnen. *Naturwissenschaften* **60**: 481.

(1975). Untersuchung zum Sexualverhalten von *Cyrtophora cicatrosa* (Araneae, Araneidae). *Zeitschrift für Tierpsychologie* **377**: 62–74.

Bristowe, W. S. and Locket, G. H. (1926). The courtship of British lycosid spiders, and its probable significance. *Proceedings of the Zoological Society, London* **1926**: 317–347.

Clark, R. J. and Jackson, R. R. (1994). Self recognition in a jumping spider: *Portia labiata* females discriminate between their own draglines and those of conspecifics. *Ethology, Ecology and Evolution* **6**: 371–375.

(1995a). Araneophagic jumping spiders discriminate between the draglines of familiar and unfamiliar conspecifics. *Ethology, Ecology and Evolution* **7**: 185–190.

(1995b). Dragline-mediated sex recognition in two species of jumping spiders (Araneae, Salticidae), *Portia labiata* and *P. fimbriata*. *Ethology, Ecology and Evolution* **7**: 73–77.

Clark, R. J., Jackson, R. R. and Waas, J. R. (1999). Draglines and assessment of fighting ability in cannibalistic jumping spiders. *Journal of Insect Behavior* **12**: 753–766.

Clark, R. J., Jackson, R. R. and Cutler, B. (2000). Chemical cues from ants influence predatory behavior in *Habrocestum pulex*, an ant-eating jumping spider (Araneae, Salticidae). *Journal of Arachnology* **28**: 309–318.

Coddington, J. A. and Levi, H. W. (1991). Systematics and evolution of spiders (Araneae). *Annual Reviews of Ecology and Systematics* **22**: 565–592.

Coyle, F. A. and Shear, W. A. (1981). Observations on the natural history of *Sphodros abboti* and *Sphodros rufipes* (Araneae, Atypidae), with evidence for a contact sex pheromone. *Journal of Arachnology* **9**: 317–326.

Crane, J. (1949). Comparative biology of salticid spiders at Rancho Grande, Venezuela, Part IV. An analysis of display. *Zoologica, New York* **34**: 159–215.
Cushing, P. E. (1997). Myrmecomorphy and myrmecophily in spiders: a review. *Florida Entomologist* **80**: 166–193.
Dettner, K. and Liepert, C. (1994). Chemical mimicry and camouflage. *Annual Review of Entomology* **39**: 129–154.
Dijkstra, H. (1976). Searching behavior and tactochemical orientation in males of the wolf spider *Pardosa amentata* (Cl.) (Araneae, Lycosidae). *Proceedings of the Koninklijke Nederlandse Akademie van Wetenschappen, Series C, Biological and Medical Science* **79**: 235–244.
Dondale, C. D. and Hegdekar, B. M. (1973). The contact sex pheromone of *Pardosa lapidicina* Emerton (Araneae: Lycosidae). *Canadian Journal of Zoology* **51**: 400–401.
Dumais, J., Perron, J. M. and Dondale, C. D. (1973). Eléments du comportement sexuel chez *Pardosa xerampelina* (Keyserling) (Araneidae: Lycosidae). *Canadian Journal of Zoology* **51**: 265–271.
Dumpert, K. (1978). Spider odor receptor: electrophysiological proof. *Experientia* **34**: 754–755.
Eberhard, W. G. (1977). Agressive chemical mimicry by bolas spiders. *Science* **198**: 1173–1175.
(1980). The natural history and behavior of the bolas spider *Mastophora dizzydeani* sp. n. (Araneidae). *Psyche* **87**: 143–169.
Enders, F. (1975). Airborne pheromone probable in orb web spider *Argiope aurantia* (Araneidae). *British Arachnological Society News* **13**: 5–6.
Escoubas, P., Diochot, S. and Corzo, G. (2000). Structure and pharmacology of spider venom neurotoxins. *Biochimie* **82**: 893–907.
Evans, T. A. and Main, B. Y. (1993). Attraction between social crab spiders: silk pheromones in *Diaea socialis*. *Behavioral Ecology* **4**: 99–105.
Farley, C. and Shear, W. A. (1973). Observations on the courtship behaviour of *Lycosa carolinensis*. *Bulletin of the British Arachnological Society* **2**: 153–158.
Foelix, R. F. (1996). *Biology of Spiders*, 2nd edn, Oxford: Oxford University Press.
Foelix, R. F. and Chu-Wang, I. W. (1973). The morphology of spider sensilla. II. Chemoreceptors. *Tissue Cell* **5**: 451–460.
Francke, W. and Schulz, S. (1999). Pheromones. In *Comprehensive Natural Products Chemistry*, eds. D. Barton, K. Nakanishi, O. Meth-Cohn and K. Mori, vol. 8, pp. 197–261. Amsterdam: Elsevier.
Gemeno, C., Yeargan, K. V. and Haynes, K. F. (2000). Aggressive chemical mimicry by the bolas spider *Mastophora hutchinsoni*: identification and quantification of a major prey's sex pheromone components in the spider's volatile emissions. *Journal of Chemical Ecology* **26**: 1235–1243.
Gertsch, W. J. (1947). Spiders that lasso their prey. *Natural History* **56**: 152–158.
Grishin, E. V. (1996). Neurotoxin from black widow spider venom. Structure and function. *Advances in Experimental Medicine and Biology* **391**: 231–236.
Guggisberg, A. and Hesse, M. (1998). Natural polyamine derivatives: new aspects of their isolation, structure elucidation, and synthesis. In *The Alkaloids*, vol. 50, eds. R. H. F. Manske, R. G. A. Rodrigo, A. Brossi and G. A. Cordell, pp. 219–256. New York: Academic Press.
Gwinner-Hanke, H. (1970). Zum Verhalten zweier stridulierender Spinnen *Steatoda bipunctata* Linné und *Teutana grossa* Koch (Theridiidae, Araneae), unter besonderer Berücksichtigung ihres Fortpflanzungsverhaltens. *Zeitschrift für Tierpsychologie* **27**: 649–678.

Hamilton, J. G. C., Brazil, R. P., Morgan, E. D. and Alexander, B. (1999a). Chemical analysis of oxygenated homosesquiterpenes: a putative sex pheromone from *Lutzomyia lichyi* (Diptera: Psychodidae). *Bulletin of Entomological Research* **89**: 139–145.

Hamilton, J. G. C., Hooper, A. M., Ibbotson, H. C. *et al.* (1999b). 9-Methylgermacrene-B is confirmed as the sex pheromone of the sandfly *Lutzomyia longipalpis* from Lapinha, Brazil, and the absolute stereochemistry defined as S. *Chemical Communications* 2335–2336.

Hamilton, J. G. C., Hooper, A. M., Mori, K., Pickett, J. A. and Sano, S. (1999c). 3-Methyl-α-himachalene is confirmed, and the relative stereochemistry defined, by synthesis as the sex pheromone of the sandfly *Lutzomyia longipalpis* from Jacobina, Brazil. *Chemical Communications* 335–356.

Haynes, K. F., Yeargan, K. V., Millar, J. G. and Chastan, B. B. (1996). Identification of sex pheromone of *Tetanolita myenesalis* (Lepidoptera: Noctuidae), a prey species of bolas spiders, *Mastophora hutchinsoni*. *Journal of Chemical Ecology* **22**: 75–89.

Hegdekar, B. M. and Dondale, C. D. (1969). A contact sex pheromone and some response parameters in lycosid spiders. *Canadian Journal of Zoology* **47**: 1–4.

Hickman, V. V. (1964). On *Atrax infensus* sp. n. (Araneida: Dipluridae), its habits and a method of trapping the males. *Papers and Proceedings of the Royal Society of Tasmania* **98**: 107–112.

Higgins, L. E., Townley, M. A., Tillinghast, E. K. and Rankin, M. A. (2001). Variation in the chemical composition of orb webs built by the spider *Nephila clavipes* (Araneae, Tetragnathidae). *Journal of Arachnology* **29**: 82–94.

Horton, C. C. (1979). Apparent attraction of moths by the webs of araneid spiders. *Journal of Arachnology* **7**: 88.

Hutchinson, D. E. (1903). A bolas throwing spider. *Scientific American* 89–172.

Jackson, R. R. (1978). Male mating strategies of dictynid spiders with differing types of social organization. *Symposia of the Zoological Society, London* **42**: 79–88.

(1982). Comparative study of *Dictyna* and *Mallos* (Araneae: Dictynidae): IV. Silk mediated interattraction. *Insectes Sociaux* **29**: 15–24.

(1986). Use of pheromones by males of *Phidippus johnsoni* (Araneae, Salticidae) to detect subadult females that are about to molt. *Journal of Arachnology* **14**: 137–139.

(1987). Comparative study of releaser pheromones associated with the silk of jumping spiders (Araneae, Salticidae). *New Zealand Journal of Zoology* **14**: 1–10.

Jackson, R. R. and Cooper, K. J. (1990). Variability in the responses of jumping spiders (Araneae: Salticidae) to sex pheromones. *New Zealand Journal of Zoology* **17**: 39–42.

Jocqué, R. and Billen, J. (1987). The femoral organ of the Zodariinae (Araneae, Zodariidae). *Revue de Zoologie Africaine* **101**: 165–170.

Jocqué, R. and Dippenaar-Schoeman, A. S. (1992). Two new termite-eating *Diores* species (Araneae, Zoodariidae) and some observations on unique prey immobilization. *Journal of Natural History* **26**: 1405–1412.

Kaston, B. J. (1936). The senses involved in the courtship of some vagabound spiders. *Entomologica Americana* **16**: 97–169.

Kolosvary, G. V. (1932). Neue Daten zur Lebensweise der *Trochosa* (*Hogna*) *singoriensis* (Laxm.). *Zoologischer Anzeiger* **98**: 307–317.

Kovoor, J. (1981). Une source probable de phéromones sexuelles: les glandes tégumentaires de la région génitale des femelles d'araignées. *Atti della Soccietà Toscana di Scienze Natturali* **88**: 1–15.

Krafft, B. (1970). Contribution à la biologie et à l'éthologie d'*Agelena consociata* Denis (araignée sociale du Gabon), II. *Biologia Gabonica* **4**: 308–369.

(1971). Contribution à la biologie et à l'éthologie d'*Agelena consociata* Denis (araignée sociale du Gabon), III: Etude expérimentale de certains phénomènes sociaux (suite). *Biologia Gabonica* **7**: 3–56.
(1974). La tolérance réciproque chez l'araignée sociale *Agelena consociata* Denis. *Proceedings of the 6th International Arachnological Congress* 107–112.
(1978). The recording of vibratory signals performed by spiders during courtship. *Symposia of the Zoological Society, London* **42**: 59–67.
Kronestedt, T. (1986). A presumptive pheromone-emitting structure in wolf spiders (Araneae, Lycosidae). *Psyche* **93**: 127–131.
Kullmann, E. J. (1972). Evolution of social behavior in spiders (Araneae: Eresidae and Theridiidae). *American Zoologist* **12**: 419–426.
Leborgne, R. (1981). Soie et communication chez les araignées (le rapprochement des sexes). *Atti della Soccietà Toscana di Scienze Natturali* **88**: 132–142.
Lendl, A. (1887–88). *Trochosa infernalis* Motschl. párzásáról ès párzási szerveirol. *Temr Füzetek* **11**: 30–39.
Lizotte, R. and Rovner, J. S. (1989). Water-resistant sex pheromones in lycosid spiders from a tropical wet forest. *Journal of Arachnology* **17**: 121–125.
Magazanik, L. G. (1996). Spider neurotoxins as tools for the investigation of glutamate receptors. *Journal of Toxicology, Toxin Reviews* **15**: 59–76.
McCormick, K. D. and Meinwald, J. (1993). Neurotoxic acylpolyamines from spider venoms. *Journal of Chemical Ecology* **19**: 2411–2451.
Millar, J. G. (2000). Polyene hydrocarbons and epoxides: a second major class of lepidopteran sex attractant pheromones. *Annual Review of Entomology* **45**: 575–604.
Miyashita, T. and Hayashi, H. (1996). Volatile chemical cue elicits mating behavior of cohabiting males of *Nephila clavata* (Araneae, Tetragnathidae). *Journal of Arachnology* **24**: 9–15.
Mueller, A. L., Roeloffs, R. and Jackson, H. (1995). Pharmacology of polyamine toxins from spiders and wasps. In *The Alkaloids*, vol. 46, eds. R. H. F. Manske, R. G. A. Rodrigo, A. Brossi and G. A. Cordell, pp. 63–94. New York: Academic Press.
Olive, C. W. (1982). Sex pheromones in two orb weaving spiders, (Araneae, Araneidae): an experimental field study. *Journal of Arachnology* **10**: 241–245.
Papke, M. D., Schulz, S., Tichy, H., Gingl, E. and Ehn, R. (2000). Identification of a new sex pheromone from the silk dragline of the tropical wandering spider *Cupiennius salei*. *Angewandte Chemie, International Edition* **39**: 4339–4341.
Papke, M. D., Riechert, S. E. and Schulz, S. (2001). An airborne female pheromone associated with male attraction and courtship in a desert spider. *Animal Behaviour* **61**: 877–886.
Pasquet, A., Trabalon, M., Bagnères, A. G. and Leborgne, R. (1997). Does group closure exist in the social spider *Anelosimus eximus*? Behavioural and chemical approach. *Insectes Sociaux* **44**: 159–169.
Persons, M. H. and Rypstra, A. L. L. (2000). Preference for chemical cues associated with recent prey in the wolf spider *Hogna helluo* (Araneae: Lycosidae). *Ethology* **106**: 27–35.
(2001). Wolf spiders show graded antipredator behavior in the presence of chemical cues from different sized predators. *Journal of Chemical Ecology* **27**: 2493–2504.
Persons, M. H., Walker, S. E., Rypstra, A. L. and Marshall, D. S. (2001). Wolf spider predator avoidance tactics and survival in the presence of diet-associated predator cues (Araneae: Lycosidae). *Animal Behaviour* **61**: 43–51.

Platnick, N. I. (1999). Dimensions of biodiversity: targeting megadiverse groups. In *The Living Planet in Crisis. Biodiversity Science and Policy*, eds. J. Cracraft and F. T. Grifo, pp. 33–52. New York: Columbia University Press.
(2001). *The World Spider Catalog, Version 2.0.* American Museum of Natural History, http://research.amnh.org/entomology/ spiders/catalog81-87/index.html (version of January 2001).
Pollard, S. D., Macnab, A. M. and Jackson, R. R. (1987). Communication with chemicals: pheromones and spiders. In *Ecophysiology of Spiders*, ed. W. Nentwig, pp. 133–141. Berlin: Springer.
Pourié, G. and Trabalon, M. (2001). Plasticity of agonistic behaviour in relation to diet and contact signals in experimentally group-living of *Tegenaria atrica. Chemoecology* **11**: 175–181.
Prenter, J., Elwood, R. W. and Montgomery, W. I. (1994). Assessments and decisions in *Metellina segmentata* (Araneae: Metidae): evidence of a pheromone involved in mate guarding. *Behavioral Ecology and Sociobiology* **35**: 39–43.
Prouvost, O., Trabalon, M., Papke, M. and Schulz, S. (1999). Contact sex signals on web and cuticle of *Tegenaria atrica* (Araneae, Agenelidae). *Archives of Insect Biochemistry and Physiology* **40**: 194–202.
Richter, C. J. J., Stolting, H. C. J. and Vlijm, L. (1971). Silk production in adult females of the wolf spider *Pardosa amentata. Journal of Zoology, London* **165**: 285–290.
Riechert, S. E. and Singer, F. D. (1995). Investigation of potential male mate choice in a monogamous spider. *Animal Behaviour* **49**: 719–723.
Robert, T. and Krafft, B. (1981). Contribution à l'étude des mécanismes de la communication tacto-chimique intervenant dans le rapprochement des sexes chez *Pardosa hortensis* Thorell (Araneae, Lycosidae). *Atti della Soccietà Toscana di Scienze Natturali* **88**: 143–153.
Robinson, M. H. (1982). The ecology and biogeography of spiders in Papua New Guinea. *Monographiae Biologicae* **42**: 557–581.
Roelofs, W. L. (1995). Chemistry of sex attraction. *Proceedings of the National Academy of Sciences, USA* **92**: 44–49.
Roland, C. (1984). Chemical signals bound to the silk in spider communication (Arachnida, Araneae). *Journal of Arachnology* **11**: 309–314.
Roland, C. and Rovner, J. S. (1983). Chemical and vibratory communication in the aquatic pisaurid spider *Dolomedes triton. Journal of Arachnology* **11**: 77–85.
Ross, K. and Smith, R. L. (1979). Aspects of the courtship behavior of the black widow spider *Latrodectus hesperus* (Araneae: Theridiidae), with evidence for the existence of a contact sex pheromone. *Journal of Arachnology* **7**: 69–77.
Rovner, J. S. (1968). An analysis of display in the lycosid spider *Lycosa rabida* Walckenaer. *Animal Behaviour* **16**: 358–369.
Rovner, J. S. and Barth, F. G. (1981). Vibratory communication through living plants by a tropical wandering spider. *Science* **214**: 464–466.
Sarinana, F. O., Kittredge, J. S. and Lowrie, D. C. (1971). A preliminary investigation of the sex pheromone of *Pardosa ramulosa. Notes on the Arachnology of the Southwest* **2**: 9–11.
Schäfer, A., Benz, H., Fiedler, W., Guggisberg, A., Bienz, S. and Hesse, M. (1994). Polyamine toxins from spiders and wasps. In *The Alkaloids*, vol. 45, eds. G. Cordell and A. Brossi, pp. 1–125. New York: Academic Press.
Schuck-Paim, C. (2000). Orb-webs as extended-phenotypes: web design and size assessment in contests between *Nephilengys cruentata* females (Araneae, Tetragnathidae). *Behaviour* **137**: 1331–1347.

Schulz, S. (1997a). The chemistry of spider toxins and spider silk. *Angewandte Chemie, International Edition in English* **36**: 314–326.

(1997b). Mass spectrometric determination of methyl group positions in long chain methyl ethers and alcohols via nitriles. *Chemical Communications* 969–970.

(1999). Structural diversity of surface lipids from spiders. In *Bioorganic Chemistry: Highlights and New Aspects*, eds. U. Diederichsen, T. K. Lindhorst, B.Westermann and L. A. Wessjohann, pp. 1–7. Weinheim: Wiley-VHC.

(2001). Composition of the silk lipids of the spider *Nephila clavipes*. *Lipids* **36**: 637–647.

Schulz, S. and Toft, S. (1993a). Branched long chain alkyl methyl ethers: a new class of lipids from spider silk. *Tetrahedron* **49**: 6805–6820.

(1993b). Identification of a sex pheromone from a spider. *Science* **260**: 1635–1637.

Searcy, L. E., Rypstra, A. L. and Persons, M. H. (1999). Airborne chemical communication in the wolf spider *Pardosa milvina*. *Journal of Chemical Ecology* **25**: 2527–2533.

Seibt, U. and Wickler, W. (1988). Interspecific tolerance in social *Stegodyphus* spiders (Eresidae, Araneae). *Journal of Arachnology* **16**: 35–39.

Shao, Z. and Vollrath, F. (1999). The effect of solvents on the contraction and mechanical properties of spider silk. *Polymer* **40**: 1799–1806.

Stewart, D. M. (1988). Endocrinology of arachnids. In *Endrocrinology of Selected Invertebrate Types*, eds. H. Laufer and G. H. Downer, pp. 415–428. New York: Alan R. Liss.

Stowe, M. K. (1986). Prey specialization in the Araneidae. In *Spiders: Webs, Behavior, and Evolution*, ed. W. A. Shear, pp. 101–131. Stanford, CT: Stanford University Press.

(1988). Chemical mimicry. In *Chemical Mediation of Coevolution*, ed. K. C. Spencer, pp. 513–580. San Diego, CA: Academic Press.

Stowe, M. K., Tumlinson, J. H. and Heath, R. R. (1987). Chemical mimicry: bolas spiders emit components of moth prey species sex pheromones. *Science* **236**: 964–967.

Suter, R. B. and Hirscheimer, A. J. (1986). Multiple web-borne pheromones in a spider *Frontinella pyramitela* (Araneae: Linyphiidae). *Animal Behaviour* **34**: 748–753.

Suter, R. B. and Renkes, G. (1982). Linyphiid spider courtship: releaser and attractant functions of a contact sex pheromone. *Animal Behaviour* **30**: 714–718.

Suter, R. B., Shane, C. M. and Hirscheimer, A. J. (1987). Communication by cuticular pheromones in a linyphiid spider. *Journal of Arachnology* **15**: 157–162.

(1989). *Frontinella pyramitela* detects *Argyrodes trigonum* via cuticular chemicals. *Journal of Arachnology* **17**: 237–240.

Taylor, P. W. (1998). Dragline-mediated mate-searching in *Trite planiceps* (Araneae, Salticidae). *Journal of Arachnology* **26**: 330–334.

Tichy, H. and Ehn, R. (1994). Hygro- and thermoreceptive tarsal organ in the spider *Cupiennius salei*. *Journal of Comparative Physiology A* **174**: 345–350.

Tichy, H., Gingl, E., Ehn, R., Papke, M. and Schulz, S. (2001). Female sex pheromone of a wandering spider: identification and sensory reception. *Journal of Comparative Physiology A* **187**: 75–78.

Tietjen, W. J. (1977). Dragline-following by male lycosid spiders. *Psyche* **84**: 165–178.

(1979a). Is the sex pheromone of *Lycosa rabida* (Araneae: Lycosidae) deposited on a substratum? *Journal of Arachnology* **6**: 207–212.

(1979b). Tests for olfactory communication in four species of wolf spiders (Araneae, Lycosidae). *Journal of Arachnology* **6**: 197–206.

Tietjen, W. J. and Rovner, J. S. (1980). Trail-following behaviour in two species of wolf spiders: sensory and etho-ecological concomitants. *Animal Behaviour* **28**: 735–741.

(1982). Chemical communication in lycosids and other spiders. In *Spider Communication. Mechanisms and Ecological Significance*, eds. P. N. Witt and J. S. Rovner, pp. 249–279. Princeton, NJ: Princeton University Press.

Tietjen, W. J., Ayyagari, L. R. and Uetz, G. W. (1987). Symbiosis between social spiders and yeast: the role in prey attraction. *Psyche* **94**: 151–158.

Toft, S. (1987). Microhabitat identity of two species of sheet-web spiders: field experimental demonstration. *Oecologica* **72**: 216–220.

(1989). Mate guarding in two *Linyphia* species (Araneae: Linyphiidae). *Bulletin of the British Arachnological Society* **8**: 33–37.

Trabalon, M., Bagnères, A. G., Hartmann, N. and Vallet, A. M. (1996). Changes in cuticular compounds composition during the gregarious period and after dispersal of the young in *Tegenaria atrica* (Araneae, Agelenidae). *Insect Biochemistry and Molecular Biology* **26**: 77–84.

Trabalon, M., Bagnéres, A. G. and Roland, C. (1997). Contact sex signals in two sympatric spider species, *Tegenaria domestica* and *Tegenaria pagana*. *Journal of Chemical Ecology* **23**: 747–758.

Usherwood, P. N. R. and Blagbrough, I. S. (1991). Spider toxins affecting glutamate receptors: polyamines in therapeutical neurochemistry. *Pharmacological Therapeutics* **52**: 245–268.

van Helsdingen, P. J. (1965). Sexual behaviour of *Lepthyphantes leprosus* (Ohlert) (Araneida, Linyphiidae), with notes on the function of the genital organs. *Zoologische Mededelingen* **41**: 15–42.

Vander Meer, R. K., Breed, M. D., Winston, M. L. and Espelie, K. E. (1998). *Pheromone Communication in Social Insects*. Boulder, CO: Westview Press.

Vollrath, F., Fairbrother, W. J., Williams, R. J. P. *et al.* (1990). Compounds in the droplets of the orb spider's viscid spiral. *Nature* **345**: 526–528.

Watson, P. J. (1986). Transmission of a female sex pheromone thwarted by males in the spider *Linyphia litigosa* (Linyphiidae). *Science* **233**: 219–221.

Willey, M. B. and Jackson, R. R. (1993). Olfactory cues from conspecifics inhibit the web-invasion behavior of *Portia*, web invading araneophagic jumping spiders (Araneae: Salticidae). *Canadian Journal of Zoology* **71**: 1415–1420.

Witt, P. N. (1975). The web as a means of communication. *Bioscience Communications* **1**: 7–23.

Witte, V., Hänel, H., Weissflog, A., Rosli, H. and Maschwitz, U. (1999). Social integration of the myrmecophilic spider *Gamasomorpha maschwitzi* (Araneae: Oonopidae) in colonies of the south east Asian army ant, *Leptogenys distinguenda* (Formicidae: Ponerinae). *Sociobiology* **34**: 145–159.

Yeargan, K. V. (1994). Biology of bolas spiders. *Annual Review of Entomology* **39**: 81–99.

Yeargan, K. V. and Quate, L. W. (1996). Juvenile bolas spiders attract psychodid flies. *Oecologia* **106**: 266–271.

Yoshida, H. and Suzuki, Y. (1981). Silk as a cue for mate location in the jumping spider, *Carrhotus xanthogramma* (Latreille) (Araneae: Salticidae). *Applied Entomology and Zoology* **16**: 315–317.

5

Why do flowers smell? The chemical ecology of fragrance-driven pollination

Robert A. Raguso

Department of Biological Sciences,
University of South Carolina at Columbia, USA

Introduction

Animal-assisted sexual reproduction in flowering plants–pollination – is a phenomenon in which volatile signal production and chemical communication play important and diverse roles. Plant–pollinator interactions are of paramount importance in terrestrial biology, because they bind together food webs within complex ecosystems (Gilbert, 1980), drive co-adaptive evolution among hundreds of thousands of plant and animal species (Feinsinger, 1983; Williams, 1983), and frequently determine agricultural productivity (Metcalf, 1987; Robacker *et al.*, 1988). Most plant–pollinator relationships are considered to be mutually beneficial, such that plants derive reproductive benefits (pollen export and deposition, fertilization) in exchange for resources (nectar, pollen, oils) that directly or indirectly enhance the pollinator's fitness (Heinrich and Raven, 1972; Proctor *et al.* 1996). In this context, floral scent functions alone or in conjunction with visual cues (e.g., Ômura *et al.*, 1999a) to attract pollinators, induce them to land, indicate a reward's presence and location, and teach pollinators to associate the reward with specific flowers (reviews: Dobson, 1994; Raguso, 2001).

Recent investigations have uncovered a rich panorama of odor-mediated–plant-pollinator interactions, including cheating and exploitation by either party, as well as third-party interventions by predators of the plant or its pollinator (Nishida *et al.*, 1997; Pellmyr, 1997; Gibernau *et al.*, 1998). Therefore, for modern chemical ecologists, the simple question "why do flowers smell?" has given way to more beguiling questions, such as "why do flowers have different odors?", "why do flowers change their scents?", and "why don't flowers have stronger odors?". In

This chapter is dedicated to the memory of Bastiaan J. D. Meeuse, a pioneer in the study of plant odors and chemical communication,

Advances in Insect Chemical Ecology, ed. R. T. Cardé and J. G. Millar. Published by Cambridge University Press.

this chapter, I explore the implications of these three questions for the chemical ecology of flowering plants and their pollinators, particularly in the light of recent technological and conceptual advances. In keeping with the ultimate (rather than proximate) nature of "why" questions (Alcock, 1998), I will focus more on evolutionary consequences than on physiological mechanisms. The "how" questions concerning fragrance biosynthesis and pollinator olfaction are equally compelling, and the reader is directed to recent reviews by Hildebrand and Shepard (1997), Hansson (1999), Raguso and Pichersky (1999), Dudareva and Pichersky (2000), Hansson and Anton (2000), Galizia and Menzel (2000), and Raguso (2001).

Improvements in chemical analysis and neuroethological and phylogenetic methods have provided the tools to investigate rigorously pattern and process in plant–pollinator dynamics (Armbruster, 1997; Schiestl *et al.*, 1999). The current empirical reassessment of pollination mutualisms, both in terms of their putative specificity and their inherent conflict of interest among parties, has led to new insights about the evolutionary forces shaping floral cues and pollinators' responses to them (Bronstein, 1994; Ollerton, 1996; Johnson and Steiner, 2000). I will conclude by discussing the remaining technical and intellectual barriers to the attainment of a more predictive understanding of "better pollination through chemistry."

Why do flowers have different odors?

Chemical characterization of fragrance diversity

Flower fragrances are remarkably complex: one of the greatest challenges to a rigorous understanding of fragrance is the full elucidation of its composition, both instantaneously and as it changes over time. The floral scent of a given plant species may vary in the number and relative abundance of the constituent volatiles, in the temporal emission patterns over the course of floral anthesis, and in the spatial emission patterns from different floral organs. Fragrance composition may vary among populations of the same species, and, despite considerable environmental plasticity, some of this variation is heritable (Raguso and Pichersky, 1999; Raguso, 2001). Consequently, floral scent qualifies as a form of phenotype upon which natural selection can act (Dodson *et al.*, 1969).

At a finer level of resolution, compounds with similar hydrocarbon skeletons may vary in functional group, degree of unsaturation, or oxidative state. For example, the volatile methyl esters of benzoic, salicylic and anthranilic acids (**1**; Fig. 5.1) differ only in *ortho*-position functionality, but they are olfactorily distinct to humans and at least to some phytophagous insects (Raguso *et al.*, 1996; Maekawa *et al.*, 1999). Furthermore, many scent compounds have positional isomers and enantiomers that elicit different behavioral responses from flower visitors (Williams and Whitten, 1983; Hick *et al.*, 1999). The refinement of capillary gas chromatography–mass

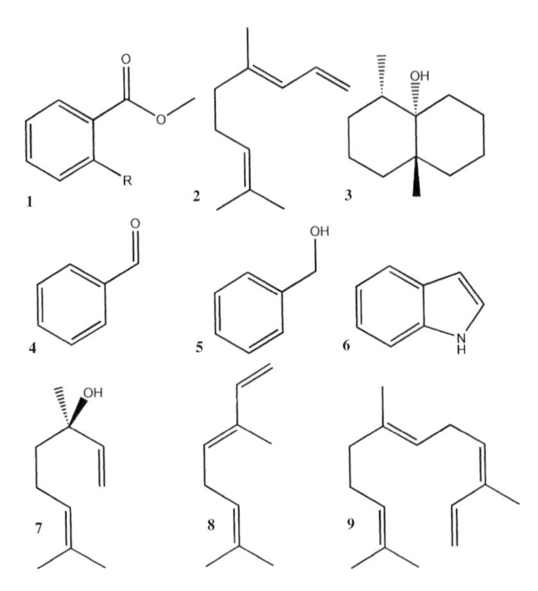

Fig. 5.1. A sample of floral scent compounds. (**1**) Substituted methyl esters of benzoic acid. The ester smells unpleasantly sweet when R = H (methyl benzoate), of wintergreen when R = OH (methyl salicylate), and of concord grape when R = NH_2 (methyl anthranylate). Odorants with unusual origins and biological functions include the homoterpene 4,8-dimethyl-1,3,7-nonatriene (**2**) and the "wet-earth" compound geosmin (**3**). Some ubiquitous fragrance compounds are benzaldehyde (**4**), benzyl alcohol (**5**), indole (**6**), (*S*)-linalool (**7**), (*E*)-β-ocimene (**8**), and α-farnesene (**9**).

spectrometry (GC–MS), automated, controlled-environment collection chambers, solid-phase micro-extraction (SPME), and other thermal desorption-based methods have made it possible to collect, separate, and identify hundreds of floral scent components from the "headspace" of living plants (Heath and Manukian, 1994; Jakobsen and Olsen, 1994; Zhang *et al.*, 1994; Agelopoulos and Pickett, 1998; Fäldt *et al.*, 2000). The importance of these sensitive non-invasive methods, both in terms of understanding the physiology of plant volatile emissions and in identifying relevant odor blends for behavioral assays, cannot be overemphasized. Collecting, separating, and identifying the diverse array of compounds making up a floral scent often requires complementary analytical approaches (Kaiser, 1991; Heath and Dueben, 1998). For example, polar compounds such as vanillin and indole (**6**), each of which provides distinctive olfactory notes to fragrances from diverse plants (Knudsen *et al.*, 1993), are collected at low efficiency using standard adsorbents (Tenax, Porapak) and elute poorly when hexane is used as a solvent (Raguso and Pellmyr, 1998). SPME is useful as a qualitative supplement to dynamic headspace sampling but, as a static headspace technique, does not provide quantitative data on emission rates nor a sample that can be reinjected on another column (Agelopoulos and Pickett, 1998). Also, it is not always possible to identify unknowns simply by comparing mass spectra and retention times on several columns with those of known standards. In such cases, the researcher may use a fraction collector to accumulate material for further analysis (e.g., by nuclear magnetic resonance or infrared spectroscopy) or use organic synthesis and further analysis to identify the compound and its exact configuration (see Kaiser, 1993). Therefore, several forms of chemical analysis often are needed for rigorous identification of the full spectrum of floral volatiles.

By definition, plant odors comprise heterogeneous groups of organic compounds whose low molecular weights and high vapor pressures determine their volatility. At the low-molecular-weight extreme, sapromyophilous flowers emit fecal or cadaverous odors formed mostly by small molecules (C_2–C_5, amines, and sulfides) that may be too volatile, too polar/hydrophilic, or too chemically unstable to be reliably analyzed without cryogenic trapping and cold on-column GC injection (Skubatz *et al.*, 1996; da Silva *et al.*, 1999; Stránský and Valterová, 1999). At the high-molecular-weight extreme, some deceptive orchids produce what humans perceive as faint odors; these mimic insect cuticular semiochemicals (Schiestl *et al.*, 2000). These large (C_{21}–C_{29}) hydrocarbons have low vapor pressure under ambient conditions and in many cases are more easily extracted with solvent or collected by contact SPME than collected from floral headspace (Schiestl and Ayasse, 2001). Interestingly, each of these examples represents a mode of pollinator deception, in which plant odors elicit oviposition and copulatory behaviors, respectively (Vogel, 1978; Dafni, 1984). Between these extremes lies the spectrum of "typical" floral scents, most of which are "honest" indicators of floral nectar, pollen, or oils and consist of heterogeneous blends of plant volatiles (Knudsen *et al.*, 1993; Dudareva and Pichersky, 2000). The major classes of scent compound are the products of anabolic or catabolic biosynthetic pathways and include the C_{10} (mono) and C_{15} (sesqui) terpenoids, aromatics, short-chain (C_4–C_{18}) aliphatics, and a variety of compounds containing heteroatoms such as nitrogen, sulfur, or oxygen (Schreier, 1984; Croteau and Karp, 1991). Scent biosynthesis incurs tangible metabolic costs; for example, aromatic volatiles are produced at the expense of photosynthate (through phosphoenolpyruvate) and amino acids (Schmid and Amrhein, 1995). Additional costs accrue when oxidative modification (Schnitzler *et al.*, 1992; Gershenzon and Croteau, 1993) or elevated respiratory activity (Vogel, 1963, 1983; Meeuse and Raskin, 1988) accompanies scent production.

Huge gaps remain in our understanding of odor chemical diversity across most plant families. Novel fragrance compounds are being discovered continuously (Kaiser, 1991, 1995; Patt *et al.*, 1992), some of which have behavioral relevance in a floral foraging context. For example, Kaiser (1991) and Gäbler *et al.* (1991) have characterized 4,8-dimethyl-1,3-(*E*)-7-nonatriene (**2**; Fig. 5.1) as an unusual but widespread component of floral scents, particularly of *Yucca filamentosa*.

Subsequent research has shown that a distinctively "floral" blend of volatiles including this irregular homoterpenoid is synthesized *de novo* and emitted systemically from the vegetation of a variety of unrelated crop plants in response to herbivore attack (review: Paré and Tumlinson, 1998). Furthermore, this blend is attractive to spider mites and parasitic wasps, whose recruitment can be thought of as an induced antiherbivore defense (Dicke, 1994; Karban and Baldwin, 1997; De Moraes *et al.*, 1998). In light of this and other examples (Galen, 1983; Vogel, 1983; Beker *et al.*, 1989), the distinction between "floral" and "vegetative" odors can be artificial and underrepresentative of the multitude of mechanisms by which plants attract animal mutualists. Two compounds not previously isolated from flowers, geosmin (**3**; Fig. 5.1) and dehydrogeosmin (named for their organoleptic similarity to the odor of wet earth), have recently been identified in several cactus flowers (Kaiser and Nussbaumer, 1990; Feng *et al.*, 1993). These flowers are pollinated by pollen-collecting, earth-nesting bees (Kaiser, 1991; Kaiser and Tollsten, 1995), for whom the geosmins may function as innate attractants (Schlumpberger, 2002). Such biosynthetic diversity and its associated physiological costs could result in different floral odors through genetic drift, even if fragrance had no adaptive value whatsoever. I now examine the extent to which scent chemistry is *not* random, with respect to phylogenetic distribution and pollinator affinities.

Distributional patterns of fragrance chemistry

The lack of systematic chemical analytical sampling, combined with the extraordinary diversity of flowering plants, makes any broad phylogenetic or ecological characterization of scent chemistry premature. However, certain compounds have been gained and lost independently in the evolution of diverse plant families (homoplasy) (Bohlmann *et al.*, 1998; Barkman, 2001; Levin *et al.*, 2001), whereas others are restricted to specific lineages or modes of pollination (Knudsen *et al.*, 1993; Kaiser and Tollsten, 1995). Furthermore, not all fragrance components occur independently, because related metabolites of the same biosynthetic pathway branches (e.g., phenylpropanoids) often are emitted together (Steele *et al.*, 1998; Barkman, 2001). Compounds such as benzaldehyde (**4**), benzyl alcohol (**5**), indole (**6**), linalool (**7**), and isomers of ocimene (**8**) and farnesene (**9**) (Fig. 5.1) are nearly ubiquitous in floral scents (Knudsen *et al.*, 1993) and are unlikely of themselves to indicate highly specific pollinator interactions. However, the combinatorial potential among floral scent components and the variation in their relative abundance present flower-visiting animals with a virtually unlimited universe of species-specific odor blends or phenotypes (Hills *et al.*, 1972; Laurent, 1999).

The earliest pollination biologists (Delpino, 1874; Kerner von Marilaum, 1895; Knuth, 1906) described striking differences in fragrance quality (by human perception) associated with different floral morphologies and putative pollinator classes. Such patterns, elaborated upon by subsequent authors (Vogel, 1954; Faegri and van der Pijl, 1979), fostered a widespread belief in highly specific "pollination syndromes," of which fragrance was a key component, and by which floral phenotypes could selectively attract appropriate and exclude inappropriate pollen vectors, thus maximizing reproductive fitness. This concept underestimates spatial and temporal variation in the spectrum of pollinators for any one plant, their foraging breadth, abundance, and effectiveness as pollen vectors; to some extent, it implies exclusivity in plant–pollinator mutualisms (Schemske and Horvitz, 1984; Waser *et al.*, 1996).

The hypothesis of convergent evolution for pollinator-specific odors is difficult to test because of technical difficulties and the need for phylogenetic controls and an appropriate null model with which to define odor convergence. Few studies have explicitly addressed the ecological or evolutionary roles for floral scent in the alternative paradigm, in which flowering plants are recognized to have mixed mating systems and more generalized pollinator spectra (Herrera, 1996; Ollerton, 1996; Waser *et al.*, 1996). However, non-random patterns of fragrance composition exist for plants with such specialized pollinators as male euglossine bees (Whitten *et al.*, 1986; Williams and Whitten, 1999), glossophagine bats (review: Winter and von Helversen, 2001), and perhaps dynastine beetles (Schatz, 1990; Gottsberger, 1999), each of which show behavioral preferences for specific odors (reviews: Dobson, 1994; Raguso, 2001). These modes of insect-mediated plant reproduction are prevalent among species-rich neotropical plant lineages (Feinsinger, 1983; Schatz 1990), and floral phenotypes (including scent) in these groups may be thought of as hyperdimensional peaks in an adaptive landscape (Williams and Whitten, 1983; Armbruster, 1990; Cresswell and Galen, 1991).

Therefore, pollinator affinity and phylogeny are to some extent correlated with differences in floral scent. The specialist–generalist debate notwithstanding, the question of why flowers produce different odors, or why odor blends tend to be species-specific, will require complex functional analyses apportioning fragrance variation to phylogenetic, physiological, and ecological influences (Azuma *et al.*, 1997). When insects that visit flowers for their own reproduction (e.g., fig wasps and male euglossine bees) are pollinators, variation in floral odor chemistry and animal behavior should be subject to the forces of sexual selection. Alternatively, when animals visit flowers as potential food sources, at least two outcomes are plausible. Plants pollinated by generalist pollinators which visit a broad spectrum of flowers and utilize associative learning, should produce odors that feature salient (easily learned and remembered) conditioning stimuli but also encode unique, species-specific entities, in order to enhance associative learning and floral constancy (Roy and Raguso, 1997; Dornhaus and Chittka, 1999). Tollsten *et al.* (1994) found low chemical similarity and high complexity among odors of three *Angelica* species with generalized pollination and suggested that such fragrances may contain nested attractants suitable for different classes of flower visitor, which differ in their relative efficiency as pollinators. Many more such studies are needed. Specialist pollinators with limited learning abilities and narrow olfactory tuning (e.g., some scarab beetles; Hansson *et al.*, 1999) could exert directional selection on fragrance variation if they are the most abundant, effective, or exclusive pollinators. This phenomenon would be comparable to patterns of "sensory bias" driving signal/receiver evolution in animal courtship, in which signal evolution is constrained by the sensory capabilities of the receiver organism (Ryan and Rand, 1993; Basolo, 1995). Adaptive radiation within plant lineages could enhance the prevalence of any of these patterns (Meeuse and Raskin, 1988; Williams and Whitten, 1999).

Functional dissection of odor blends

The chemical analysis of floral volatiles for any one species can easily reveal over 100 identified components (e.g., sunflower; Thièry *et al.*, 1990), which often vary by orders of magnitude in their relative abundances. The central challenge to the chemical ecologist is to determine the function of such a complex blend to the extent that the identity and abundance of its components are relevant to the plant's reproductive biology or to the olfactory physiology and foraging behavior of its pollinators. This challenge is especially difficult when odor blends trigger different behaviors at different spatial scales (Brantjes, 1976; Robacker *et al.*, 1988). A common approach to olfactory–behavioral analysis of odor blends is to record olfactory receptor neuron (ORN) action potentials elicited by specific compounds, either from whole antennae (electroantennogram (EAG) detection (EAD)) or from single ORNs (single cells) (review: Marion-Poll and Thièry, 1996; Hallberg and Hansson, 1999). Positive results from either assay indicate only that the stimulus was detected at its test concentration (Mayer *et al.*, 1984), whereas negative results suggest (but do not absolutely prove) that an undetected compound may be omitted from further consideration as a mediator of behaviors. A critical assumption, based on pheromone-type assays, is that odors that are potent olfactory stimulants at very low concentrations should correlate with a behavioral function(s) as an attractant, repellent, or synergist (Ômura *et al.*, 2000; Schiestl and Marrion-Poll, 2002). One rapid way to screen complex floral blends for potentially stimulatory components is through combined GC–EAD. EAG or single-cell responses to individual fragrance components are recorded as each compound elutes from the GC column (Struble and Arn, 1984; Wadhams *et al.*, 1994; review: Schiestl and Marrion-Poll, 2002). Figure 5.2 provides an example of this strategy, plotting the antennal responses of a *Manduca sexta* moth to the components of a simple floral blend that elicits

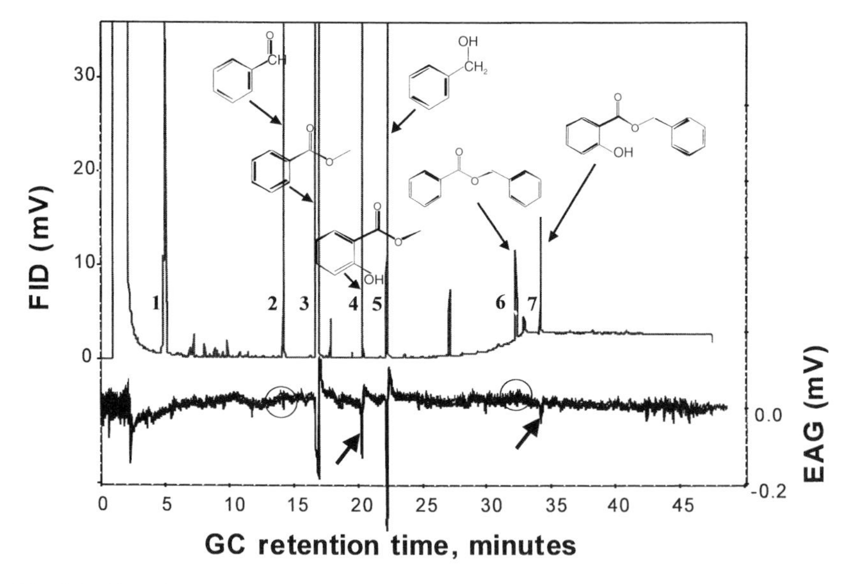

Fig. 5.2. Gas chromatography (GC) and electroantennography (EAG) analysis of male *Manduca sexta* antennal responses to floral volatiles from the night blooming cactus *Peniocereus greggii*. The upper trace is a flame ionization detection (FID) chromatogram of floral headspace odors separated on a carbowax GC column, while the lower trace is a simultaneous recording of summed antennal action potentials elicited by individual compounds as they elute. The largest absolute responses followed methyl benzoate, methyl salicylate, and benzyl alcohol (peaks 3–5, respectively). Note the poor responses (circled) to benzaldehyde and benzyl benzoate (peaks 2, 6) and the disproportionately higher responses (bold arrows) to methyl salicylate and benzyl salicylate (peak 7) relative to their peak areas. Peak 1 is the internal standard (toluene); remaining unnumbered peaks are ambient contaminants.

upwind flight in wind tunnel assays (Raguso and Willis, 2003). Although action potentials are elicited by most components in this sample, the disproportionately large response to benzyl salicylate (peak 7) relative to its small peak area (low concentration) merits further exploration; indeed, this and similar compounds have been used to trap *M. sexta* (Morgan and Lyon, 1928). However, such analyses are best employed to complement behavioral assays, not replace them. Proboscis extension and choice assays with honeybees (Henning and Teuber, 1992; Henning *et al.*, 1992; Le Mètayer *et al.*, 1997) and butterflies (Honda *et al.*, 1998; Ômura *et al.*, 1999b) have shown that an odor constituent's effectiveness as a feeding stimulant or attractant is not always positively correlated with EAG response.

In sum, both rigorous analyses of floral odors and olfactory/behavioral assays will be required to tease apart the critical factors that underlie pollinator–flower interactions. Recent work from independent sources indicates progress in this area. For example, studies using GC–EAD and conditioned proboscis extension (Wadhams *et al.*, 1994; Blight *et al.*, 1997; Pham-Delègue *et al.*, 1997) have identified a core group of compounds – 2-phenyl-ethanol, phenylacetaldehyde, benzyl alcohol and linalool – that are salient conditioning stimuli for honeybees. In studies on two widespread flower-feeding lepidopterans, *Pieris rapae* (Honda *et al.*, 1998; Ômura *et al.*, 1999a) and *Trichoplusia ni* (Haynes *et al.*, 1991; Heath *et al.*, 1992), floral scents from several plant species were functionally analyzed using EAG, proboscis extension, wind-tunnel, and feeding assays. Interestingly, 2-phenyl-ethanol, phenylacetaldehyde, benzyl acetate, and benzaldehyde were highly effective attractants and feeding stimulants in these studies. Moreover, rust fungi that mimic generalist-pollinated alpine flowers produce a powerful fragrance featuring these same compounds and their metabolites, which attract diverse flies and bees (Roy and Raguso, 1997; Raguso and Roy, 1998). If behavioral assays using comprehensive arrays of odorants continue to support the above patterns, this core blend of compounds might profitably be considered a broad peak or plateau in an adaptive landscape of floral phenotypes. Although these compounds are ubiquitous in the fragrances of generalist-pollinated plants (Dobson *et al.*, 1990; Connick and French, 1991; Erhardt, 1993; Neilsen *et al.*, 1995), it is unclear whether this reflects pollinator-mediated selection for such signals or, conversely, a metabolic predisposition of angiosperms to synthesize and emit them. Factors related to biosynthesis, plant defense, local pollinator adaptation, phylogenetic history, and genetic drift should dictate the identity of the remaining scent components in the blend for each species.

Why do flowers change their scents?

Periodicity in fragrance emissions

Botanists and perfumers alike have long recognized that fragrances change quantitatively and qualitatively over time (Robacker *et al.*, 1988; Kaiser, 1993). There are at least three biologically relevant patterns of temporal variation in floral scent. The first is the diel periodicity of odor production and/or emission (Hansted *et al.*, 1994), in some cases known to be regulated by circadian rhythms (Matile and Altenburger, 1988; review: Dudareva *et al.*, 1999). Fragrance emission may be synchronized with pollinator activity cycles, resulting in nocturnal bursts of fragrance in moth-pollinated *Cestrum nocturnum* (Overland, 1960), several *Nicotiana* spp. (Loughrin *et al.*, 1991; Euler and Baldwin, 1996), and in *Lonicera* spp. (Miyake *et al.*, 1998), and diurnal peaks in bee-pollinated orchids (Hills, 1989; Schiestl *et al.*, 1997) and *Antirrhinum majus* (Dudareva *et al.*, 2000). The systemic emissions of wound induced volatiles from foliage (many of them characteristic of flower scents) show a similar diurnal rhythm, putatively coincident with the activity of parasitoid wasps (Loughrin *et al.*, 1994; Paré and Tumlinson, 1999). Indeed, inducible vegetative volatile emissions represent rhythmicity (or at least plasticity) on a larger temporal scale (Baldwin, 1999). These patterns imply that scent emission incurs costs, both in metabolic currency or allocation (see p. 154) and in exploitation by enemies (Nielsen *et al.*, 1995; Galen, 1999), that provide a check to directional selection for pollinator attraction (see p. 159). This issue is essentially identical to the larger questions of how showy flowers should be (Strauss *et al.*, 1997; Mothershead and Marquis, 2000) and how long flowers should remain open and functional (Ashman and Schoen, 1994; Miyake and Yahara, 1998). In an extreme case, female flowers of *Catasetum maculatum*, a monoecious, epiphytic orchid, may remain open for weeks, emitting prodigious amounts of fragrance on a diurnal cycle, ceasing abruptly when a euglossine bee successfully deposits a pollinarium (Vogel, 1963; Hills *et al.*, 1972; Janzen, 1981; see below). Although such physiological investment in floral emissions must consume a significant fraction of floral mass and metabolism, these costs are essentially halved by restricting emission to discrete daily intervals.

The broad distribution of fragrance periodicity among angiosperms suggests multiple origins and mechanisms of control. The *sine qua non* of an endogenous plant rhythm is its continuation, with a slightly displaced period, under invariant ("free-running") environmental conditions (Piechulla, 1993). Experiments in constant light or darkness reveal different patterns of rhythmicity for individual compounds: some are endogenous; others are enhanced by light intensity and/or temperature, and many are out of phase with each other (Jakobsen and Olsen, 1994; Helsper *et al.*, 1998; MacTavish *et al.*, 2000). This pattern is intriguing in its implication that scent-emission rhythms are both variable and homoplaseous, and that selection for odor/pollinator synchrony is constrained by the underlying physiology and evolutionary history of a given plant. Meeuse and Raskin (1988) have suggested that shifts in fragrance chemistry and periodicity among sympatric Araceae (e.g., *Philodendron* and *Arum* spp.), with inflorescences that often grow to unusual sizes and emit heat and strong, sometimes quite foul, odors have resulted from reproductive character displacement and historical competition for pollinators. Mechanistic studies of scent emission, thermogenesis, and insect attraction to these bizarre plants (Raskin *et al.*, 1989; Gottsberger and Silberbauer-Gottsberger, 1991; Skubatz *et al.*, 1995, 1996; Kite and Hetterschieid, 1997; Seymour and Schultze-Motel, 1999), combined with advances in phylogenetic analysis, should make it possible to test this hypothesis rigorously.

Postpollination odor change: flowers as billboards

The second pattern of temporal fragrance variation is the change in odor quality or quantity following pollination. Many plants entirely cease odor production within minutes (*Catasetum*; see above) to hours (*Nicotiana attenuata*; Euler and Baldwin, 1996) to days (*Platanthera bifolia*; Tollsten and Bergström, 1989) after pollination, presumably to conserve resources and direct subsequent visitors toward receptive and/or rewarding flowers. Again, the loss of scent could also conceal fertilized, developing embryos from floral predators (Baldwin *et al.*, 1997). Cessation of odor production commonly accompanies general physiological collapse in short-lived or senescent flowers (Eisikowitch and Lazar, 1987; Schade *et al.*, 2001). However, postpollination flowers that remain turgid and are retained to contribute to distance attraction may modify odor in parallel to changes in visual cues, depending upon the scale at which floral cues interact in plant–pollinator communication (Lex, 1954; Eisikowitch and Rotem, 1987; Weiss, 1991; Miyake and Yahara, 1998). Under this scenario, change or loss of scent is an honest signal of a flower's sexual state (irrelevant to most pollinators) and, more importantly, of the absence of floral rewards, particularly when pollen (review: Dobson and Bergström, 2000) and nectar (review: Raguso, 2004) are scented.

In an optimal foraging context, age-specific floral odor changes should allow pollinators to avoid empty flowers and/or reduce handling times, perhaps analogous to the way that bumblebees use scent marks to avoid depleted flowers (Goulson *et al.*, 1998, 2000, and references therein), and thus enhance floral constancy. However, there are lamentably few published data bearing on this prediction, besides the purely behavioral work of Lex (1954). The flowers of *Lantana montevidense* (Verbenaceae) exhibit age-specific color changes (loss of yellow and white centers on purple background) correlated with two kinds of odor change (M. R. Weiss and R. A. Raguso, unpublished data). Terpenoid compounds (especially linalool and its oxides) are reduced manyfold in emission rate with age, while aromatic alcohols, aldehydes, and esters (including 2-phenylethanol and phenylacetaldehyde) are completely lost (Fig. 5.3). That these specific compounds are among the most attractive and easily learned floral stimuli (see p. 159) suggests that they are components of an integrated signal promoting assortative visitation to rewarding, sexually functional flowers (review: Weiss, 1995). The multivariate methods used by Tollsten (1993) and Schiestl *et al.* (1997) to track odor changes, combined with behavioral assays, should facilitate manipulative tests of this hypothesis in diverse plant–pollinator systems.

Fragrance shuffling in food-deceptive orchids: a mechanism for learning disruption?

The third pattern of temporal fragrance variation centers on the role of odor changes in deceptive flowers, such as nectarless orchids (reviews: Dafni, 1984; Ackerman, 1986). Hundreds of deceptive orchid species use odor and other cues to mimic general and specific floral food sources, oviposition sites, and female insects (Gill, 1989; Nilsson, 1992). Moya and Ackerman (1993) observed species-level qualitative and

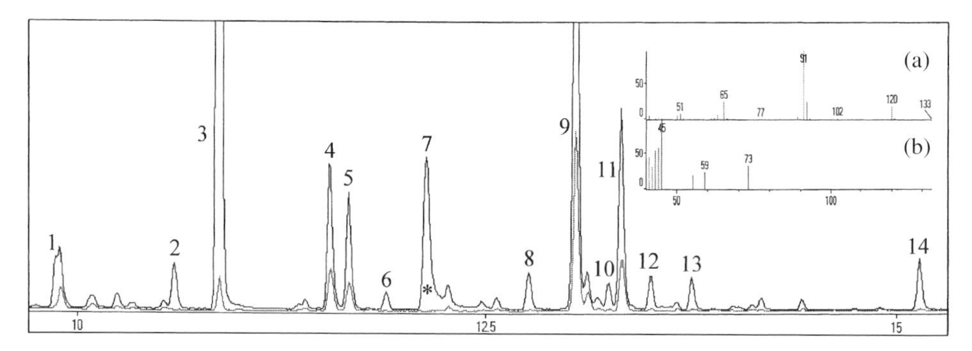

Fig. 5.3. Odor changes that track floral color changes. Gas chromatography–mass spectrometry total ion chromatograms of floral headspace collected from young (upper trace) and old (lower trace) flowers of *Lantana montevidense*. Peaks 1, 8, 10, and 12 are metabolites of linalool (peak 3), all of which decrease dramatically with floral age and color change. Sesquiterpene hydrocarbons (peaks 4, 5, 9, and 11) show comparable decreases over time. Peaks 2, 6, 7, 13, and 14 are oxygenated aromatics and are present only in newly opened, rewarding flowers. Insert mass spectra highlight loss of phenylacetaldehyde (*peak 7) taken from young (a) and old (b) flowers. (M. R. Weiss and R. A. Raguso, unpublished data.)

quantitative variation in the fragrance of individual flowers of *Epidendrum ciliare*, a nectarless, hawkmoth-pollinated orchid, over the course of a week. This unusual pattern suggests that *E. ciliare* modifies odor composition to disrupt negative associative learning by moths seeking profitable nectar rewards. Rather than mimic a specific model, *E. ciliare* and similar frauds present the highly visible floral display and nocturnal fragrance common to guilds of nectar-rich hawkmoth-pollinated flowers (Haber, 1984; Nilsson *et al.*, 1985; Haber and Frankie, 1989; Murren and Ellison, 1996). However, because fragrance chemistry is only loosely convergent among such guilds (Knudsen and Tollsten, 1993; Miyake *et al.*, 1998), moths may be equally attracted by broad variations on a fragrance theme. High population-level fragrance variation also occurs among deceptive *Cypripedium* orchids (Barkman *et al.*, 1997) and is thought to contribute to pollinator deception by inhibiting associative learning (review: Nilsson, 1992). However, fragrance cycling by individual flowers awaits further examination in these and other general floral mimics.

Pre- and postpollination odors in sex-deceptive orchids

Orchids of the Mediterranean genus *Ophrys* and several Australian genera, including *Cryptostylis* and *Caladenia*, have converged upon a remarkable category of deception – "pseudocopulation" – because their flowers use odor, visual, and tactile cues to mimic female hymenoptera (reviews: Staudamire, 1983; Borg-Karlson, 1990). After decades of intensive study (Priesner, 1973; Kullenberg and Bergström,

1976; Paulus and Gack, 1990), the subtle mechanisms by which *Ophrys* flowers manipulate the behavior of male bees and wasps through sexual mimicry have recently been elucidated. Flowers of *Ophrys sphegodes* use a blend of low-volatility alkanes to solicit copulatory behavior and incidental pollination by male *Andrena nigroaenea*; the same blend is produced as a sex pheromone by the female bees (Schiestl *et al.*, 1999, 2000). Once a flower has been pollinated, alkane production decreases markedly (Schiestl *et al.*, 1997), with a concomitant increase in a novel ester, farnesyl hexanoate, which also is a postcopulatory repellent produced by female *A. nigroaenea* (Schiestl and Ayasse, 2001). The combined decrease in sex pheromone and increase in repellent result in close-range avoidance of post-pollination flowers by male bees, just as they avoid non-virgin females, and their redirection to receptive flowers, sometimes on the same inflorescence (Schiestl *et al.*, 1997). In contrast, Australian male thynnine wasps simply leave the area after legitimate matings or attempted copulations with *Caladenia* and *Drakea* orchids, which occur as single flowers and for whom successful pollination events are rare (Peakall, 1990; Peakall and Beattie, 1996). Thus, postpollination repellents would be superfluous among thynnine-pollinated orchids. The large orchid genus *Caladenia* offers a unique opportunity to track the events associated with the evolution of pseudocopulation to and from legitimate, food-based pollination systems (Staudamire, 1983).

Why don't flowers have stronger odors?

Plant defense and the pollinator–attraction bias

Historical views of the origins of angiosperms offered an image of voluptuous, thermogenic magnoliid flowers wafting powerful fragrances into Cretaceous evenings, with beetles gnawing away at their fleshy petals (van der Pijl, 1960; Crepet, 1983; Endress, 1990; Azuma *et al.*, 1999). Pellmyr *et al.* (1991) persuasively argue the case for allomonal (deterrent) origins of "floral" fragrances in cycads and ancestral angiosperms from the volatile defensive compounds ubiquitous in terrestrial plants. In this scenario, herbivorous insects not deterred by these odors evolved to use them as kairomonal attractants and visited plant reproductive structures to feed and/or mate upon them. Selection may have favored variants with strong odor production or floral localization of emissions if incidental pollen movement by these animals enhanced fruit set or reduced inbreeding depression (Pellmyr and Thien, 1986). When increases in reproductive fitness outweigh tissue loss to herbivory or florivory, odor-mediated attraction can be considered synomonal (mutualistic). If scented flowers represent the *de facto* ancestral condition among angiosperms, then the spatial and temporal modifications of fragrance emissions presented above must be derived traits. Further, if pollinator attraction were the only function of floral scent, the following patterns should be observed:

1. Intense fragrances (> 0.1 mg/h per g flower emission rates; see Kaiser, 1997) should persist in solitary or low-density flowers that recruit euglossine bees, hawkmoths, and beetles as long-distance pollen-dispersal agents (Janzen, 1971; Haber, 1984; Young, 1988; Gottsberger and Silberbauer-Gottsberger, 1991).
2. More subtle, organ-specific odors should predominate when long-range attraction (> 10 m) is primarily visual but scent is required for landing, feeding, or associative learning (Pellmyr Then, 1986; Lunau, 1992; Dobson *et al.*, 1996).
3. Fragrance (as well as colorful visual display) would be superfluous in autogamous and wind-pollinated flowers, and, because of its metabolic costs, should be lost entirely.

These predictions are only partially supported by available data. Although many night-blooming, hawkmoth-pollinated flowers emit intense fragrances, emission rates for the entire scent blend vary at least three orders of magnitude among guilds of such plants (Haber and Frankie, 1989; Knudsen and Tollsten, 1993; Raguso and Willis, 2003). It remains to be tested whether specific subsets of these blends are emitted at comparable rates. The presence of mixed mating systems and alternative pollinators for some hawkmoth-pollinated plants (see Motten and Antonovics, 1992; Barthell and Knops, 1997) may introduce further variation in scent chemistry and emission rate, and the degree to which selection may act upon such variation. Phylogenetic or developmental constraints also may prevent some plants from emitting odors from floral tissues. For example, in the Chihuahuan Desert, *Acleisanthes acutifolia* (Nyctaginaceae), *Calylophus hartweggii* (Onagraceae), and *Ipomopsis longiflora* (Polemoniaceae) appear to compensate for meager floral scents with complex vegetative emissions, which are functionally sufficient as hawkmoth attractants and also contribute to plant defense (Jungers and Bergelson, 1997; Levin *et al.*, 2001; Raguso and Willis, 2003). Indeed, vegetative volatiles may also serve as close orientation and learning cues for honeybees, especially among plants of the mint family (Vogel, 1983; Beker *et al.*, 1989). Fragrance compounds clearly are labile among angiosperm lineages, but the biochemical mechanisms of odor gain and loss remain largely unknown and vary among pathways and compound classes (Raguso and Pichersky, 1999). Consequently, scent components may be retained in plants in which they are not pollinator attractants if they deter or are not attractive to floral predators (Ackerman *et al.*, 1997; Levin *et al.*, 2001).

Unbidden visitors and conspiratorial whispers: the perils of advertisement

Predator-mediated stabilizing selection on sexual signals has long been a major theme of animal behavior research (Endler, 1987; Ryan, 1990; Zuk *et al.*, 1998), but this rich body of theory has only recently been applied to plant reproductive biology (Strauss, 1997, and references therein). One idea emerging from this trend is that the physiological costs of fragrance production, however tangible, may be trivial in comparison to the ecological costs of apparency to enemies (Metcalf and Metcalf, 1992; Baldwin, 1999). Baldwin *et al.* (1997) experimentally augmented benzyl acetone emissions among populations of night-blooming *N. attenuata* (Solanaceae). Greater fragrance emissions did not enhance seed set but instead were correlated with increased browsing damage by mammals and seed predation by *Cormelina* bugs. Because the manipulation negated the flowers' nocturnal fragrance rhythms *and* amplified benzyl acetone emissions six-fold, it is difficult to determine whether constant odor or too much odor was most responsible for the observed plant damage. Further, *N. attenuata* is self-compatible, so the benefits of additional pollinator recruitment might not have been apparent in seed set alone (e.g., Waser and Price, 1994). However, this study clearly demonstrates directional selection *against* intense, constant scent production through reduced female fitness as a result of exploitation of floral scents by herbivores and seed predators.

Galen's long-term study of *Polemonium viscosum* (Polemoniaceae) paints a similar picture, with the added twist of floral variation and scent polymorphism over an altitudinal gradient. In the alpine tundra, flowers with large nectar volumes, flared corollas, and a "sweet" but as yet uncharacterized fragrance are preferentially visited and pollinated, nearly exclusively, by *Bombus* bees (Galen and Kevan, 1983; Galen and Newport, 1988). Assortative bumblebee visitation to such flowers can drive rapid adaptive shifts in correlated floral traits within a few generations (Cresswell and Galen, 1991; Galen, 1996). Below the timberline, smaller flowers lacking the sweet fragrance but secreting viscid, skunky exudates from calyx trichomes are more successful, although the pollinator guild of flies and small bees exhibit no clear odor preferences (Galen, 1985; Galen *et al.*, 1987). Here, *Formica* ants are the key selective force, exerting directional selection against large, sweet-smelling flowers, which they pillage for nectar, detaching styles and negating seed set in the process (Galen, 1983, 1999). The ants are physically deterred by the calyx exudates and repelled by their skunky odor; the rarity of this form above the timberline is inversely correlated with the frequency and reproductive advantage of sweet-smelling variants.

Therefore, in *P. viscosum*, altitudinal variation in fragrance intensity and composition is maintained by the opposing influences of pollinator attraction and herbivore avoidance. In an similar way, stabilizing selection has been shown to dictate the frequency and intensity of frog songs, the breadth and heft of antlers, and the brightness of breeding coloration (Endler, 1992). In animals, there are two extreme responses to conflicting selective pressures that could modulate fragrance intensity. The first extreme is runaway sexual selection. The progeny of an individual with an exaggerated signal trait obtains a dual fitness advantage – dominant mating status to the signal-bearing sex and preference for such a signal among the signal-receiving sex – that statistically overwhelms the negative impact of signal-exploiting predators (Lande, 1981; Kirkpatrick and Ryan, 1991). Although their reproductive biology remains mysterious, euglossine bees and *Stanhopea* orchids are good candidates for this model, with the powerful fragrances collected by male bees and thought to be used in female attraction (reviews: Dressler, 1982; Lunau, 1992). The second and opposite extreme is a concerted diminution of signal intensity by the producer and enhancement of signal detection and amplification by the recipient, resulting in what Johnstone (1998) terms "conspiratorial whispers." This model would apply to plants with specialized or obligate pollinators, including deceptive species whose fragrances mimic subtle oviposition cues (e.g., fungal odors; Kaiser, 1993) or, particularly, sex pheromones (see above). Further progress in understanding the selective pressures that act upon scent variation will require both the rigor and subtlety of Galen's field experiments and the sophistication of modern chemical analytical methods, ideally in a model system that can be genetically manipulated. The transformable mustard *Brassica napus*, with chemically characterized genetic lines and well-documented olfactory–behavioral responses from its honeybee pollinators (see above) and coleopteran seed predators (Evans and Allen-Williams, 1992; Blight *et al.*, 1995; Smart and Blight, 2000), would seem an ideal choice for such an initiative.

Conclusions

This chapter has examined three major evolutionary questions in the chemical ecology and evolution of floral scent: why fragrance composition is so variable between plant species; why patterns of scent emission change over time; and why odor intensity appears to vary so widely. The cases presented above suggest that any serious study of fragrance evolution must take into account both the phylogenetic history of the focal species and the potential alternative roles played by its volatiles for functions other than pollination (Farmer, 2001). Despite 30 years of technological innovation, interdisciplinary collaboration, and attempts to develop model systems, nearly every question in fragrance research remains open to further inquiry. Fragrance remains difficult to analyze and discuss in any comparative context, and further exploration of multivariate graphical methods and statistical testing of null models is sorely needed. In addition, the literature on insect olfactory physiology and behavior is so heavily biased toward economically important herbivores and agricultural systems that a careful survey of the responses of *any* genus or family of insects to *any* flower odors would constitute a major contribution. Finally, there remains no graphical method by which to visualize how insects perceive flower odors, in the way that flowers' visual spectra can be plotted in the trichromatic visual space of bees (Chittka, 1992). The approach used by Galizia and co-workers (Galizia *et al.*, 1999, 2000; Sachse *et al.* 1999) to visualize topological glomerular maps of neural odor-processing in bees and moths has provided the first glimpse of a non-pheromonal odor code in insects, and a giant step toward defining perceptual odor space.

Acknowledgements

Thanks to Ian Baldwin, Todd Barkman, Heidi Dobson, Jette Knudsen, Olle Pellmyr, Eran Pichersky, and Mark Whitten for a decade of ideas and encouragement, and to Laurel Hester, Ring Cardé and Jocelyn Millar for editorial suggestions. Figures 5.2 and 5.3 were made possible by the generosity and inspiration of Ann Fraser, Cyndi Henzel, John Hildebrand, and Martha Weiss. This chapter was prepared with support from National Science Foundation grant DEB-9806840.

References

Ackerman, J. D. (1986). Mechanisms and evolution of food deceptive pollination systems in orchids. *Lindleyana* **1**: 108–113.

Ackerman, J. D., Melendez-Ackerman, E. J. and Salguero-Faria, J. (1997). Variation in pollinator abundance and selection on fragrance phenotypes in an epiphytic orchid. *American Journal of Botany* **84**: 1383–1390.

Agelopoulos, N. G. and Pickett, J. A. (1998). Headspace analysis in chemical ecology: effects of different sampling methods on ratios of volatile compounds present in headspace samples. *Journal of Chemical Ecology* **24**: 1161–1172.

Alcock, J. (1998). *Animal Behavior: An Evolutionary Approach*, 6th edn. Sunderland, MA: Sinauer.

Armbruster, W. S. (1990). Estimating and testing adaptive surfaces: the morphology and pollination of *Dalechampia* blossoms. *American Naturalist* **135**: 14–31.

(1997). Exaptations link evolution of plant–herbivore and plant–pollinator interactions: a phylogenetic inquiry. *Ecology* **78**: 1661–1672.

Ashman, T.-L. and Schoen, D. J. (1994). How long should flowers live? *Nature* **371**: 788–791.

Azuma, H., Toyota, M., Asakawa, Y. *et al.* (1997). Chemical divergence in floral scents of *Magnolia* and allied genera (Magnoliaceae). *Plant Species Biology* **12**: 69–83.

Baldwin, I. T. (1999). Inducible nicotine production in native *Nicotiana* as an example of adaptive phenotypic plasticity. *Journal of Chemical Ecology* **25**: 3–30.

Baldwin, I. T., Preston, C., Euler, M. and Gorham, D. (1997). Patterns and consequences of benzyl acetone floral emissions from *Nicotiana attenuata* plants. *Journal of Chemical Ecology* **23**: 2327–2343.

Barkman, T. J. (2001). Character coding of secondary chemical variation for use in phylogenetic analyses. *Biochemical Systematics and Ecology* **29**: 1–20.

Barkman, T. J., Beaman, J. H. and Gage, D. A. (1997). Floral fragrance variation in *Cypripedium*: implications for evolutionary and ecological studies. *Phytochemistry* **44**: 875–882.

Barthell, J. F. and Knops, J. M. H. (1997). Visitation of evening primrose by carpenter bees: evidence of a "mixed" pollination syndrome. *Southwestern Naturalist* **42**: 86–93.

Basolo, A. L. (1995). Phylogenetic evidence for the role of a pre-existing bias in sexual selection. *Proceedings of the Royal Society of London, Series B* **259**: 307–311.

Beker, R., Dafni, A., Eisikowitch, D. and Ravid, U. (1989). Volatiles of two chemotypes of *Majorana syriaca* L. (Labiatae) as olfactory cues for the honeybee. *Oecologia* **79**: 446–451.

Blight, M. M., Pickett, J. A., Wadhams, L. J. and Woodcock, C. M. (1995). Antennal perception of oilseed rape, *Brassica napus* (Brassicaceae), volatiles by the cabbage seed weevil *Ceutorhynchus assimilis* (Coleoptera, Curculionidae). *Journal of Chemical Ecology* **21**: 1649–1664.

Blight, M. M., Le Métayer, M., Pham-Delègue, M.-H., Pickett, J. A., Marion-Poll, F. and Wadhams, L. J. (1997). Identification of floral volatiles involved in recognition of oilseed rape flowers, *Brassica napus*, by honeybees, *Apis mellifera*. *Journal of Chemical Ecology* **23**: 1715–1727.

Bohlmann, J., Meyer-Gauen, G. and Croteau, R. (1998). Plant terpenoid synthases: molecular biology and phylogenetic analysis. *Proceedings of the National Academy of Sciences, USA*. **95**: 4126–4133.

Borg-Karlson, A.-K. (1990). Chemical and ethological studies of pollination in the genus *Ophrys* (Orchidaceae). *Phytochemistry* **29**: 1359–1387.

Brantjes, N. B. M. (1976). Senses involved in the visiting of flowers by *Cucullia umbratica* (Noctuidae: Lepidoptera). *Entomologia Experimentalis et Applicata* **20**: 1–7.

Bronstein, J. L. (1994). Conditional outcomes in mutualistic interactions. *Trends in Ecology and Evolution* **9**: 214–217.

Chittka, L. (1992). The colour hexagon: a chromaticity diagram based on photoreceptor excitations as a generalized representation of colour opponency. *Journal of Comparative Physiology A* **170**: 533–543.

Connick, W. J. and French, R. C. (1991). Volatiles emitted during the sexual stage of the Canada thistle rust fungus and by thistle flowers. *Journal of Agricultural and Food Chemistry* **39**: 185–188.

Crepet, W. L. (1983). The role of insect pollination in the evolution of the angiosperms. In *Pollination Biology*, ed. L. A. Real, pp. 29–50. Orlando, FL: Academic Press.

Cresswell, J. E. and Galen, C. (1991). Frequency-dependent selection and adaptive surfaces for floral character combinations: the pollination of *Polemonium viscosum*. *American Naturalist* **138**: 1342–1353.

Croteau, R. and Karp, F. (1991). Origin of natural odorants. In *Perfumes: Art, Science and Technology*, eds. P. M. Müller and D. Lamparsky, pp. 101–126. New York: Elsevier.

Dafni, A. (1984). Mimicry and deception in pollination. *Annual Review of Ecology and Systematics* **15**: 259–278.

da Silva, U. F., Borba, E. L., Semir, J. and Marsaioli, A. J. (1999). A simple solid injection device for the analyses of *Bulbophyllum* (Orchidaceae) volatiles. *Phytochemistry* **50**: 31–34.

Delpino, F. (1874). Ulteriori osservazioni e considerazioni sulla dicogamia nel regno vegetale. 2 (IV). Delle piante zoidifile. *Atti della Societa Italiana Scientifica Natura* **16**: 151–349.

De Moraes, C. M., Lewis, W. D., Paré, P. W., Alborn, H. T. and Tumlinson, J. H. (1998). Herbivore-infested plants selectively attract parasitoids. *Nature* **393**: 570–573.

Dicke, M. (1994). Local and systemic production of volatile herbivore-induced terpenoids: their role in plant–carnivore mutualism. *Journal of Plant Physiology* **143**: 465–472.

Dobson, H. E. M. (1994). Floral volatiles in insect biology. In *Insect–Plant Interactions*, vol. 5, ed. E. Bernays, pp. 47–81. Boca Raton, FL: CRC Press.
Dobson, H. E. M. and Bergström, L. G. (2000). The ecology and evolution of pollen odors. *Plant Systematics and Evolution* **222**: 63–87.
Dobson, H. E. M., Bergström, L. G. and Groth, I. (1990). Differences in fragrance chemistry between flower parts of *Rosa rugosa* Thunb. (Rosaceae). *Israel Journal of Botany* **39**: 143–156.
Dobson, H. E. M., Groth, I. and Bergström, L. G. (1996). Pollen advertisement: chemical contrasts between whole-flower and pollen odors. *American Journal of Botany* **83**: 877–885.
Dodson, C., Dressler, R., Hills, H., Adams, R. and Williams, N. (1969). Biologically active compounds in orchid fragrances. *Science* **164**: 1243–1249.
Dornhaus, A. and Chittka, L. (1999). Evolutionary origins of bee dances. *Nature* **401**: 38.
Dressler, R. L. (1982). Biology of the orchid bees (Euglossini). *Annual Review of Ecology and Systematics* **13**: 373–394.
Dudareva, N. and Pichersky, E. (2000). Biochemical and molecular aspects of floral scents. *Plant Physiology* **122**: 627–634.
Dudareva, N., Piechulla, B. and Pichersky, E. (1999). Biogenesis of floral scent. *Horticultural Review* **24**: 31–54.
Dudareva, N., Murfitt, L. M., Mann, C. J. *et al.* (2000). Developmental regulation of methyl benzoate biosynthesis and emission in snapdragon flowers. *Plant Cell* **12**: 949–961.
Eisikowitch, D. and Lazar, Z. (1987). Flower change in *Oenothera drummondii* Hooker as a response to pollinators' visits. *Botanical Journal of the Linnean Society* **95**: 101–111.
Eisikowitch, D. and Rotem, R. (1987). Flower orientation and color change in *Quisqualis indica* and their possible role in pollinator partitioning. *Botanical Gazette* **148**: 175–179.
Endler, J. A. (1987). Predation, light intensity and courtship behaviour in *Poecilia reticulata* (Pisces: Poeciliidae). *Animal Behaviour* **35**: 1376–1385.
(1992). Signals, signal conditions, and the direction of evolution. *American Naturalist* **139**(suppl.): S125–S153.
Endress, P. K. (1990). *Diversity and Evolutionary Biology of Tropical Flowers*. Cambridge: Cambridge University Press.
Erhardt, A. (1993). Pollination of the edelweiss, *Leontopodium alpinum*. *Botanical Journal of the Linnaean Society* **111**: 229–240.
Euler, M. and Baldwin, I. T. (1996). The chemistry of defense and apparency in the corollas of *Nicotiana attenuata*. *Oecologia* **107**: 102–112.
Evans, K. A. and Allen-Williams, L. J. (1992). Electroantennogram responses of the cabbage seed weevil, *Ceutorrhynchus assimilis*, to oilseed rape, *Brassica napus* ssp. *oleifera*, volatiles. *Journal of Chemical Ecology* **18**: 1641–1659.
Faegri, K. and van der Pijl, L. (1979). *The Principles of Pollination Ecology*, 3rd edn. Oxford: Pergamon Press.
Fäldt, J., Eriksson, M., Valterová, I. and Borg-Karlson, A.-K. (2000). Comparison of headspace techniques for sampling volatile natural products in a dynamic system. *Verlag der Zeitschrift für Naturforschung* **55c**: 180–188.
Farmer, E. E. (2001). Surface to air signals. *Nature* **411**: 854–856.
Feinsinger, P. (1983). Coevolution and pollination. In *Coevolution*, eds. D. J. Futuyma and M. Slatkin, pp. 282–310. Sunderland, MA: Sinauer.

Feng, Z. F., Huber, U. and Boland, W. (1993). Biosynthesis of the irregular C-12-terpenoiddehydrogeosmin in flower heads of *Rebutia marsoneri* Werd (Cactaceae). *Helvetica Chimica Acta* **76**: 2547–2552.

Gäbler, A., Boland, W., Preiss, U. and Simon, H. (1991). Stereochemical studies on homoterpene biosynthesis in higher plants; mechanistic, phylogenetic and ecological aspects. *Helvetica Chimica Acta* **74**: 1773–1789.

Galen, C. (1983). The effects of nectar thieving ants on seedset in floral scent morphs of *Polemonium viscosum*. *Oikos* **41**: 245–249.

(1985). Regulation of seed set in *Polemonium viscosum*: floral scents, pollination and resources. *Ecology* **6**: 792–797.

(1996). Rates of floral evolution: adaptation to bumblebee pollination in an alpine wildflower, *Polemonium viscosum*. *Evolution* **50**: 120–125.

(1999). Flowers and enemies: predation by nectar thieving ants in relation to variation in floral form of an alpine wildflower, *Polemonium viscosum*. *Oikos* **85**: 426–434.

Galen, C. and Kevan, P. G. (1983). Bumblebee foraging and floral scent dimorphism: *Bombus kirbyellus* Curtis (Hymenoptera: Apidae) and *Polemonium viscosum* Nutt. (Polemoniaceae). *Canadian Journal of Zoology* **61**: 1207–1213.

Galen, C. and Newport, M. E. (1988). Pollination quality, seed set and flower traits in *Polemonium viscosum*: complementary effects of variation in flower scent and size. *American Journal of Botany* **75**: 900–905.

Galen, C., Zimmer, K. A. and Newport, M. E. (1987). Pollination in floral scent morphs of *Polemonium viscosum*: a mechanism for disruptive selection on flower size. *Evolution* **41**: 599–606.

Galizia, C. G. and Menzel, R. (2000). Odour perception in honeybees: coding information in glomerular patterns. *Current Opinion in Neurobiology* **10**: 504–510.

Galizia, C. G., Sachse, S., Rappert, A. and Menzel, R. (1999). The glomerular code for odor representation is species specific in the honeybee *Apis mellifera*. *Nature Neuroscience* **2**: 473–478.

Galizia, C. G., Sachse, S. and Mustaparta, H. (2000). Calcium responses to pheromones and plant odours in the antennal lobe of the male and female moth *Heliothis virescens*. *Journal of Comparative Physiology* A **186**: 1049–1063.

Gershenzon, J. and Croteau, R. (1993). Terpenoid biosynthesis: the basic pathway and formation of monoterpenes, sesquiterpenes and diterpenes. In *Lipid Metabolism in Plants*, ed. T. S. Moore Jr, pp. 339–388. Boca Raton, FL: CRC Press.

Gibernau, M., Hossaert-McKey, M., Frey, J. and Kjellberg, F. (1998). Are olfactory signals sufficient to attract fig pollinators? *Ecoscience* **5**: 306–311.

Gilbert, L. E. (1980). Food web organization and the conservation of Neotropical diversity. In *Conservation Biology*, eds. M. E. Soulé and B. A. Wilcox, pp. 11–33. Sunderland, MA: Sinauer.

Gill, D. E. (1989). Fruiting failure, pollinator inefficiency and speciation in orchids. In *Speciation and its Consequences*, eds. D. Otte and J. A. Endler. pp. 458–481. Sunderland MA: Sinauer.

Gottsberger, G. (1999). Pollination and evolution in neotropical Annonaceae. *Plant Species Biology* **14**: 143–152.

Gottsberger, G. and Silberbauer-Gottsberger, I. (1991). Olfactory and visual attraction of *Erioscelis emarginata* (Cyclocephalini, Dynastinae) to the inflorescences of *Philodendron selloum* (Araceae). *Biotropica* **23**: 23–28.

Goulson, D., Hawson, S. A. and Stout, J. C. (1998). Foraging bumblebees avoid flowers already visited by conspecifics or by other bumblebee species. *Animal Behaviour* **55**: 199–206.

Goulson, D., Stout, J. C., Langley, J. and Hughes, W. O. H. (2000). Identity and function of scent marks deposited by foraging bumblebees. *Journal of Chemical Ecology* **26**: 2897–2911.

Haber, W. A. (1984). Pollination by deceit in a mass-flowering tropical tree *Plumeria rubra* L. (Apocynaceae). *Biotropica* **16**: 269–275.

Haber, W. A. and Frankie, G. W. (1989). A tropical hawkmoth community: Costa Rican dry forest Sphingidae. *Biotropica* **21**: 155–172.

Hallberg, E. and Hansson, B. S. (1999). Arthroped sensilla: morphology and phylogenetic considerations. *Microscopy Research and Technique* 47: 428–439.

Hansson, B. S. (ed.) (1999). *Insect Olfaction*. Berlin: Springer.

Hansson, B. S. and Anton, S. (2000). Function and morphology of the antennal lobe: new developments. *Annual Review of Entomology* **45**: 203–231.

Hansson, B. S., Larsson, M. D. and Leal, W. S. (1999). Green leaf volatile-detecting olfactory receptor neurones display very high sensitivity and specificity in a scarab beetle. *Physiological Entomology* **24**: 121–126.

Hansted, L., Jakobsen, H. B. and Olsen, C. E. (1994). Influence of temperature on the rhythmic emission of volatiles from *Ribes nigrum* flowers in situ. *Plant, Cell and Environment* **17**: 1069–1072.

Haynes, K. F., Zhao, J. Z. and Latif, A. (1991). Identification of floral compounds from *Abelia grandiflora* that stimulate upwind flight in cabbage looper moths. *Journal of Chemical Ecology* **17**: 637–646.

Heath, R. R. and Dueben, B. D. (1998). Analytical and preparative gas chromatography. In *Methods in Chemical Ecology*, vol. 1: *Chemical Methods*, eds. J. G., Millar and K. F. Haynes, pp. 85–126. New York: Chapman & Hall.

Heath, R. R. and Manukian, A. (1994). An automated system for use in collecting volatile chemicals released from plants. *Journal of Chemical Ecology* **20**: 593–608.

Heath, R. R., Landolt, P. J., Dueben, B. and Senczewski, B. (1992). Identification of floral compounds of night-blooming jessamine attractive to cabbage looper moths. *Environmental Entomology* **21**: 854–859.

Heinrich, B. and Raven, P. H. (1972). Energetics and pollination ecology. *Science* **176**: 597–602.

Helsper, J. P. F. G., Davies, J. A., Bouwmeester, H. J., Krol, A. F and van Kampen, M. V. (1998). Circadian rhythmicity in emission of volatile compounds by flowers of *Rosa hybrida* L. cv. Honesty. *Planta* **207**: 88–95.

Henning, J. A. and Teuber, L. R. (1992). Combined gas chromatography-electro-antennogram characterization of alfalfa floral volatiles recognized by honey bees (Hymenoptera: Apidae). *Journal of Economic Entomology* **85**: 226–232.

Henning, J. A., Peng, Y.-S., Montague, M. A. and Teuber, L. R. (1992). Honey bee (Hymenoptera: Apidae) behavioral response to primary alfalfa (Rosales: Fabaceae) floral volatiles. *Journal of Economic Entomology* **85**: 233–239.

Herrera, C. M. (1996). Floral traits and plant adaptation to insect pollinators: a devil's advocate approach. In *Floral Biology*, eds. S. C. H. Barrett and D. G. Lloyd, pp. 65–87. New York: Chapman & Hall.

Hick, A. J., Luszniak, M. C. and Pickett, J. A. (1999). Volatile isoprenoids that control insect behaviour and development. *Natural Product Reports* **16**: 39–54.

Hildebrand, J. G. and Shepherd, G. M. (1997). Mechanisms of olfactory discrimination: converging evidence for common principles across phyla. *Annual Review of Neuroscience* **20**: 595–631.

Hills, H. G. (1989). Fragrance cycling in *Stanhopea pulla* (Orchidaceae, Stanhopeinae) and identification of *trans*-limonene oxide as a major fragrance component. *Lindleyana* **4**: 61–67.

Hills, H. G., Williams, N. H. and Dodson, C. H. (1972). Floral fragrances and isolating mechanisms in the genus *Catasetum* (Orchidaceae). *Biotropica* **4**: 61–76.

Honda, K, Ômura, H. and Hayashi, N. (1998). Identification of floral volatiles from *Ligustrum japonicum* that stimulate flower visiting by cabbage butterfly, *Pieris rapae*. *Journal of Chemical Ecology* **24**: 2167–2180.

Jakobsen, H. B. and Olsen, C. E. (1994). Influence of climatic factors on emission of flower volatiles in situ. *Planta* **192**: 365–371.

Janzen, D. H. (1971). Euglossine bees as long distance pollinators of tropical plants. *Science* **171**:203–205.

(1981). Visitor and pollinator abundance at two Costa Rican female *Catasetum* orchid inflorescences. *Oikos* **36**: 177–183.

Johnson, S. D. and Steiner, K. E. (2000). Generalization versus specialization in plant pollination systems. *Trends in Ecology and Evolution* **15**: 140–143.

Johnstone, R. A. (1998). Conspiratorial whispers and conspicuous displays: games of signal detection. *Evolution* **52**: 1554–1563.

Jungers, T. and Bergelson, J. (1997). Pollen and resource limitation of compensation to herbivory in scarlet gilia, *Ipomopsis aggregata*. *Ecology* **78**: 1684–1695.

Kaiser, R. (1991). Trapping, investigation and reconstitution of flower scents. In *Perfumes: Art, Science and Technology*, eds. P. M. Müller and D. Lamparsky, pp. 213–250. London: Elsevier Applied Science.

(1993). *The Scent of Orchids*. Amsterdam: Elsevier Science.

(1995). New or uncommon volatile compounds in floral scents. *Proceedings of the 13th International Congress of Flavours, Fragrances and Essential Oils*, pp. 135–168.

(1997). Environmental scents at the Ligurian coast. *Perfumer and Flavorist* **22**: 7–18.

Kaiser, R. and Nussbaumer, C. (1990). 1,2,3,4,4A,5,8,8A-Octahydro-4-β,8A-α-dimethylnaphthalen-4A-β-O (= dehydrogeosmin), a novel compound occurring in the flower scent of various species of Cactaceae. *Helvetica Chimica Acta* **73**: 133–139.

Kaiser, R. and Tollsten, L. (1995). An introduction to the scent of cacti. *Flavour and Fragrance Journal* **10**: 153–164.

Karban, R. and Baldwin, I. T. (1997). *Induced Responses to Herbivory*. Chicago, IL: Chicago University Press.

Kerner von Marilaum, A. (1895). *The Natural History of Plants: Their Forms, Growth, Reproduction and Distribution*. London: Blackie and Son.

Kirkpatrick, M. and Ryan, M. A. (1991). The evolution of mating preferences and the paradox of the lek. *Nature* **350**: 33–38.

Kite, G. C. and Hetterschieid, W. L. A. (1997). Inflorescence odours of *Amorphophallus* and *Pseudodracontium* (Araceae). *Phytochemistry* **46**: 71–75.

Knudsen, J. T. and Tollsten, L. (1993). Trends in floral scent chemistry in pollination syndromes: floral scent composition in moth-pollinated taxa. *Botanical Journal of the Linnaean Society* **113**: 263–284.

Knudsen, J. T., Tollsten, L and Bergström, L. G. (1993). Floral scents: a check list of volatile compounds isolated by head-space techniques. *Phytochemistry* **33**: 253–280.

Knuth, P. (1906). *Handbook of Flower Pollination*, vol. 1. [Based upon Hermann Müller's work *The Fertilization of Flowers by Insects*.] Oxford: Clarendon Press.

Kullenberg, B. and Bergström, L. G. (1976). The pollination of *Ophrys* orchids. *Botaniska Notiser* **129**: 11–19.

Lande, R. (1981). Models of speciation by sexual selection of polygenic traits. *Proceedings of the National Academy of Sciences, USA*. **78**: 3721–3725.
Laurent, G. (1999). A systems perspective on early olfactory coding *Science* 286: 723–726.
Le Mètayer, M., Marion-Poll, F., Sandoz, J. C. *et al.* (1997). Effect of conditioning on discrimination of oilseed rape volatiles by the honeybee: use of a combined gas chromatography-proboscis extension behavioral assay. *Chemical Senses* **22**: 391–398.
Levin, R. D, Raguso, R. A. and McDade, L. A. (2001). Fragrance chemistry and pollinator affinities in Nyctaginaceae. *Phytochemistry* **58**: 429–440.
Lex, T. (1954). Duftmale an blüten. *Zeitschrift für Vergleichende Physiologie* **36**: 212–234.
Loughrin, J. H., Hamilton-Kemp, T. D, Andersen, R. A. and Hildebrand, D. F. (1991). Circadian rhythm of volatile emission from flowers of *Nicotiana sylvestris* and *N. suaveolens*. *Physiologia Plantarum* **83**: 492–496.
Loughrin, J. H., Manukian, A., Heath, R. R., Turlings, T. C. J. and Tumlinson, J. H. (1994). Diurnal cycle of emission of induced volatile terpenoids by herbivore-injured cotton plants. *Proceedings of the National Academy of Sciences, USA* **91**: 11836–11840.
Lunau, K. (1992). Evolutionary aspects of perfume collection in male euglossine bees (Hymenoptera) and of nest deception in bee-pollinated flowers. *Chemoecology* **3**: 65–73.
MacTavish, H. S., Davies, N. W. and Menary, R. C. (2000). Emission of volatiles from brown *Boronia* flowers: some comparative observations. *Annals of Botany* **86**: 347–354.
Maekawa, M., Imai, T., Tsuchiya, S., Fujimori, T. and Leal, W. S. (1999). Behavioral and electrophysiological responses of the soybean beetle, *Anomala rufocuprea* Motschulsky (Coleoptera: Scarabaeidae) to methyl anthranilate and its related compounds. *Applied Entomology and Zoology* **34**: 99–103.
Marion-Poll, F. and Thièry, D. (1996). Dynamics of EAG responses to host plant volatiles delivered by gas-chromatograph. *Entomologia Experimentalis et Applicata* **80**: 120–123.
Matile, P. and Altenburger, R. (1988). Rhythms of fragrance emission in flowers. *Planta* **174**: 242–247.
Mayer, M. S., Mankin, R. W. and Lemire, G. F. (1984). Quantitation of the insect electro-antennogram: measurement of sensillar contributions, elimination of background potentials and relationship to olfactory sensation. *Journal of Insect Physiology* **30**: 757–763.
Meeuse, B. J. D. and Raskin, I. (1988). Sexual reproduction in the arum lily family, with emphasis on thermogenicity. *Sexual Plant Reproduction* **1**: 3–15.
Metcalf, R. L. (1987). Plant volatiles as insect attractants. CRC *Critical Reviews in Plant Sciences* **5**: 251–301.
Metcalf, R. L. and Metcalf, E. R. (1992). *Plant Kairomones in Insect Ecology and Control*. New York: Chapman & Hall.
Miyake, T. and Yahara, T. (1998). Why does the flower of *Lonicera japonica* open at dusk? *Canadian Journal of Botany* **76**: 1806–1811.
Miyake, T., Yamaoka, R. and Yahara, T. (1998). Floral scents of hawkmoth-pollinated flowers in Japan. *Journal of Plant Research* **111**: 199–205.
Morgan, A. and Lyon, S. (1928). Notes on armyl salicylate as an attractant to the tobacco hornworm moth. *Journal of Economic Entomology* **21**: 189–191.

Mothershead, K. and Marquis, R. J. (2000). Fitness impacts of herbivory through indirect effects on plant–pollinator interactions in *Oenothera macrocarpa*. *Ecology* **81**: 30–40.

Motten, A. F. and Antonovics, J. (1992). Determinants of outcrossing rate in a predominantly self-fertilizing weed, *Datura stramonium* (Solanaceae). *American Journal of Botany* **79**: 419–427.

Moya, S. and Ackerman, J. D. (1993). Variation in the floral fragrance of *Epidendrum ciliare* (Orchidaceae). *Nordic Journal of Botany* **13**: 41–47.

Murren, C. J. and Ellison, A. M. (1996). Effects of habitat, plant size and floral display on male and female reproductive success of the Neotropical orchid, *Brassavola nodosa*. *Biotropica* **28**: 30–40.

Nielsen, J. K., Jakobsen, H. B., Friis, P., Hansen, K., Møller, J. and Olsen, C. E. (1995). Asynchronous rhythms in the emission of volatiles from *Hesperis matronalis* flowers. *Phytochemistry* **38**: 847–851.

Nilsson, L. A. (1992). Orchid pollination biology. *Trends in Ecology and Evolution*, **7**: 255–259.

Nilsson, L. A., Jonsson, L., Rason, L. and Randrianjohany, E. (1985). Monophily and pollination mechanisms in *Angraecum arachnites* Schltr. (Orchidaceae) in a guild of long-tongued hawk-moths (Sphingidae) in Madagascar. *Biological Journal of the Linnaean Society* **26**: 1–19.

Nishida, R., Shelly, T. E. and Kaneshiro, K. Y. (1997). Acquisition of female-attracting fragrance by males of oriental fruit fly from a Hawaiian lei flower, *Fagraea berteriana*. *Journal of Chemical Ecology* **23**: 2275–2285.

Ollerton, J. (1996). Reconciling ecological processes with phylogenetic patterns: the apparent paradox of plant–pollinator systems. *Journal of Ecology* **84**: 767–769.

Ômura, H., Honda, K. and Hayashi, N. (1999a). Chemical and chromatic bases for preferential visiting by the cabbage butterfly, *Pieris rapae*, to rape flowers. *Journal of Chemical Ecology* **25**: 1895–1906.

Ômura, H., Honda, K., Nakagawa, A., and Hayashi, N. (1999b). The role of floral scent of the cherry tree, *Prunus yedoensis*, in the foraging behavior of *Luehdorfia japonica* (Lepidoptera; Papilionidae). *Applied Entomology and Zoology* **34**: 309–313.

Ômura, H., Honda, K. and Hayashi, N. (2000). Floral scent of *Osmanthus fragrans* discourages foraging behavior of cabbage butterfly, *Pieris rapae*. *Journal of Chemical Ecology* **26**: 655–666.

Overland, L. (1960). Endogenous rhythm in opening and odor of flowers of *Cestrum nocturnum*. *American Journal of Botany* **47**: 378–382.

Paré, P. W. and. Tumlinson, J. H. (1999). Plant volatiles as a defense against insect herbivores. *Plant Physiology* **121**: 325–331.

Patt, J. M., Hartman, T. G., Creekmore, R. W. *et al.* (1992). The floral odour of *Peltandra virginica* contains novel trimethyl-2,5-dioxabicyclo[3.2.1.]nonanes. *Phytochemistry* **31**: 487–491.

Paulus, H. F. and Gack, C. (1990). Pollinators as prepollinating isolation factors: evolution and speciation in *Ophrys* (Orchidaceae). *Israel Journal of Botany* **39**: 43–79.

Peakall, R. (1990). Responses of male *Zaspilothynnus trilobatus* Turner wasps to females and the sexually deceptive orchid it pollinates. *Functional Ecology* **4**: 159–167.

Peakall, R. and Beattie, A. J. (1996). Ecological and genetic consequences of pollination by sexual deception in the orchid *Caladenia tentactulata*. *Evolution* **50**: 2207–2220.

Pellmyr, O. (1997). Stability of plant–animal mutualisms: keeping the benefactors at bay. *Trends in Plant Science* **2**: 408–409.

Pellmyr, O. and Thien, L. B. (1986). Insect reproduction and floral fragrances: keys to the evolution of the angiosperms? *Taxon* **35**: 76–85.

Pellmyr, O., Tang, W., Groth, I., Bergström, L. G. and Thien, L. B. (1991). Cycad cone and angiosperm volatiles: inferences for the evolution of insect pollination. *Biochemical Systematics and Ecology* **19**: 623–627.

Pham-Delègue, M.-H., Blight, M. M., Kerguelen, V. *et al.* (1997). Discrimination of oilseed rape volatiles by the honeybee: combined chemical and biological approaches. *Entomologia Experimentalis et Applicata* **83**: 87–92.

Piechulla, B. (1993). Circadian clock directs the expression of plant genes. *Plant Molecular Biology* **22**: 533–542.

Priesner, E. (1973). Reaktionen von Riechrezeptoren männlicher Solitärbienen (Hymenoptera, Apoidea) auf Inhaltsstoffe von *Ophrys*-Blüten. *Zoon Supplement* **1**: 3–54.

Proctor, M., Yeo, P. and Lack, A. (1996). *The Natural History of Pollination*. Portland, OR: Timber Press.

Raguso, R. A. (2001). Floral scent, olfaction and scent-driven foraging behavior. In *Cognitive Ecology of Pollination; Animal Behavior and Floral Evolution*, eds. L. Chittka and J. D. Thomson, pp. 83–105, Cambridge: Cambridge University Press.

Raguso, R. A. (2004). Why are some floral nectars scented? *Ecology*, in press.

Raguso, R. A. and Pichersky, E. (1999). A day in the life of a linalool molecule: chemical communication in a plant–pollinator system. Part 1: Linalool biosynthesis in flowering plants. *Plant Species Biology* **14**: 95–120.

Raguso, R. A. and Roy, B. A. (1998). "Floral" scent production by *Puccinia* rust fungi that mimic flowers. *Molecular Ecology* **7**: 1127–1136.

Raguso, R. A. and Willis, M. A. (2003). Hawkmoth pollination in Arizona's Sonoran Desert: behavioral responses to floral traits. In *Evolution and Ecology Taking Flight: Butterflies as Model Systems*, ch. 3, eds. C. L. Boggs, W. B. Watt. and P. R. Ehrlich. Rocky Mountain Biological Laboratory Symposium Series. Chicago, IL: University of Chicago Press.

Raguso, R. A., Light, D. M. and Pichersky, E. (1996). Electroantennogram responses of *Hyles lineata* (Sphingidae: Lepidoptera) to floral volatile compounds from *Clarkia breweri* (Onagraceae) and other moth-pollinated flowers. *Journal of Chemical Ecology* **22**: 1735–1766.

Raskin, I., Turner, I. and Melander, W. R. (1989). Regulation of heat production in the inflorescences of an *Arum* lily by endogenous salicylic acid. *Proceedings of the National Academy of Sciences, USA* **86**: 2214–2218.

Robacker, D. C., Meeuse, B. J. D. and Erickson, E. H. (1988). Floral aroma: how far will plants go to attract pollinators? *BioScience* **38**: 390–398.

Roy, B. A. and Raguso, R. A. (1997). Olfactory vs. visual cues in a floral mimicry system. *Oecologia* **109**: 414–426.

Ryan, M. J. (1990). Sexual selection, sensory systems and sensory exploitation. *Oxford Surveys in Evolutionary Biology* **7**: 156–195.

Ryan, M. J. and Rand, A. S. (1993). Sexual selection and signal evolution: the ghost of biases past. *Philosophical Transactions of the Royal Society, Series B* **340**: 187–195.

Sachse, S., Rappert, A. and Galizia, C. G. (1999). The spatial representation of chemical structures in the antennal lobe of honeybees: steps towards the olfactory code. *European Journal of Neuroscience* **11**: 3970–3982.

Schade, F., Legge, R. L. and Thompson, J. E. (2001). Fragrance volatiles of developing and senescing carnation flowers. *Phytochemistry* **56**: 703–710.

Schatz, G. E. (1990). Some aspects of pollination biology in Central American forests. In *Reproductive Ecology of Tropical Forest Plants*, eds. K. S. Bawa, and M. Hadley, pp. 69–84. Paris: UNESCO/Parthenon Publishing.

Schemske, D. W. and Horvitz, C. C. (1984). Variation among floral visitors in pollination ability: a precondition for mutualism specialization. *Science* **225**: 519–521.

Schiestl, F. P. and Ayasse, M. (2001). Post-pollination emission of a repellent compound in a sexually deceptive orchid: a new mechanism for maximizing reproductive success? *Oecologia* **126**: 531–534.

Schiestl, F. P. and Marrion-Poll, F. (2002). Detection of physiologically active flower volatiles using gas chromatography coupled with electroantennography. In *Molecular Methods of Plant Analysis*, vol. 21, *Analysis of Taste and Aroma*, eds. J. F. Jackson, H. F. Linskens, and R. Inman, pp. 173–198. Berlin: Springer.

Schiestl, F. P., Ayasse, M., Paulus, H. F., Erdmann, D. and Francke, W. (1997). Variation of floral scent emission and post-pollination changes in individual flowers of *Ophrys sphegodes*. *Journal of Chemical Ecology* **23**: 2881–2895.

Schiestl, F. P., Ayasse, M., Paulus, H. D. *et al.* (1999). Orchid pollination by sexual swindle. *Nature* **399**: 421–422.

Schiestl, F. P., Ayasse, M., Paulus, H. D. *et al.* (2000). Sex pheromone mimicry in the early spider orchid (*Ophrys sphegodes*): patterns of hydrocarbons as the key mechanism for pollination by sexual deception. *Journal of Comparative Physiology A* **186**: 567–574.

Schlumpberger, B. O. (2002). Dehydrogeosmin produzierende kakfeen: Untersuchungen zur Verbreitung, Duftstoff-produktion und Bestänbung. PhD Thesis, University of Bonn, Germany.

Schmid, J. and Amrhein, N. (1995). Molecular organization of the shikimate pathway in higher plants. *Phytochemistry* **39**: 737–749.

Schnitzler, J.-P., Madlung, J., Rose, A. and Seitz, H. U. (1992). Biosynthesis of *p*-hydroxybenzoic acid in elicitor-treated carrot cell cultures. *Planta* **188**: 594–600.

Schreier, P. (1984). *Chromatographic Studies of Biogenesis of Plant Volatiles*. Heidelberg: Alfred Hüthig Verlag.

Seymour, R. S. and Schultze-Motel, P. (1999). Respiration, temperature regulation and energetics of thermogenic inflorescences of the dragon lily *Dracunculus vulgaris* (Araceae). *Proceedings of the Royal Society of London, Series B* **266**: 1975–1983.

Skubatz, H., Kunkel, D. D., Patt, J. M., Howald, W. N., Hartman, T. G. and Meeuse, B. J. D. (1995). Pathway of terpene excretion by the appendix of *Sauromatum guttatum*. *Proceedings of the National Academy of Sciences, USA*. **92**: 10084–10088.

Skubatz, H., Kunkel, D. D., Howald, W. N., Trenkle, R and Mookherjee, B. (1996). The *Sauromatum guttatum* appendix as an osmophore: excretory pathways, composition of volatiles and attractiveness of insects. *New Phytologist* **134**: 631–640.

Smart, L. E. and Blight, M. M. (2000). Response of the pollen beetle, *Meligethes aeneus*, to traps baited with volatiles from oilseed rape, *Brassica napus*. *Journal of Chemical Ecology* **26**: 1051–1064.

Staudamire, W. P. (1983). Wasp-pollinated species of *Caladenia* (Orchidaceae) in Southwestern Australia. *Australian Journal of Botany* **31**: 383–394.

Steele, C. L., Crock, J., Bohlmann, J. and Croteau, R. (1998). Sesquiterpene synthases from grand fir (*Abies grandis*): comparison of constitutive and wound-induced activities, and cDNA isolation, characterization and bacterial expression of δ-selinene synthase and γ-humulene synthase. *Journal of Biological Chemistry* **273**: 2078–2089.

Stránský, K. and Valterová, I. (1999). Release of volatiles during the flowering period of *Hydrosme rivieri* (Araceae). *Phytochemistry* **52**: 1387–1390.

Strauss, S. Y. (1997). Floral characters link herbivores, pollinators and plant fitness. *Ecology* **78**: 1640–1645.
Struble, D. L. and Arn, H. (1984). Combined gas chromatography and electroantennogram recording of insect olfactory responses. In *Techniques in Pheromone Research*, eds. H. E. Hummel and T. A. Miller, pp. 161–178. New York: Springer-Verlag.
Thièry, D., Bluet, J. M., Pham-Delègue, M.-H., Etiévant, P. and Masson, C. (1990). Sunflower aroma detection by the honeybee: study by coupling gas chromatography and electroantennography. *Journal of Chemical Ecology* **16**: 701–711.
Tollsten, L. (1993). A multivariate approach to post-pollination changes in the floral scent of *Platanthera bifolia* (Orchidaceae). *Nordic Journal of Botany* **13**: 495–499.
Tollsten, L. and Bergström, L. G. (1989). Variation and post-pollination changes in floral odours released by *Platanthera bifolia* (Orchidaceae). *Nordic Journal of Botany* **9**: 359–362.
Tollsten, L., Knudsen, J. T. and Bergström, L. G. (1994). Floral scent in generalistic *Angelica* (Apiaceae): an adaptive character? *Biochemical Systematics and Ecology* **22**: 161–169.
van der Pijl, L. (1960). Ecological aspects of flower evolution. I. *Evolution* **14**: 403–416.
Vogel, S. (1954). *Blütenbiologische Typen als Elemente der Sippengliederung*. Jena: Fischer.
(1963). *The Role of Scent Glands in Pollination*. Rotterdam: A. A. Balkema.
(1978). Evolutionary shifts from reward to deception in pollen flowers. In *The Pollination of Flowers by Insects*, ed. A. J. Richards, pp. 89–96. London: Academic Press.
(1983). Ecophysiology of zoophilic pollination. In *Encyclopedia of Plant Physiology; Physiological Plant Ecology III*, eds. O. L. Lang, P. S. Nobel, C. B. Osmond and H. Ziegler, pp. 559–624. Berlin: Springer.
Wadhams, L. J., Blight, M. M., Kerguelen, V. *et al.* (1994). Discrimination of oilseed rape volatiles by honey bee: novel combined gas chromatographic-electrophysiological behavioral assay. *Journal of Chemical Ecology* **20**: 3221–3231.
Waser, N. M and Price, M. V. (1994). Crossing-distance effects in *Delphinium nelsonii*: outbreeding and inbreeding depression in progeny fitness. *Evolution* **48**: 842–852.
Waser, N. M., Chittka, L., Price, M. V., Williams, N. M. and Ollerton, J. (1996). Generalization in pollinator systems, and why it matters. *Ecology* **77**: 1043–1060.
Weiss, M. R. (1991). Floral colour changes as cues for pollinators. *Nature* **354**: 227–229.
(1995). Floral color change: a widespread functional convergence. *American Journal of Botany* **82**: 167–185.
Whitten, W. M., Williams, N. H., Armbruster, W. S., Battiste, M. A., Strekowski, L. and Lindquist, N. (1986). Carvone oxide: an example of convergent evolution in euglossine pollinated plants. *Systematic Botany* **11**: 222–228.
Williams, N. H. (1983). Floral fragrances as cues in animal behavior. In *Handbook of Experimental Pollination Biology*, eds. C. E. Jones, and R. J. Little, pp. 51–69. New York: Van Nostrand-Reinhold.
Williams, N. H. and Whitten, W. M. (1983). Orchid floral fragrances and male euglossine bees: methods and advances in the last sesquidecade. *Biological Bulletin* **164**: 355–395.
(1999). Molecular phylogeny and floral fragrances of male euglossine bee-pollinated orchids: a study of *Stanhopea* (Orchidaceae). *Plant Species Biology* **14**: 129–136.
Winter, Y. and von Helversen, O. (2001). Bats as pollinators: foraging energetics and floral adaptations. In *Cognitive Ecology of Pollination; Animal Behavior and Floral*

Evolution, eds. L. Chittka and J. D. Thomson, pp. 148–170. Cambridge: Cambridge University Press.
Young, H. J. (1988). Differential importance of beetle species pollinating *Dieffenbachia longispatha* (Araceae). *Ecology* **69**: 832–844.
Zhang, Z., Yang, M. and Pawliszyn, J. (1994). Solid phase microextraction, a solvent-free alternative for sample preparation. *Analytical Chemistry* **66**: 844A.
Zuk, M., Rotenberry, J. T. and Simmons, L. W. (1998). Calling songs of field crickets with and without phonotactic parasitoid infection. *Evolution* **52**: 166–171.

6

Sex pheromones of cockroaches

César Gemeno and Coby Schal

Department of Entomology, North Carolina State University at Raleigh, USA

Introduction

The association of several cockroach species with humans is "rivaled perhaps only by lice and fleas" (Cornwell, 1968). Indeed, some cockroaches, including the American cockroach, *Periplaneta americana* (L.), and the German cockroach, *Blattella germanica* L., have become so intimately associated with humans that natural populations can no longer be found easily. Pest cockroach species are ubiquitous and difficult to eliminate; they form large aggregations that have unpleasant aesthetics and odor, generate allergens that can trigger severe allergic responses including asthma, and can transmit human and animal pathogens (Brenner, 1995). It is not surprising, therefore, that there is an abundance of studies that focus on tactics to eliminate the few cockroach species that are pests to humans ($\sim 1\%$ of the 4000+ described species). However, as will become apparent from reading this chapter, cockroaches are also excellent model organisms in studies of chemical ecology.

In contrast to the closely related grasshoppers and crickets (Orthoptera), cockroaches do not rely on sound as their primary communication modality. Instead, these mainly nocturnal insects use olfactory and tactile cues in their social behavior. Chemical signals are used in attraction to and arrestment at aggregation sites. Cockroach species that can fly, and some of the non-flying species also, rely on long-range volatile pheromones for finding mates. At close range, female cuticular contact pheromones act in concert with short-range male volatile pheromones to bring the sexes together and facilitate recognition. Male dominance hierarchies and territoriality, such as those observed in the lobster cockroach, *Nauphoeta cinerea* (Olivier), are also mediated by chemical signals, and these same signals are involved in female mate choice. Chemical signals also may be involved in parent–offspring communication, stage and population recognition, trail-following behavior, and as

Advances in Insect Chemical Ecology, ed. R. T. Cardé and J. G. Millar. Published by Cambridge University

epidiectic pheromones that mediate dispersion behavior. In addition, cockroaches use irritants, repellents, and sticky excretions as chemical defenses against natural enemies.

Besides their critical reliance on pheromones in social behavior, there are other reasons why cockroaches are excellent models for studies of chemical ecology. This highly diverse group exhibits a variety of reproductive strategies, including parthenogenesis, oviparity, ovoviviparity, and viviparity. Their social organization ranges from solitary individuals to genetically related families with monogamous parents. They live in temperate as well as tropical habitats; deserts, caves, and hollow trees; bromeliad pools; and in the nests of birds and social insects (Schal *et al.*, 1984). The pest cockroach species, and some non-pest species too, can be easily maintained in the laboratory on ordinary rodent or dog food; they are relatively long lived, and their large size facilitates behavioral observations, as well as physiological and biochemical experimentation. Cockroaches have served as excellent model organisms for invertebrate endocrinology and neurobiology, and, as will become apparent in this chapter, such insight has made possible a clearer understanding of the coordination of physiology and behavior. Unfortunately, however, study of the molecular genetics of cockroaches lags significantly behind that of other model insects.

At this point, few generalizations can be made about cockroach sex pheromones. In this chapter, we summarize and analyze the information available and provide suggestions for future research. First, we outline the main aspects of the reproductive behavior and biology of cockroaches as they relate to chemical communication. We then describe the mating behavior and sex pheromones of four cockroach species that have been studied in detail (*P. americana, B. germanica, N. cinerea*, and the brownbanded cockroach, *Supella longipalpa* (F.)), and of several other species for which some experimental studies are available. For each species, we discuss sex pheromone production, reception, and the function of pheromones in mating. Last, we put forward suggestions on themes of research that will likely provide important insights into the chemical ecology of cockroaches.

The general subject of sex pheromones of cockroaches was last reviewed by Schal and Smith (1990). Reviews dealing with particular species are cited in the appropriate section.

Taxonomy, reproduction, and mating behavior

Taxonomy

Extensive reviews on the biology and taxonomy of cockroaches can be found in the literature (Roth and Willis, 1960; Cornwell, 1968; Guthrie and Tindall, 1968; Bell and Adiyodi, 1981; Schal *et al.*, 1984; Gautier *et al.*, 1988; Rust *et al.*, 1995). The most widely accepted taxonomy of cockroaches is that of McKittrick (1964), based on morphological as well as biological data, such as the reproductive mode. However, their close phylogenetic relationship to termites, and the existence of extant, primitive, and apparently related cockroach and termite taxa, have renewed interest in cockroach phylogenetics and fueled extensive debate and alternative phylogenies (Kambhampati, 1995, 1996; Grandcolas, 1996; Klass, 1997, 1998; Nalepa and Bandi, 1999; Mukha *et al.*, 2002). Nevertheless, according to McKittrick (1964) (and supported by both morphometric and molecular analyses), Blattodea (Order Dictyoptera) is divided into five families containing 20 subfamilies. Three of the families – Blattidae, Blattellidae, and Blaberidae – contain most of the cockroach species, and all the species for which sex pheromones have been identified belong to these three families (Table 6.1). The other two families are the most primitive living cockroaches: Cryptocercidae, which is monogeneric and includes wood-eating species of the genus *Cryptocercus*, and Polyphagidae, which is mostly distributed in deserts. Nothing has been reported about the use of sex pheromones by cryptocercids and polyphagids, and so these families are not discussed further in this chapter.

Reproduction

Some cockroach species exhibit obligatory or facultative parthenogenesis, but most species reproduce sexually with reproductive modes that include oviparity, ovoviviparity, and viviparity (Roth, 1970). Oviparous females oviposit their eggs inside a protective egg case (ootheca), which is either deposited soon after being produced or carried by the mother until the nymphs hatch. In ovoviviparous and viviparous species, the egg case is reduced and remains unsclerotized; it is incubated within a brood sac or uterus and live nymphs hatch from the female. At least one species, *Diploptera punctata* (Eschscholtz), is viviparous and the embryos receive nutrients directly from the mother in the form of milk (Evans and Stay, 1995).

The gonadotrophic cycle in cockroaches is regulated by several lipid and peptide hormones (Scharrer, 1987; Stay, 2000). Prominent among these is juvenile hormone (JH), an adult gonadotropic hormone. It stimulates production by the fat body of vitellogenin – a yolk protein precursor – and its uptake by the oocytes; it stimulates the production of oothecal proteins (oothecins) by accessory glands, and it elevates food intake in vitellogenic females. No wonder, then, that behavioral and physiological events related to mate finding and sexual receptivity are also regulated in a coordinated fashion by this vital hormone. *B. germanica* represents an unusual condition and an excellent example of this coordination. It is an oviparous cockroach but functionally exhibits ovoviviparous reproduction (Schal *et al.*, 1997). The

Table 6.1. *Cockroach species in which presence of sex pheromones has been demonstrated, indicating their taxonomic relationships and references to the original studies*

Superfamily	Family	Subfamily	Species	References
Blattoidea	Blattidae	Blattinae	*Periplaneta americana*	Persoons *et al.*, 1979, 1990; Still, 1979; Hauptmann *et al.*, 1986; Kuwahara and Mori, 1990
			Periplaneta australasiae	Waldow and Sass, 1984; Nishino and Manabe, 1985a
			Periplaneta brunnea	Nishino and Manabe, 1985b; Ho *et al.*, 1992
			Periplaneta fuliginosa	Takahashi *et al.*, 1995; Nishii *et al.*, 1997
			Periplaneta japonica	Takegawa and Takahashi, 1989; Nishii *et al.*, 1997
			Blatta orientalis	Warthen *et al.*, 1983; Abed *et al.*, 1993a
		Polyzosteriinae	*Eurycotis floridana*	Farine *et al.*, 1994, 1996
Blaberoidea	Blattellidae	Plectopterinae	*Supella longipalpa*	Schal *et al.*, 1992; Charlton *et al.*, 1993; Leal *et al.*, 1995; Gemeno *et al.*, 2003a
		Blattellinae	*Blattella germanica*	Nishida *et al.*, 1974, 1976; Brossut *et al.*, 1975; Nishida and Fukami, 1983; Schal *et al.*, 1990a; Abed *et al.*, 1993b; Liang and Schal, 1993a; Tokro *et al.*, 1993; Nojima *et al.*, 1999a
			Parcoblatta lata	Gemeno *et al.*, 2003b
			Parcoblatta caudelli	Gemeno *et al.*, 2003b
	Blaberidae	Oxyhaloinae	*Nauphoeta cinerea*	Fukui and Takahahi, 1983; Sreng, 1990; Siruge *et al.*, 1992
			Leucophaea maderae	Sreng, 1993
		Blaberinae	*Blaberus craniifer*	Abed *et al.*, 1993b
			Byrsotria fumigata	Moore and Barth, 1976
		Pycnoscelinae	*Pycnoscelus surinamensis*	Barth, 1965

female undergoes 2–3 days of sexual maturation after the imaginal molt; she feeds and drinks extensively as her corpora allata (CA) synthesize and secrete JH, and the fat body responds by synthesizing massive amounts of vitellogenin. In response to JH, the virgin female also produces and releases sex pheromones and mates (Schal and Chiang, 1995). The basal oocytes continue to take up yolk protein precursors and grow, and as the JH titer dramatically declines, partly in response to allatostatic neuropeptides (Vilaplana *et al.*, 1999), hemolymph ecdysteroids peak, chorionation occurs (Bellés *et al.*, 1993), and the oocytes are ovulated, fertilized, and packaged into an egg case. Most oviparous cockroaches carry the egg case for just one to a few days, but *Blattella* initiates a 3-week 'pregnancy' after oviposition, during which the egg case is incubated while still attached to the female. The egg case inhibits feeding and vitellogenesis in the gravid female through mechanoreceptors in the female's genital vestibulum. As the neonates synchronously emerge from the egg case, the female's CA are released from brain inhibition and a second gonotropic cycle ensues (Roth and Stay, 1962; Schal *et al.*, 1997).

Mating behavior

An accurate description of the mating behavior of cockroaches is essential in order to understand the specific functions of their sex pheromones. Although there are marked differences in mating behavior among species, a few generalizations can be made. The mating sequence in most cockroaches consists of (i) mate finding, (ii) contact, (iii) male release of tergal volatiles, (iv) female feeding on male tergal secretions, (v) copulation, and (vi) postcopulation behaviors (Fig. 6.1). Following is a generalized description of the role of sex pheromones in each of these steps.

Mate finding

Under laboratory conditions, some cockroach species find their mates without the aid of sex pheromones, but it is not known to what extent this occurs in nature. In the majority of the cockroach species that have been studied, one of the sexes, typically the female, releases a volatile pheromone that, when perceived by the opposite sex, elicits orientation toward the emitting individual. Some of the volatile pheromones are extremely effective at eliciting behavioral responses at very low concentrations and, therefore, may be functional over long distances (e.g., > 2 m), whereas other volatile pheromones seem active only over considerably shorter distances. The modality of locomotion (i.e., walk or fly), together with the degree of gregariousness, foraging strategy, and other spatial characteristics of the population, may determine whether long-range or close-range pheromones are used in attraction by any given species.

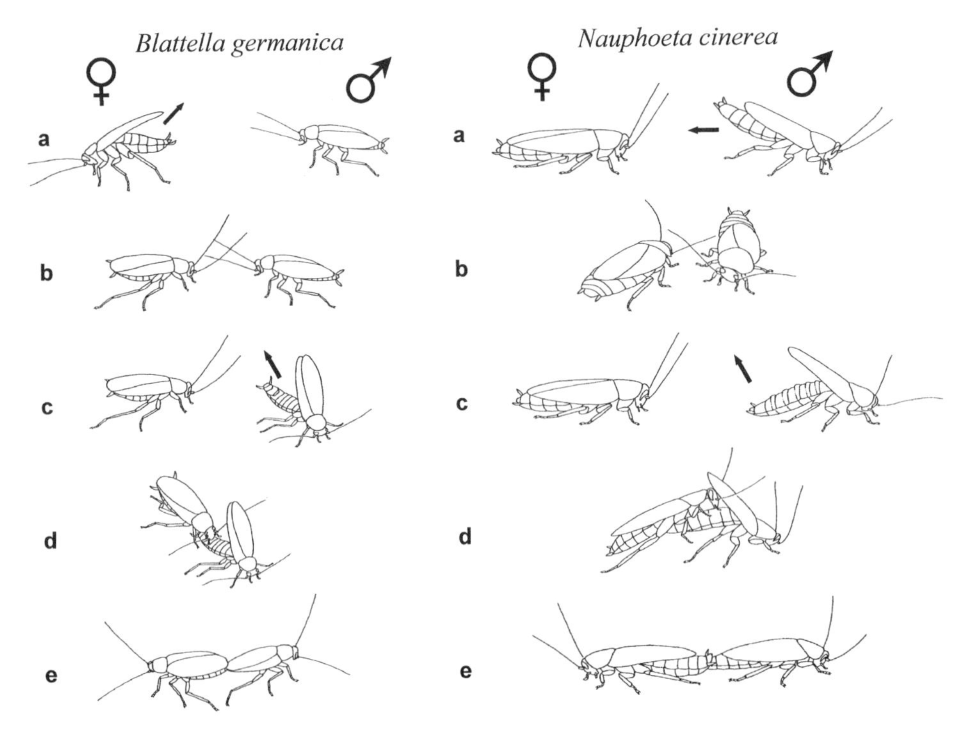

Fig. 6.1. Pheromone-mediated courtship sequences in *Blattella germanica* (Blattellidae) and *Nauphoeta cinerea* (Blaberidae). In *B. germanica*, as in most blattellids and blattids, the female calls and releases a long-range attractant (a) and after contact (b) the male raises his wings and emits volatiles from his tergites (c), which attract the female to feed on the male's dorsum (d) and position the sexes for copulation (e). In *N. cinerea* and some other blaberids, the male releases a long-range attractant from his sternites (a) and after contact with the female (b) the male raises his wings and releases volatile attractants from his tergites (c), which attract the female to feed on his tergites (d) and then mating ensues (e). Arrows indicate release of volatile pheromones from tergite 10 in *B. germanica* females, glands in tergites 7 and 8 in *B. germanica* males, and sternites 3–7 (a) and tergites 2–8 (b) in *N. cinerea* males. Drawings are based in part on photographs in Sreng (1979a, 1993) and Nojima *et al.* (1999b), and our observations. Species are not drawn to scale with respect to each other.

Flying cockroaches will probably tend to use long-range pheromones, whereas cockroaches incapable of flight are more likely to rely on short-range pheromones. In some cockroach species, both males and females have fully developed wings; in other species, the females have reduced wings. In still other species, both males and females have reduced wings. Males and females of *P. americana* have full-length wings and both sexes can fly. In *S. longipalpa* and *Parcoblatta* sp. (wood cockroach) the males have normal wings and fly, whereas females have reduced wings and cannot fly. In the latter three species, females release long-range volatile sex pheromones. *B. germanica* females also release a volatile attractant, but, as will be shown below, this pheromone does not seem to have as prominent a role in mate finding as in the previous three species, perhaps because neither male nor female *B. germanica* can fly.

The pattern that seems to emerge is that in those species in which the males, or both sexes, are good flyers, the females release a long-range volatile sex pheromone. In those species in which neither males nor females are able to fly, the males may be the releasers of sex pheromones in a reversal of sex roles. Males of the oriental cockroach, *Blatta orientalis* L., have longer wings than the females but neither sex flies. In this species Simon and Barth (1977a) showed that only males release sex attractants; however, later work by Abed *et al.* (1993a) indicated that both males and females of *B. orientalis* release sex attractants. In species of the subfamily Oxyhaloinae (Blaberidae), *N. cinerea* for example, neither sex can fly and males are the releasers of the sex attractant (Sreng, 1993). It appears that members of the Blaberidae tend to be more gregarious than blattellids or blattids (Schal *et al.*, 1984). It remains to be demonstrated in cockroaches whether aggregations in caves, hollow trees, or the leaf-litter provide for short-range chemical – or even tactile – communication or make long-range pheromones imperative to avoid inbreeding. From the scant field-based observations, it appears that receptive females of highly gregarious species (e.g., *Blaberus* spp., *Capucina patula* (Walker)) position themselves outside the aggregation and emit volatile pheromones to attract males (Schal and Bell, 1985).

Long-range attractants produced by females are usually released by means of characteristic calling postures or behaviors (Schal and Bell, 1985). Typically, the abdomen is lowered toward the substrate and this may be accompanied by periodic exposure of the genitalia (Fig. 6.1a). The calling period is species specific. Although there are few experiments that directly link female calling to the release of the sex pheromone (Smith and Schal, 1990a; Gemeno *et al.*, 2003b), the available evidence suggests that when such specific postures occur in sexually receptive female cockroaches, and not in mated females of the same stage of the reproductive cycle, then they represent calling and the release of pheromones. Males also have characteristic postures to release long-range attractants. For example, males of the subfamily Oxyhaloinae extend and elevate their abdomens with the tip pointing upward to expose sternal glands, which are normally hidden between the sternites. Here, we define calling, in either females or males, as pheromone emission aimed at attracting potential mates. We further stipulate that calling is an early act of mate finding, to distinguish it from subsequent pheromone emission during courtship.

Contact

Volatile sex pheromones provide the first specific information about mating partners. Cuticular contact pheromones provide a second step in species and sex recognition and, in most cases, they function as courtship-inducing pheromones

(Fig. 6.1b). Although they are probably present in most cockroach species, sex-specific contact pheromones have been identified in only a few species. They are thought to be distributed throughout the epicuticular surface and are perceived by means of antennal contact and with the mouthparts. Female-produced contact pheromones elicit courtship responses in males; however, in some cockroaches, stridulation and hissing may combine with, or operate in place of, contact chemoreception. In addition, male contact pheromones may function to inhibit courtship in other males.

Male release of tergal volatiles

After the sexes contact each other, the next step in mating behavior of cockroaches is male courtship and concomitant release of a volatile pheromone from the male's tergal (i.e., dorsal) abdominal glands (Fig. 6.1c). These glands may be highly specialized macroscopic structures readily visible to the naked eye, as in *B. germanica* and *S. longipalpa*, or they may consist of microscopic secretory cells distributed in the epithelium beneath the cuticle. Microscopic glands may also be present in species bearing specialized tergal structures (Brossut and Roth, 1977; Sreng, 1984). In species possessing long wings, the tergal glands are normally hidden and courtship entails wing-raising to expose them. Courting males may also bend their abdomen, either facilitating release of pheromone or directing it toward the female. Moreover, fanning of the glands with the wings occurs in some species, suggesting that the male is wafting odorants toward the female. Nymphs and other males are also attracted to male tergal gland odors, suggesting that general odorants released from these glands mediate the orientation responses. The distance over which male tergal gland pheromones attract females has not been studied empirically. Nevertheless, courted females are in close proximity to males, and it appears that male-emitted courtship pheromones operate over relatively short distances. It is possible that female mounting behavior may not be chemically mediated in some species, or this behavior may be released by signals other than chemicals from the tergal glands. In some species, the courting male actively maneuvers his abdomen under the female.

Roth (1969) surveyed the presence of specialized tergal gland structures in cockroaches. He found large variation among species, ranging from complete absence (although microscopic glands may be present) to presence in five abdominal segments. The positions of the glands on the abdomen were also highly variable. Roth (1969) proposed that in the course of evolution there has been a reduction in the number of segments that are specialized, and a change in the location of the glands from a posterior to a more anterior position on the abdomen. Despite marked anatomical diversity of the glands, only a few basic types of cell are present: type 1 are in direct contact with and secrete to the cuticle; type 2 are large, without ducts, and not in contact with the cuticle; and type 3 secretory cells are large with a single cuticular duct, formed by a canal cell that connects the glandular cell with the outer cuticle (Noirot and Quennedey, 1974; Brossut and Roth, 1977).

All seven species of the subfamily Oxyhaloinae examined by Sreng (1984) possess tergal and sternal microscopic glands, but in different arrangements. This variation is apparently correlated with the mating system. Sreng (1984) proposed that species with well-developed tergal and sternal glands, like *N. cinerea*, in which the female mounts the male to feed from his tergal secretions, represent the primitive condition, which evolved toward a reduction in the amount of glandular tissue and no mounting and feeding by the females. Sternal glands play a defensive role in other species (Brossut and Roth, 1977).

Female feeding on male tergal secretions

Females attracted to courting males mount their abdomen and nibble or feed on their tergal secretions (Fig. 6.1d). If no specialized tergal gland is present, the female seems to nibble throughout the tergites. The more specialized tergal glands posses bristles, which appear to function as mechanoreceptors, possibly indicating to the male that the female is in the correct position for copulation. However, in those species that lack specialized tergal glands, the males orient just as well to the position of the mounting female. It is possible that these mechanoreceptors may play other roles, including triggering the release of pheromone and nuptial secretions.

Copulation

With the female positioned over the male and feeding on the tergal secretions, the male telescopes his abdomen under the female's abdomen, engages her subgenital plate with his hooked left phallomere, and attempts copulation. If successful, the couple immediately turns around and face in opposite directions (Fig. 6.1e). This mating sequence (type A) is observed in most cockroach species, including *N. cinerea* and the Madeira cockroach, *Leucophaea maderae* (Fabricius) within the Oxyhaloinae. Sreng (1993) described two other types of mating behavior in the Oxyhaloinae, both characterized by lack of a male wing-raising display or female feeding behavior. In type B mating, represented by *Jagrehnia madecassa*, after the sexes contact, the male, without raising his wings, mounts the female's dorsum, curves his abdomen under the female's abdomen, and copulates. In type C mating, which is characteristic of *Gomphrolita* species, after the sexes contact, the male positions himself in the opposite direction to the motionless female (heads directed away from each other). He then moves backward and copulation follows (Sreng, 1993). The length of copulation is highly variable in cockroaches. In *B. germanica*, copulation averages 95 min under laboratory conditions (Schal and Chiang, 1995), whereas in *Xestoblatta hamata* (Giglio-Tos), copulation in the tropical rainforest may last for up to 5 h (Schal and Bell, 1982).

Postcopulation behaviors

In most cockroaches, the male and female disengage and do not interact further after copulation. In some species, however, gift giving and postcopulatory guarding may be chemically mediated. In many blattellids, the male offers a postnuptial gift of uric acid to his mate. In some (e.g., *X. hamata*) the female is attracted to the male's genital area, the source of urate exudates (Schal and Bell, 1982).

Sex pheromones of Blattidae

Periplaneta americana *(L.) (American cockroach)*

P. americana is a large cockroach, 28–44 mm long, shiny red-brown with a paler yellow area around the edge of the pronotum. The fully developed wings extend beyond the tip of the abdomen in males but only just overlap the abdomen in females. Both sexes can fly, and like most cockroach pest species it probably originated in Africa. It prefers warm, moist environments and is a common outdoor urban pest in tropical and subtropical regions.

Mating behavior

The mating behavior of *P. americana* has been described in detail (Roth and Willis, 1952; Barth, 1970; Abed *et al.*, 1993c). Females exhibit a calling behavior in which the tip of the abdomen is lowered while the extended legs raise the anterior part of the body (Abed *et al.*, 1993c). Calling results in pheromone release and attraction of males from a distance (Seelinger, 1984; Abed *et al.*, 1993c); in the field, attracted males fly to the vicinity of the pheromone source and orient to the female by walking (Waldow and Sass, 1984). Upon contacting the female, the male proceeds with courtship, which consists of raising both pairs of wings and turning 180° from his previous position of facing the female. Males exposed to the volatile female pheromone may engage in wing-raising displays without prior direct contact with the female. While the male is raising his wings, he releases an attractant pheromone that directs the female to his dorsum. He generally also flutters his wings, presumably directing the pheromone toward the female. Attraction to this male pheromone is not sex specific because males and nymphs are also attracted to and feed upon the tergal secretions. The female mounts the male and palpates his back, licking the exposed tergal glands on the anterior tergites 2–4 (Abed *et al.*, 1993c), while gradually moving forward on the male. The male may move his antennae backward and contact the female. While in this position, the male, probably in response to tactile stimulation from the female, telescopes his abdomen and attempts copulation. If the female is receptive they copulate in the opposed position.

Female volatile pheromone: (−)-periplanone-B

The identification of the sex pheromone of *P. americana* has not been straightforward. A brief historical account provides a good case study of the possible problems associated with pheromone identification (Stinson, 1979; Schal and Smith, 1990). Roth and Willis (1952) discovered that females release an odorant, minute amounts of which stimulate male courtship. This pheromone could be extracted with ether from filter paper taken from containers that housed virgin females. Wharton *et al.* (1962) made the first attempt to isolate and chemically identify the sex pheromone. They isolated a chromatographic peak that retained 75% of the behavioral activity of the crude extract and preliminarily identified it as an aliphatic compound containing an ester carbonyl. Using volatile collections from 10 000 females (Yamamoto, 1963), accumulated over a period of 9 months, and a different purification procedure, Jacobson *et al.* (1963) reported the identification of the sex pheromone of *P. americana*. This identification received wide press coverage (Stinson, 1979), but its validity was soon questioned by Wharton *et al.* (1963) based on, among other concerns, the enormous difference in gas chromatographic (GC) retention times between their active compound (Wharton *et al.*, 1962) and that of Jacobson *et al.* (1963). Day and Whiting (1964, 1966) synthesized the compound that Jacobson *et al.* (1963) proposed as the pheromone, but neither its chemical characteristics nor its biological activity corresponded with those of the putative pheromone. Jacobson and Beroza (1965) retracted their original identification and proposed a new configuration, which was not followed up by further research. Using the purification method of Jacobson *et al.* (1963), the antennae of males, females, and nymphs produced electroantennogram (EAG) responses to the putative sex pheromone (Boeckh *et al.*, 1963). Jacobson's compound finally was shown to be an artifact of their purification method. In a later study, Boeckh *et al.* (1970), using the purification method of Wharton *et al.* (1963), showed that only the adult male antennae produced strong EAG responses to the true pheromone component, but those of nymphs and adult females – except for late instar male nymphs – produced no EAG responses.

An important step toward the correct identification of the sex pheromone of *P. americana* was the identification of several phytochemicals, such as bornyl acetate and santalol, that stimulated courtship behavior in males (Bowers and Bodenstein, 1971; see p. 196). Some years later, Tahara *et al.* (1975) and Kitamuta *et al.* (1976) reported that germacrene-D, extracted from composite plant species, stimulated sexual behavior in males of *P. americana* and the Japanese cockroach *Periplaneta japonica* Karny. Persoons *et al.* (1974, 1976) determined, on the basis of spectroscopic data, that the sex pheromone (periplanone-B) had a germacranoid structure of molecular weight 248, but the definitive structure was not elucidated. Kitamura and Takahashi (1976) also isolated an active component of molecular weight 248, supporting Persoons *et al.* (1976). Chow *et al.* (1976), independently from Persoons *et al.* (1976), identified two active peaks, one of which provided further support for this candidate pheromone. Periplanone-B was identified as **1** (Table 6.2), and its structure was confirmed by comparison with a synthetic, biologically active epimer of periplanone-B, but its stereochemistry was not resolved (Persoons *et al.*, 1979).

The tentative structure identification of Persoons *et al.* (1979) required an unambiguous synthesis to confirm the gross structure and determine the stereochemistry of periplanone-B. The elegant total synthesis of periplanone-B by Still (1979; see Nicolaou and Sorensen, 1996) allowed the preparation of three different diastereoisomers and the assignment of **1** (Table 6.2) as the natural sex pheromone (as proposed by Persoons *et al.* (1979)). The relative configuration was unambiguously established by an X-ray crystallographic study (Adams *et al.*, 1979; but see Seybold, 1993). Kitahara *et al.* (1986) synthesized (−)-periplanone-B and found it to cause the characteristic sexual excitement in males. Okada *et al.* (1990a,b) found a behavioral threshold of 10^{-5} ng to (−)-periplanone-B. Behavioral tests, using different assay procedures, showed that threshold responses to both synthetic and natural periplanone-B were similar (10^{-3} to 10^{-4} ng) (Adams *et al.*, 1979), which was later confirmed by Persoons *et al.* (1982). The behavioral responses of *P. americana* to a range of concentrations of synthetic periplanone-B were comparable to those to extracts of papers contaminated by virgin females (Tobin *et al.*, 1981).

Female volatile pheromone: (−)-periplanone-A

The identification of a second component of the female's pheromone, periplanone-A, also was not straightforward (Kuwahara and Mori, 1990; Persoons *et al.*, 1990). In their original purification, Persoons *et al.* (1974, 1976) identified several active peaks, one of which was named periplanone-A (Persoon's periplanone-A is isoperiplanone-A) (Persoons *et al.*, 1990), but two of the four possible diastereomers of isoperiplanone-A had different spectral data from isoperiplanone-A (Shizuri *et al.*, 1987a,b) and one of them was tested and found to be behaviorally inactive (Persoons *et al.*, 1990). Several other compounds were proposed, synthesized, and shown to be inactive (Shizuri *et al.*, 1987c; Macdonald *et al.*, 1987). However, an active compound (Hauptmann's periplanone-A), similar to but with different spectral data from Persoons' periplanone-A, was isolated earlier (**2**; Table 6.2) (Hauptmann *et al.*, 1986; Nishino *et al.*, 1988). Both enantiomers of Hauptmann's periplanone-A were synthesized and only (–)-periplanone-A was found to be highly active (Kuwahara and Mori, 1990; Okada *et al.*, 1990a). Persoons' isoperiplanone-A

Table 6.2. *Chemical structures of the principal identified volatile sex pheromones of cockroaches, listed by the species and the sex that produces it; for most pheromones common names are also listed.*

	Structure	Common name	Species/sex	Source
1	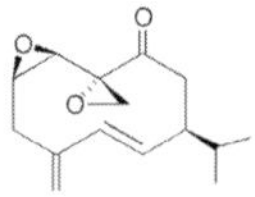	Periplanone-B	*Periplaneta americana* ♀	Digestive tract[a]
2	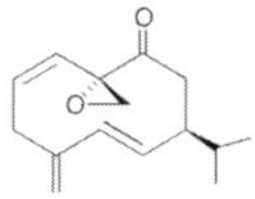	Periplanone-A	*Periplaneta americana* ♀	Digestive tract[a]
3	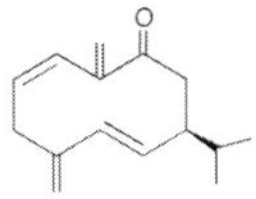	Periplanone-C	*Periplaneta americana* ♀	Digestive tract[a]
4	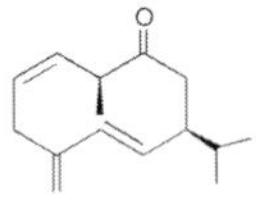	Periplanone-D	*Periplaneta americana* ♀ *Periplaneta fuliginosa* ♀	Digestive tract[a]
5	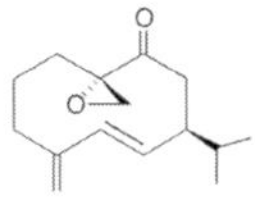	Periplanone-J	*Periplaneta japonica* ♀	Digestive tract[a]
6	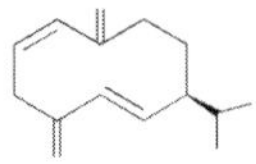	Periplanene-Br	*Periplaneta brunnea* ♀	Digestive tract[a]
7	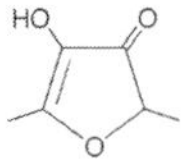 HO O O		*Eurycotis floridana* ♂	Tergites
12	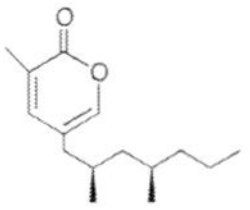 O O	Supellapyrone	*Supella longipalpa* ♀	Tergites
13	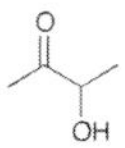 O OH	Seducin (acetoin)	*Nauphoeta cinerea* ♂	Sternites[b]
14	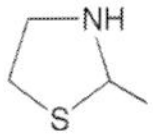 NH S	Seducin	*Nauphoeta cinerea* ♂	Sternites[b]
15	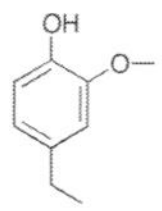 OH O–	Seducin	*Nauphoeta cinerea* ♂	Sternites[b]

[a]Other tissues may be involved.
[b]Also found in the tergites, but in lower quantities.

(Persoons *et al.*, 1982) appears to have been a biologically inactive derivative of periplanone-A, possibly generated under the harsh conditions of GC (Kuwahara and Mori, 1990). These findings required a revision of the entire nomenclature of the periplanones (Persoons *et al.*, 1990).

Functions of periplanone-A and periplanone-B

In the initial stages of identification, it was suggested that periplanone-A was a byproduct of periplanone-B, or vice versa (Talman *et al.*, 1978; Persoons *et al.*, 1979), but behavioral and electrophysiological studies have demonstrated that both compounds are used in sexual communication of *P. americana* and related species, and that each compound has a specific role in male orientation behavior. Sass (1983) discovered two types of antennal receptor cell, one for each of the periplanones. This finding was later confirmed by differential adaptation EAG to pheromone extracts (Nishino and Manabe, 1983). Males showed similar amplitude EAG responses to periplanone-A and periplanone-B (Nishino and Kuwahara, 1983; Yang *et al.*, 1992), and both compounds elicited running, upwind orientation, and wing-raising in a wind tunnel, with increasing running at higher concentrations (Seelinger, 1985a; Seelinger and Gagel, 1985). However, in behavioral tests using the same concentrations of periplanones, males were more responsive to periplanone-B than to periplanone-A (Seelinger, 1985a; Seelinger and Gagel, 1985; Okada *et al.*, 1990b). In addition, the response of males to periplanone-B declined faster and the insects were more easily adapted by repeated stimulation with periplanone-B than with periplanone-A (Seelinger, 1985a; Seelinger and Gagel, 1985). All this suggests that periplanone-B is used for long-distance attraction, whereas periplanone-A influences male behavior closer to the female (Seelinger, 1985a). In fact, odor source location is more precise when the natural blend is presented than with periplanone-B alone (Seelinger and Gagel, 1985).

In addition to their role in orientation behavior, the periplanones may participate in reproductive isolation among sympatric *Periplaneta* species. Periplanone-A from *P. americana* attracts *Periplaneta australasiae*. Periplanone-B, and possibly some other components, inhibits attraction of male *P. australasiae* (Seelinger, 1985b). Two other periplanones, periplanone-C (**3**; Table 6.2) and periplanone-D (**4**; Table 6.2), have been isolated from *P. americana* (Biendl *et al.*, 1989; Takahashi *et al.*, 1995) and the pure stereoisomers ((–)-periplanones) have been synthesized (Nishii *et al.*, 1997). Three sympatric *Periplaneta* species respond differently to blends of these compounds (Takahashi *et al.*, 1988b; Nishii *et al.*, 1997), suggesting that they may contribute to species-specific blends and, hence, reproductive isolation.

Sources of periplanones

The sex pheromone of *P. americana* is highly volatile and it readily impregnates the substrate or cage in which virgin females are kept (Roth and Willis, 1952;

Wharton *et al.*, 1962). Therefore, the identification of its tissue source has been confounded by adsorption of the pheromone to females and their feces, as well as by the extraordinary sensitivity of the males. The first attempt at identifying the source of pheromone concluded that it was produced in the head (Stürckow and Bodenstein, 1966), but later studies disproved this unusual finding and now it is generally accepted that the pheromone occurs mainly in the abdomen, specifically in the digestive system, feces, or in glandular tissues in the posterior abdominal segments.

The exact anatomical site of production of the pheromone in the abdomen has yet to be resolved. Bodenstein (1970) showed that the greatest amount of pheromone activity was in the crop (foregut) and midgut, but he also found it in the caeca and feces. Takahashi *et al.* (1976) also found most of the pheromone activity in the midgut. However, Persoons *et al.* (1979) extracted the foregut, hindgut, and midgut (cut into five sections) and recovered pheromone only from a single section of the midgut containing the proventriculus and caeca. The main difficulty in establishing that the source of the pheromone is in the digestive system is that the pheromone can be easily transported in contaminated food and feces. Because feeding and digestion show diel periodicities in cockroaches (Cornwell, 1968), the time of the photocycle at which the pheromone is extracted could affect the location of the pheromone in the digestive system and be responsible for the differences observed in various studies. Moreover, release of pheromone with the feces is an unlikely mate-finding strategy in highly mobile insects (Cardé and Baker, 1984), such as *P. americana* and other cockroaches, which frequently change calling location (Schal *et al.*, 1984; Schal and Bell, 1985).

The calling behavior of *P. americana* females involves opening the genital vestibulum (posterior end of the abdomen), without excretion of feces (Abed *et al.*, 1993c), and results in the attraction of males (Seelinger, 1984). Abed *et al.* (1993c) contend that the sex pheromone of *P. americana* is produced by the atrial glands, a pair of glandular areas in the genital vestibulum that are everted during calling by pressure of the hemolymph, and that the feces may become contaminated with pheromone from the atrial gland. Analysis by GC–mass spectrometry (GC–MS) of extracts from the alimentary canal, the last two abdominal segments (which contain the atrial glands), and the anterior part of the abdomen indicate that the atrial glands contain most of the periplanone-B, up to 60 ng (Abed *et al.*, 1993c). In contrast, Yang *et al.* (1998) extracted several abdominal tissues, including the atrial glands, and concluded that periplanone-A and periplanone-B were most abundant in the colon (0.34 and 8.31 ng, respectively), and that this tissue produced the strongest EAG responses. Moreover, feces extracted from the colon induced strong sexual responses in males, showing that fecal material contained pheromone before it could contact the atrial glands. This is also true of other studies (Bodenstein, 1970; Takahashi *et al.*, 1976; Persoons *et al.*, 1979). It is unclear how calling behavior figures in this scenario.

Female contact pheromone

Mated *P. americana* females do not elicit courtship in males. The observation that in the presence of periplanone-B males direct courtship toward mated females, and not males, prompted Seelinger and Schuderer (1985) to investigate a female-specific contact pheromone in *P. americana*. The contact stimulus can be extracted from the cuticle with non-polar solvents and when transferred to glass dummies it elicits courtship wing-raising responses in males as do live females. However, the contact pheromone is ineffective without the female volatile pheromone (periplanone-B), suggesting that the former is used to discriminate between males and females only in the presence of a calling female. The contact pheromone is also species specific, facilitating close-range discrimination of species that release similar volatile pheromones. The chemical identity of this pheromone remains unknown.

Factors affecting pheromone production and response

Females begin to emit pheromone 9 or more days after the adult molt (Bodenstein, 1970; Takahashi *et al.*, 1976; Hawkins and Rust, 1977), but clearly, this is variable and temperature dependent. The attractancy of gut extracts made on the first day after the imaginal molt corresponds to that of 0.1 ng (±)-periplanone-B (Sass, 1983). During the next 20 days, the effectiveness of both fractions of the sex pheromone (periplanone-A and periplanone-B) in behavioral assays increases 100-fold and remains high for at least the next 45 days. Collection of airborne pheromone with Tenax followed by behavioral assays showed that periplanone-A and periplanone-B were released by 10–25-day-old females in equal amounts, equivalent to ~0.6 ng periplanone-B per female per day (Sass, 1983). Yang *et al.* (1998) confirmed an increase in pheromone activity in the early adult but showed a decline in pheromone between days 20 and 30.

There are no published reports on the neuroendocrine control of pheromone production in female *P. americana*, but preliminary observations (in Barth and Lester, 1973; C. Schal, unpublished data) indicated that removal of the CA completely abolished pheromone production. Pheromone production resumed in allatectomized females rescued with either implantation of active CA or topical application of a JH analog. Successful copulation also suppressed the production of the volatile pheromone (Wharton and Wharton, 1957), as in some other cockroaches (see below). However, Sass (1983) stated that virgin females carrying parthenogenetically produced egg cases maintained the same level of pheromone as females without egg cases.

Virgins, unlike mated females, spend significantly more time outside the shelter and engage in extensive calling behavior (Seelinger, 1984). Calling commenced in vitellogenic females and occurred during the early scotophase (Seelinger, 1984) or during the second half of the scotophase (Abed *et al.*, 1993c). Yet, Sass (1983) determined that equal amounts of pheromone were produced in the photophase, when females did not call, and scotophase, when calling took place. This is clearly inconsistent with the diel periodicity of calling and the data of Abed *et al.* (1993c), which showed that males exhibited no sexual reaction when placed with a non-calling female. There are no observations on the effect of mating on calling behavior in *P. americana*.

P. americana males are exquisitely sensitive to stimulation with periplanone-B. As little as 10^{-3} ng of purified extract loaded on a dispenser and puffed into a cage can elicit a motor response in males (Seelinger, 1985). Because only a fraction of the pheromone is actually emitted from the dispenser, and only a fraction of that is trapped by and enters the sensilla, it appears that the cockroach rivals the male silkmoth (Kaissling and Priesner, 1970) in its ability to respond to minute quantities of pheromone. The sequence of male behaviors in response to purified sex pheromone includes rapid antennation, standing erect, increased locomotion, wing-raising and fluttering, and backing toward the pheromone source (Rust, 1976). Running activity and courting displays by males have been major bioassay criteria used by many investigators, and both increase with dose of the sex pheromone (Rust, 1976).

Detailed observations of male behavior in response to pheromone extracts indicate the existence of positive chemotaxis and positive chemoanemotaxis, and probably chemoklinotaxis, chemotropotaxis, and anemotaxis (Rust and Bell, 1976; Bell, 1981; Bell and Tobin, 1981; Tobin, 1981). In response to the sex pheromone, time spent immobile and frequency of pauses decreased, whereas running speed, time between pauses, and distance traveled increased (Hawkins, 1978). Bell and Kramer (1980) used a servosphere (two-dimensional locomotion compensator ("treadmill") for walking animals) and determined that straightness and direction of movement are concentration dependent and that the cockroaches turn back toward the plume if they depart its edge, all the while moving upwind. Source location, once close to the pheromone source, is achieved by increased overall levels of activity and initiation of local search, which results in coverage of a large area in a short period of time (Hawkins and Gorton, 1982). Observations in the field generally have confirmed laboratory studies. However, an important component of male response in the field is an oriented flight of 30 m or more to the vicinity of the pheromone source, followed by a local search while walking (Sass, 1983; Seelinger, 1984).

The sexual response of adult males depends on both intrinsic and environmental factors. Males become responsive to the pheromone 6–9 days after adult emergence (Wharton *et al.*, 1954; Hawkins and Rust, 1977). Antennation and erect body posture appear at the same time during sexual maturation, followed by, in this order, increased locomotion, running, wing-raising, and abdominal extension (Silverman,

1977). The first time that males wing-raise and extend the abdomen is on day 9 (Silverman, 1977). Male sexual activity is mainly nocturnal (Hawkins and Rust, 1977; Seelinger, 1984), and males respond to lower concentrations of pheromone at night than during the day (Hawkins and Rust, 1977). Zhukovskaya (1995) confirmed that the sexual wing-raising response and its latency exhibited a diel periodicity (not circadian, as stated in the paper) under a light:dark cycle coupled with a diel temperature cycle.

Males respond better at intermediate (25 °C) than at more extreme temperatures (10 °C, 30 °C) (Appel and Rust, 1983). Fecal extracts from females reared at low temperatures elicited significantly less male courtship, suggesting that females produce less pheromone at the lower temperatures than at the higher temperatures (Appel and Rust, 1983). However, because females at low temperatures also defecate less, it is possible that while the total amount of pheromone was lower, the amount of pheromone per unit mass of feces may be the same as at 25 °C or 30 °C. Wharton *et al.* (1954) indicated that isolated males were much less responsive than grouped males, but Takahashi and Kitamura (1972) demonstrated that isolation from females for more than 1 month caused the males to be highly responsive. Adult males can survive at least a month without food. When food and pheromone (extract of filter paper that had been contaminated by virgin females) were presented separately to starved males, equal numbers of males responded to both stimuli, but when they were offered simultaneously, males preferred the pheromone (McCluskey *et al.*, 1969).

Pheromone analogs and structure–activity studies

Several phytochemicals have been shown to elicit male sexual behavior in *Periplaneta* spp., and their identification has played an important role in the elucidation of the sex pheromone of *P. americana*. The first compounds identified were (+)-bornyl acetate, a monoterpenoid constituent of conifer trees, the sesquiterpenoids α- and β-santalol, and a sesquiterpene hydrocarbon, later identified as germacrene-D (Bowers and Bodenstein, 1971). These compounds were extracted from 19 plant species in the families Simaroubaceae, Araliaceae, Labiatae, and Compositae. Extensive screening of analogs followed discovery of these pheromone mimics. Nishino and colleagues (Washio and Nishino, 1976; Nishino and Washio, 1976) tested the antennal responses of *P. americana* to 30 different monoterpenoids and 43 straight-chain compounds and concluded that C_6 compounds having carbonyl and hydroxyl groups elicited the largest EAG responses. Although the relative activity of the test compounds was the same in both male and female antennae, most compounds elicited greater EAG responses from male antennae than from the female antennae. Similar results have been extended to other *Periplaneta* spp., including *P. australasiae, Periplaneta fuliginosa*, and *Periplaneta brunea* (Nishino *et al.*, 1980; Nishino and Usui, 1985). Manabe and Nishino (1983) tested several compounds related to bornyl acetate by EAG and behavior assays in *P. americana*. All the alcohols were inactive and the elimination of the C-1 methyl group caused a complete loss of pheromone activity. Although (+)-bornyl acetate elicited the same sexual display from males as did the sex pheromone, its structure is quite different from that of the sex pheromone. Moreover, very high doses and a longer time were required to induce sexual display, and the effect of bornyl acetate persisted for a much shorter period than with the sex pheromone (30 s versus 5–7 min) (Manabe and Nishino, 1983).

Nishino *et al.* (1977b) tested several compounds related to (+)-bornyl acetate and found that (+)-verbanyl acetate was as behaviorally stimulatory as (+)-bornyl acetate, which suggested that the functional groups (ester carbonyl and methyl groups) are essential to the sex pheromone activity of bicyclic monoterpenoids. Manabe *et al.* (1983) determined that the C-9 methyl group of (+)-verbanyl acetate was essential for sex pheromone activity. Using differential adaptation EAG, Nishino and Manabe (1983) demonstrated cross-adaptation between (+)-verbanyl propanoate (VaP), which was more active than the acetate (Nishino *et al.*, 1982), and periplanone-B, indicating that VaP and periplanone-B stimulated the same receptor neuron. Adaptation with periplanone-A resulted in no decrease in EAG response to periplanone-B or VaP, indicating that there are two receptor types, one for periplanone-B and VaP and one for periplanone-A. Behavioral assays were performed to determine the specific molecular activity of verbanyl analogs (Manabe *et al.*, 1985). Electron density of the carbonyl oxygen atom in the substituent, its length, and the position of the carbonyl group were all important in response, with the strongest response elicited by (+)-verbanyl methylcarbonate, which possesses the highest electron density on the carbonyl oxygen atom of all analogs tested (Manabe *et al.*, 1985).

The sesquiterpene hydrocarbon that was first isolated by Bowers and Bodenstein (1971) has been found in several plants, including *Erigeron annus* (L.), *Eupatorium chinense* L., and *Chrysanthemum morifolium* Ramat, and it was identified as germacrene-D, a compound rather similar in structure to the later identified periplanones (Tahara *et al.*, 1975). Moreover, it stimulated sexual behavior in males of *P. americana* and *P. japonica*; in addition, unlike (+)-bornyl acetate, germacrene-D elicited EAG responses from male antennae but not from female antennae (Washio *et al.*, 1976). Nevertheless, high concentrations of germacrene-D were required. Several germacrene analogs were tested by EAG and behavioral assays on males and females and only germacrene-D produced greater EAG responses in males than in females, and the typical sexual behavior of males (Nishino *et al.*, 1977b). Germacrene-D and periplanone-B both possess a conjugated diene system and an isopropyl group with the same relative structural positions, suggesting that these two might be important in pheromone recognition.

Another phytochemical that stimulated males and females differentially was T-cadinol, extracted from *Solidago altissima* L. It stimulated male and female antennae, but in comparison with germacrene-D, the responses were neither as sex specific nor as strong (Nishino *et al.*, 1977c).

Sex pheromone reception and processing

P. americana is one of just a few species of insects in which both peripheral and central olfactory processing have been studied. In contrast to many short-lived lepidopterans, in which the male antenna is highly specialized for sex pheromone reception, the antennae of male cockroaches contain numerous food-responsive sensilla. In addition to olfactory sensilla, the antennae also house mechano-, hygro- and thermoreceptors, as well as contact chemoreceptors (Schaller, 1978; review: Boeckh *et al.*, 1984). Extensive ultrastructural and electrophysiological evidence has demonstrated that morphologically defined sensillum types house receptor cells of specific functional types (Sass, 1976, 1978, 1983; Schaller, 1978; Selzer, 1981, 1984; review: Boeckh and Ernst, 1987). Boeckh and Ernst (1987) defined 25 types of cell according to their odor spectra, but of the 65 500 chemo- and mechanosensory sensilla on the antenna of adult male *P. americana*, an estimated 37 000 house cells that respond to periplanone-A and periplanone-B.

Sexual dimorphism of antenna sensillum types does not become morphologically apparent before the adult stage. Antennal segments increase in length approximately three-fold during postembryonic development in both males and females (Schafer and Sanchez, 1976). In the female, the sensillar population increases 7.5-fold, whereas adult males have 12 times more sensilla than first instars; the difference results from a significant proliferation of olfactory sensilla in males.

The density of antennal sensilla in males rises sharply away from the basal segment for about 1 cm then declines over the next 4 cm to the tip of the antenna (Schaller, 1978; Hösl, 1990). The two receptor cells that are tuned to each of the two periplanones are housed within the same sensillum, the basiconic single-walled type B, along with two other cells that respond to terpenes and alcohols (Boeckh and Ernst, 1987). However, unlike the highly specialized receptor cells of male moths, the periplanone-A and periplanone-B cells have overlapping response spectra to these two compounds. Also, it is not known how responsiveness of pheromone-sensitive sensilla to food odorants (terpenes and alcohols) affects behavior of the male cockroach.

There has been no research reported on the interaction of periplanones with pheromone-binding proteins or dendritic receptors. Nevertheless, two studies, one empirical and one theoretical, report on intrasensillar events in *P. americana*. Picimbon and Leal (1999) determined the N-terminal amino acid sequences of two soluble proteins (putative odorant-binding proteins) that are specifically expressed in male antennae, and several other proteins that are expressed in antennae of both males and females. It is tempting to speculate that the male antenna-specific proteins may act as pheromone-binding proteins playing roles in solubilization of pheromone molecules within the perilymph of the olfactory sensilla. However, confirmation will have to await localization of the proteins to olfactory sensilla, cloning and expression of the proteins, and demonstration of specific binding of one or more of the periplanones.

To gain insight into ligand–receptor interactions, Bykhovskaia and Zhorov (1996) examined structure–activity relationships for periplanone-A, periplanone-B, and various structural analogs including agonists, antagonists, and related but inactive compounds. An analysis of the minimum-energy conformation of periplanone-B was used to model the ligand-binding site of a theoretical pheromone receptor. Although numerous olfactory receptors have now been sequenced from *Drosophila melanogaster* Meigen (de Bruyne *et al.*, 2001), only one putative pheromone receptor has now been identified, which is expressed in chemosensory neurons of male-specific gustatory bristles in the forelegs (Bray and Amrein, 2003).

Axons originating in sensilla that are responsive to food odorants and those that originate in sex pheromone-responsive neurons terminate in different glomeruli, the first synaptic relay station within the antennal lobe (Boeckh and Ernst, 1987). About 120 glomeruli have been described in *P. americana*, but all sex pheromone-sensitive neurons project into a deutocerebral macroglomerulus, a large, densely packed neuropil nearest to the antennal nerve. A convergence ratio of 5000:1 has been calculated between receptor cells and output neurons within the macroglomerulus (Boeckh *et al.*, 1984). Approximately 15–20 projection neurons run from the macroglomerulus to the protocerebrum, but each then branches extensively, resulting in > 1000 postsynaptic elements (~ 70 endings for each pheromone-sensitive projection neuron) (Esslen, 1982; cited in Boeckh *et al.*, 1984).

The structuring of the olfactory neuropil into adult-like glomeruli of *P. americana* takes place in the second half of embryogenesis (Salecker and Boeckh, 1995). Removal of the antenna (and ingrowing receptor axons) reduced the volume and altered the organization of the ipsilateral antennal lobe (Salecker and Boeckh, 1996). Interestingly, however, development of the macroglomerulus in the male must be controlled differently. In contrast to holometabolous insects, the macroglomerulus of male cockroaches is not developed *de novo* in the adult brain. Rather, it is formed during the last third of postembryonic (nymphal) development, by allometric growth and fusion of several small larval glomeruli (Prillinger, 1981; Schaller-Selzer, 1984), apparently before pheromone-responsive receptor neurons form. If true, then unlike the macroglomerular complex of holometabolous insects, which is dependent on afferent input (Schneiderman and Hildebrand, 1985), the growth and organization of the cockroach macroglomerulus might be independent of receptor axon input.

Pheromone-sensitive units within the deutocerebrum respond only to stimulation of the ipsilateral antenna and exhibit a well-marked dose dependency (Burrows *et al.*, 1982). Two groups of pheromone-sensitive neurons occur in the deutocerebrum of male nymphs (from the 10th instar onward) and adult males: group I, equally responsive to periplanone-A and periplanone-B, and group II, less responsive to periplanone-B but equally as responsive to periplanone-A as group I neurons. Both, however, respond to all components of the female pheromone, including a yet to be identified compound (Burrows *et al.*, 1982; Boeckh and Selsam, 1984; Schaller-Selzer, 1984). The neurons that project from the macroglomerulus to the protocerebrum fall into two categories: (i) neurons with global receptive fields (26% of pheromone-responsive neurons), which give responses of similar, but relatively low, intensity regardless of the antennal region that is stimulated; and (ii) neurons with local receptive fields (74% of pheromone-responsive neurons), which exhibit stronger responses to stimuli applied to particular parts of the antenna (Hösl, 1990). In this way, information about the site of stimulation is available in the deutocerebrum with no involvement of mechanosensory signals.

When the antennae are completely ablated, the male cockroach still responds to the sex pheromone (Roth and Willis, 1952). This suggests that some pheromone-responsive cells are also located elsewhere, such as on the mouthparts. Olfactory receptor cells on the palps of *P. americana* project to the lobus glomeratus within the posterior region of the ventral deutocerebrum (Boeckh and Ernst, 1987). However, it is not known whether pheromone-responsive cells on the palps project to the more anterior macroglomerulus.

Sex pheromones of other blattid species

Over much of its tropical and subtropical range, *P. americana* is sympatric with other species within its genus, as well as the closely related genera *Blatta* and *Eurycotis*. Attractant female sex pheromones, and close-range male tergal pheromones, are also used by these members of the Blattidae. In a series of four comprehensively detailed papers, Simon and Barth (1977a–d) described a comparative analysis of courtship behaviors and sensory modalities of six species within these genera. The mating behaviors of *P. americana* and the other *Periplaneta* spp. are similar but differ from the mating behaviors of *Blatta* and *Eurycotis* spp., in which the males appear to emit the attractant volatile pheromone. Periplanone-B, the major component of the sex pheromone of *P. americana*, or pheromone extracts from this species, are attractive to males of some of the other *Periplaneta* spp. (Schafer, 1977a,b; Takahashi *et al.*, 1988a,b; Nishii *et al.*, 1997), suggesting that they use similar structures as their sex pheromones (Persoons *et al.*, 1979). When rejection of heterospecifics occurs, it appears to result from species-specific blends of periplanones (and possibly other female pheromone components), male tergal pheromones (Simon and Barth, 1977d), and perhaps also contact female pheromones.

Periplaneta australasiae *(F.) (Australian or Australasian cockroach)*

Waldow and Sass (1984) tested the response of *P. americana* and *P. australasiae* males in the field to crude extracts of female *P. americana* and to fractions containing either periplanone-A or periplanone-B. Males of *P. americana* were attracted only to crude extracts and to the periplanone-B fraction, whereas males of *P. australasiae* were attracted only to the periplanone-A fraction. Interestingly, addition of the periplanone-B fraction to the periplanone-A fraction eliminated the attractiveness of the latter to *P. australasiae*. Therefore, the sex pheromone of *P. australasiae* is probably similar to periplanone-A, and periplanone-B seems to act as an antagonist. Seelinger (1985b) confirmed these results under laboratory conditions. Laboratory behavioral tests and EAGs confirmed the former results and showed that males of both *P. australasiae* and *P. americana* responded to extracts of *P. australasiae* (Nishino and Manabe, 1985a). Takahashi *et al.* (1988a,b) compared the response of several *Periplaneta* spp. to the two periplanones of *P. americana* and found that *P. australasiae* did not respond to periplanone-B; although the response of all species to periplanone-A was low, *P. australasiae* showed the highest response to this compound. Similarly, Schafer (1977a) recorded very low behavioral responses of *P. australasiae* to *P. americana* pheromone extracts. Nonetheless, the EAG response was highly sex specific (Schafer, 1977b), with males more sensitive than females at the peripheral level. Lack of a behavioral response is in agreement with the hypothesis that periplanone-B may be perceived as an antagonist by *P. australasiae*.

Periplaneta fuliginosa *(Serville) (smoky brown cockroach)*

P. fuliginosa males were not attracted to extracts of *P. americana* (Schafer, 1977a; Takahashi *et al.*, 1988a) but because their antennae were more responsive to *P. americana* female extracts than to male extracts (Schafer, 1977b), components of the sex pheromone of these two species must be related. Total synthesis of the pheromone from (–)-germacrene-D and assays of male response unambiguously determined the structure of the pheromone of *P. fuliginosa* as (–)-periplanone-D (Takahashi *et al.*, 1995; Nishii *et al.*, 1997; **4**, Table 6.2). Of several species tested in bioassays, *P. fuliginosa* is the most sensitive to periplanone-D, responding to 10^{-4} ng, followed by *P. japonica, P. americana*, and *B. orientalis* (Takahashi *et al.*, 1995; Nishii *et al.*, 1997).

Periplaneta japonica *Karny (Japanese cockroach)*

Kitamura and Takahashi (1973) described the mating behavior of *P. japonica* and provided behavioral evidence for a sex pheromone. A purified active fraction from filter papers lining the cages of virgin females elicited male behavioral responses at low concentrations, but contact with the female was required to elicit the courtship wing-raising response in males. Therefore, it appears that, as in *P. americana*, a contact pheromone is necessary for close-range sexual responses in *P. japonica*. The behavioral and EAG responses of *P. japonica* to *P. americana* extracts were very low (Schafer, 1977a,b; Takahashi *et al.*, 1988a). A tentative structure for periplanone-J, the sex pheromone of *P. japonica*, was proposed by Takegawa and Takahashi (1989) as (1*R*,6*E*,8*S*)-1(14)-epoxy-5(15),6-germacradien-10-one (**5**; Table 6.2). Synthetic (–)-periplanone-C elicited behavioral responses from *P. americana* and *P. japonica* at 0.1 and 1 ng, respectively (Nishii *et al.*, 1997).

Periplaneta brunnea *Burmeister (brown cockroach)*

P. brunnea males respond to extracts of *P. americana* females with the same intensity elicited by extracts of *P. brunnea* females (Schafer, 1977a), and *P. brunnea* EAG responses are stronger to female than to male *P. americana* extracts (Schafer, 1977b). In the field, 500 ng periplanone-B attracted more males than females of both *P. americana* and *P. brunnea* (Takahashi *et al.*, 1988a). Furthermore, a purified fraction from *P. brunnea* extracts showed activity to *P. americana* (Nishino and Manabe, 1985b), which suggests that the pheromone of *P. brunnea* may be similar to periplanone-B. Indeed, in behavioral and EAG assays, synthetic periplanone-B was much more active than periplanone-A to *P. brunnea* males (Nishino and Manabe, 1985b). Also, using antennae of *P. americana* and differential adaptation, strong cross-adaptation was observed between the active fraction from extracts of *P. brunnea* and periplanone-B (Nishino and Manabe, 1985b). Midgut extracts from *P. brunnea* contained a female-specific compound that has been tentatively assigned the structure (2*Z*,6*E*,8*S*)-1(14),2,5(15),6-germacratetraene, or periplanene-Br (**6**; Table 6.2) (Ho *et al.*, 1992).

Blatta orientalis *L. (oriental cockroach)*

B. orientalis is a large, 20–24 mm long, dark reddish-brown to black cockroach. Both sexes are similar in size and the tegmina and wings are short in the male and greatly reduced in the female. Neither sex can fly, and although they probably originated in North Africa, they now occur throughout the world as residential pests, primarily in cooler basements and cellars (Cornwell, 1968). Early studies of the mating behavior of *B. orientalis* failed to demonstrate a volatile sex pheromone in either males or females (Roth and Willis, 1952; Barth, 1970; Simon and Barth, 1977b), but in recent years the mating system of this species has turned out to be more complex than that of any of the other blattids. Abed *et al.* (1993a) observed that unlike *Periplaneta* spp. and other cockroaches, in which the males are good flyers, male calling behavior appears to initiate the mate-finding process in *B. orientalis*, although occasionally females may be the first to call. The female detects the presence of the calling male, and after contacting him she initiates calling by opening her genital atrium. The male then orients toward the female, contacts her, and raises his wings. The female mounts the male to feed from his tergites and copulation follows (Abed *et al.*, 1993a).

Males call mainly during the second half of the scotophase by stretching and telescoping the abdomen, revealing the brightly colored anterior region of their tergites (Abed *et al.*, 1993a). A different glandular epithelium underlies the anterior region of each male tergite than its posterior, and the anterior portion is more effective at inducing females to expose their genital atria, a sign of sexual receptivity (Abed *et al.*, 1993a). It appears that the male pheromone is emitted from widely distributed tergal glands that terminate in orifices on the surface of all tergites. However, the evidence that male calling behavior is associated with the release of an attractant is equivocal. In a Y-tube olfactometer, females preferred calling males over females (77% versus 23%), suggesting that calling males attract females (Abed *et al.*, 1993a). However, because females did not show a preference for calling males over non-calling males, or for extracts of male tergites over solvent controls, it remains unclear whether males release an attractant at all during what is termed a calling behavior. Females spent more time on the side of the tube housing the calling males, but this does not necessarily imply orientation. Similarly, tests in rearing chambers demonstrated that females spend significantly more time on filter papers conditioned by males than on unconditioned filter papers or papers conditioned by females (Abed *et al.*, 1993a). All females were attracted to the papers in the first 20 s following exposure and were sexually excited, but only 14% exposed their genital atria.

A female-produced volatile pheromone was demonstrated by Simon and Barth (1977a), who exposed groups of males to filter papers contaminated with female odors. Warthen *et al.* (1983) extracted different parts of the female body with solvent and concluded that a compound resembling periplanone-B was present in the female crop, esophagus, and proventriculus. However, more recent studies claim that the atrial glands elicit the strongest responses in *Blatta* males, as in *P. americana*, and only the atrial glands induce any males to raise their wings (Abed *et al.*, 1993a). The epithelium of the atrial gland consists of type 1 glandular cells, in which the secretion is exuded directly onto the cuticle.

A number of reports in the 1970s and 1980s demonstrated that periplanone-B was highly active to male *B. orientalis* in both behavioral and EAG assays. Periplanone-B in slow-release formulations also attracted *Blatta* males to traps in infested swine farms (C. Schal, unpublished data). However, it remains unclear which of the periplanones is produced and emitted by female *B. orientalis* and how emission of the female pheromone is coordinated with male calling behavior.

Eurycotis floridana (Walker) (Florida cockroach)

Eurycotis floridana is a large cockroach (30–40 mm) found outdoors in tropical and subtropical areas. The tegmina are rich red-brown to black and extremely small, and the wings are absent. Adults of both sexes emit potent repellent odors when attacked. The male initiates calling by extending the abdomen, exposing the brightly colored anterior parts of tergites 2 and 8, and from time to time tergite 7 (Barth, 1968a; Farine *et al.*, 1993, 1996). Males call at all hours of the photocycle, but calling is more frequent in the scotophase and peaks toward the end of the scotophase. Females respond to male calling behavior by orienting toward the male 10 to 30 s after he starts calling. Contact is followed by antennal fencing and mutual body stroking with the antennae. Receptive males show a curious side-to-side rocking movement (lateral vibration). During courtship, females sometimes open their genital atrium, but this behavior is also observed in immature females and so its involvement in the release of sex pheromones is unclear. After contact, the female mounts the wingless male and feeds from a readily accessible tergal gland located on his first abdominal segment. The tuft of setae in the tergal gland is important because if experimentally glued the male does not respond to female feeding (Farine *et al.*, 1996). Genital connection is followed by the assumption of the opposed copulatory position. Some of the chemicals produced in the male tergal secretion have been identified. Tergite 2 produces no volatile compounds. Extracts from tergite 7 are attractive from some distance and contain 8.5 μg/male 4-hydroxy-5-methyl-3(2*H*)-furanone (**7**; Table 6.2) and 240 ng/male 4-hydroxy-2,5-dimethyl-3(2*H*)-furanone, but choice tests reveal that only the first compound is definitely involved in attraction (Farine *et al.*, 1993). Tergite 8 produces (2*R**,3*R**)-butanediol, 1-dodecanol, and benzyl-2-hydroxybenzoate, but only the former two are attractive to females while other components may act as repellents; the function of these substances in mate attraction and courtship needs further testing (Farine *et al.*, 1994).

Sex pheromones of Blattellidae

Blattella germanica L. *(German cockroach)*

B. germanica is a small cockroach, 10–15 mm long, and pale tan to brown; the females are broader than the males, and both sexes have wings covering most or all of the abdomen. It probably originated in southeast Asia, the center of diversity of the genus, and it is now one of the most common indoor pests. It prefers warm, humid environments, and there are no records of it living outdoors away from human structures.

Mating behavior

The mating behavior of *B. germanica* has been extensively described (Roth and Willis, 1952; Nishida and Fukami, 1983; Nojima *et al.*, 1999b). Virgin females exhibit a characteristic calling behavior during which they emit a volatile sex pheromone that attracts males (Fig. 6.1a) (Liang and Schal, 1993a,b). Upon antennal contact (Fig. 6.1b), the male responds to a female contact pheromone by rotating 180°, so that the tip of his abdomen points toward the female, he simultaneously raises his wings almost perpendicular to the body's longitudinal axis, thus exposing highly specialized tergal glands on the 7th and 8th abdominal segments that were previously covered by the wings (Fig. 6.1c). Exposure of the tergal glands may be enhanced by extension and downward curving of the abdomen. These glands release a blend of volatile compounds that presumably attracts females (Brossut *et al.*, 1975). The female mounts the male's abdomen and feeds on nutrients within reservoirs of the tergal glands (Fig. 6.1d), placing her into a precopulatory position (Nojima *et al.*, 1999b). The male then couples with the female, and if successful the pair faces in opposite directions (Fig. 6.1e) and remains in copula for 72–115 min (Roth and Willis, 1952; Schal and Chiang, 1995).

Female volatile sex pheromone

B. germanica is arguably the most important cockroach pest, and numerous studies have examined its biology and pest-control tactics. Surprisingly, unequivocal evidence of a volatile sex pheromone in females was demonstrated less than a decade ago and this pheromone is yet to be chemically elucidated. In their study of cockroach mating behavior, Roth and Willis (1952) found no evidence for a volatile sex pheromone in *B. germanica*, perhaps because its effect is more subtle than in *P. americana*. Presence of a volatile sex pheromone in female *B. germanica* was first demonstrated by Volkov *et al.* (1967), who showed that males were attracted to female extracts, whereas females were not. This suggested that the response to the extract was sex specific but did not determine whether the extract itself was also sex specific. Almost 30 years later, it was confirmed that females produce a sex pheromone that attracted males. In two-choice olfactometers, most of the males preferred the extracts of virgin females over extracts of males or nymphs, showing an adult female-specific pheromone (Abed *et al.*, 1993b; Liang and Schal, 1993a; Tokro *et al.*, 1993). Because extracts and headspace collections of virgin females were significantly more attractive to males than similar samples obtained from mated females, it was concluded that the pheromone emitted by virgin females was indeed a sex pheromone (Liang and Schal, 1993a).

The volatile pheromone of *B. germanica* females is emitted while the female performs a specific calling behavior, during which the wings are raised, the abdomen lowered, and the genital vestibulum occasionally exposed (Fig. 6.1a) (Liang and Schal, 1993b). Calling females have been shown to release more pheromone and they are more likely to mate than non-calling females. Calling occurs throughout the day and night in females that are not allowed to mate, but under a 12:12 (light:dark) photoregimen, it exhibits a peak before the end of the scotophase (Liang and Schal, 1993b), which corresponds well with the period of maximal mating in *B. germanica* under time-lapse videographic observations (Schal and Chiang, 1995).

Source of the female-produced attractant pheromone The pheromone is produced in the last abdominal tergite (tergite 10, or pygidium). Fewer males are attracted to the pygidium of males than to that of virgin females (Abed *et al.*, 1993b), and in two-choice assays, females whose pygidia were ablated attracted only 13% of the males, whereas sham-operated females attracted 87% of the males (Liang and Schal, 1993a). The pygidial gland was first described by Dusham (1918), but its function was not ascertained. These tergal modifications on the anterior part of the pygidium are not present in nymphs nor in males. The gland consists of three groups of cuticular depressions, one medial and two lateral, and each group contains numerous cuticular depressions, within which are 1–32 orifices (Liang and Schal, 1993c; Tokro *et al.*, 1993). Each orifice is approximately 0.5 μm in diameter and leads to a duct that penetrates through the cuticle and inserts deeply within a large secretory cell located in the modified epithelium beneath the cuticle. The secretory unit is of type 3, consisting of a large secretory cell and a much smaller duct or canal cell. Within the secretory cell, a complex end-apparatus is elaborated that comprises numerous long microvilli. The secretory cells of mature, pheromone-producing virgin females are characterized by abundant mitochondria, smooth endoplasmic reticuli, rough endoplasmic reticuli, a large nucleus, and numerous secretory vesicles. The last, presumably containing pheromone, are discharged into the end-apparatus and up through the duct to the cuticular surface. Large amounts of secreted material can be found within the ducts and in the cuticular depressions (Liang and Schal, 1993c).

Factors affecting pheromone production and response The volatile sex pheromone of *B. germanica* has yet to be identified. Nevertheless, ultrastructural studies as well as behavioral and EAG assays of both pygidium extracts and headspace collections have elucidated several mechanisms that regulate pheromone production and release. First, the pheromone gland exhibits a developmental maturation in relation to sexual maturation of the female, followed by cycles of cellular plasticity in relation to the reproductive cycle (Abed *et al.*, 1993b; Liang and Schal, 1993c; Tokro *et al.*, 1993; Schal *et al.*, 1996). The secretory cells of newly formed glands in the imaginal female are small and contain few secretory vesicles, few short microvilli around the end-apparatus, and little material within the end-apparatus. The amount of behaviorally active and EAG-active material in the gland is low on day 0 but it increases with age and peaks on day 6, corresponding to the physiological stage when females become sexually receptive (Liang and Schal, 1993c). Both the thickness of the glandular tissue and the size of each secretory cell decline markedly after mating and increase again after the release of the ootheca (Abed *et al.*, 1993b; Liang and Schal, 1993c; Tokro *et al.*, 1993; Schal *et al.*, 1996). This pattern has been related to the titer of JH in the hemolymph (Liang and Schal, 1993c; Schal *et al.*, 1996), but no experimental manipulations of hormone titers have been conducted to verify the hypothesis that JH controls the cellular plasticity of the pheromone gland. Interestingly, the pygidium becomes less attractive soon after the female mates, suggesting a cessation of pheromone production (Abed *et al.*, 1993b; Liang and Schal, 1994; Schal *et al.*, 1996). However, in mated vitellogenic females, the pheromone gland hypertrophies and produces pheromone again. It, therefore, appears that the gland cycles in relation to hormone titer, and independently of the mating status of the female.

Therefore, communication of the female's sexual receptivity is not at the level of pheromone production but rather through the behavior of calling, which, in turn, is tightly neuroendocrinologically regulated. Calling commences 5–6 days after adult emergence, when the basal oocytes are about 1.6 mm in length, and the female is sexually receptive (Liang and Schal, 1993b). If the female fails to mate, she continues to exhibit bouts of calling that are interrupted only by ovulation. Thus, a clear temporal coordination of calling and carrying an infertile ootheca is evident in virgin females, suggesting a relationship to the titer of JH. Liang and Schal (1994) tested this hypothesis by ablating the CA from young females. Allatectomized females never expressed the calling behavior, whereas treatment of allatectomized females with a JH analog restored calling. Manipulations that increase the JH titer, such as severance of the nerves between the CA and the brain or exposure to JH analogs, also accelerated the onset of calling.

However, this rather simple view of endocrine regulation of calling is appreciably complicated by events in mated females. Calling is immediately, but only transiently, suppressed by transfer of a spermatophore to the female, and it is completely suppressed after transfer of viable sperm to her spermatheca (Liang and Schal, 1994). Therefore, castrated males inhibit female calling for only a few days because of the action of the spermless spermatophore. Later, presence of an egg case suppresses calling behavior in both virgin and mated females. However, when females carrying an ootheca were treated with JH analogs, they were not induced to call (Liang and Schal, 1994). This suggested that JH must reach a certain minimum titer for calling to be expressed; however, even in the presence of this hormone, calling can be suppressed by neuronal signals from the spermatheca and genital chamber. The latter was clearly demonstrated by transecting the ventral nerve cord in mated, egg-case-carrying females that were exposed to JH analog (Liang and Schal, 1994). Normally, gravid females have low, usually undetectable, levels of JH (Gadot *et al.*, 1989; Sevala *et al.*, 1999). However, treated females resumed calling, showing that transmission of the calling-inhibiting signals from the genital vestibulum (presence of the ootheca) and the spermatheca (sperm) was interrupted.

Function of the volatile female pheromone There is some ambiguity about the role of the volatile female pheromone in *B. germanica*, mainly stemming from the temporal expression of calling relative to copulation. In *B. germanica* females, basal oocyte length is a good predictor of the JH titer (Gadot *et al.*, 1989; Sevala *et al.*, 1999) and, in turn, of sexual receptivity (Schal and Chiang, 1995). The basal oocytes develop gradually from 1.2 to 1.9 mm in length between days 4 and 7, during the time when virgin females initiate calling (Liang and Schal, 1993b). Yet, when females initiate calling for the first time, their oocytes are the same size (1.6 mm) regardless of their chronological age. This suggests that females initiate calling when they reach a specific physiological stage, represented by a certain JH titer and a minimal oocyte length. However, virgin females with constant access to sexually mature males mated on average when their oocytes were 1.35 mm long, earlier than when calling occurs.

This discrepancy, suggesting that females mate before they would normally initiate calling, is problematic because copulation terminates calling and thus would eliminate any role for calling in the mate-finding system of *B. germanica*. Two hypotheses have been proposed to account for these observations (Liang and Schal, 1993b). The first suggests that the volatile sex pheromone may serve as an alternative mate-finding mechanism employed by sexually receptive females after they fail to contact males (see also Tsai and Lee, 1997). The second proposes a primary role for the volatile pheromone: its emission may actually begin prior to the overt expression of the calling behavior, and attractants may be released well before our ability to discern visually the calling stance. Furthermore, calling and the emission of the volatile sex pheromone may serve not only to attract males from a distance but also to potentiate their responses to the contact sex pheromone in aggregations. These hypotheses have yet to be tested rigorously.

Female contact pheromone

A non-volatile contact pheromone contained in the cuticular wax of females elicits a wing-raising courtship response from males (Roth and Willis, 1952; Ishii, 1972). Nishida and co-workers obtained three active chromatographic fractions from hexane extracts of 224 000 females. The major active component was identified as

3,11-dimethylnonacosan-2-one (**8**; Table 6.3) (Nishida *et al.*, 1974), whereas the other two possess the same dimethyl ketone skeleton but with hydroxyl or carbonyl groups at the C_{29} position. The alcohol (**9**: 29-hydroxy-3,11-dimethylnonacosan-2-one; Table 6.3) is ~10-fold more active than the methyl ketone (Nishida *et al.*, 1976), whereas the activity of 29-oxo-3,11-dimethylnonacosan-2-one (**10**; Table 6.3) is intermediate between the other two components (Nishida and Fukami, 1983). A fourth pheromone component, 3,11-dimethylheptacosan-2-one (**11**; Table 6.3), is less active than its C_{29} homolog (Schal *et al.*, 1990a). Other methyl ketones with a 3,11-dimethyl branching pattern are present on the female cuticular surface, including a C_{31} and a C_{33} (C. Schal, unpublished data), but their biological activity has not been assayed. The equivalent fractions obtained from adult males do not elicit courtship responses (Nishida *et al.*, 1975).

The 3,11-branching pattern appears to be essential for activity in *B. germanica* (Sato *et al.*, 1976). The stereochemistries of the 29-methyl and 29-hydroxy components have been studied and shown to be (3*S*,11*S*) (Nishida *et al.*, 1979); the other components are likely to be the same because they appear to have a common biosynthetic origin (see below). Interestingly, all four stereochemical isomers of the 3,11-positions yield similar wing-raising responses in males, suggesting lack of stereospecificity in the pheromone receptor (Mori *et al.*, 1978). Abed *et al.* (1993b), however, have asserted that the natural (3*S*,11*S*) isomer is most active at low concentrations, but empirical evidence in support of this claim has not been reported.

Structure–activity studies on the contact pheromone of *B. germanica* have been summarized by Nishida and Fukami (1983). Dose–response studies of the four known components have yielded two hypotheses. First, biological activity of the pheromone is proportional to its polarity, because 3,11-dimethylnonacosan-2-one is less active than the 29-hydroxy-analog, and the 29-oxo-analog exhibits intermediate activity. Also, reduction of the C_2 carbonyl to a hydroxyl group increases activity by ~10-fold, whereas a methylene in place of the carbonyl group eliminates activity (Nishida and Fukami, 1983). Second, behavioral activity declines as chain length is either shortened or lengthened. Again, 3,11-dimethylnonacosan-2-one is more active than its C_{27} homolog (Schal *et al.*, 1990a), confirming the earlier structure–activity studies that Sato *et al.* (1976) conducted with synthetic pheromone analogs. The contact pheromone recognition system of male *B. germanica* thus appears to accommodate much greater variation in pheromone structure than is seen in the attractant pheromone systems of most lepidopterans and dipterans, where omission of even minor components affects male response qualitatively and quantitatively. It is also notable that some synthetic analogs are more active than the natural pheromone, another departure from lepidopteran chemical communication systems.

Table 6.3. *Chemical structures of the principal identified contact sex pheromones of cockroaches, listed by the species and sex that produces it; contact pheromones are localized on the cuticular surface*

	Structure	Species/sex
8	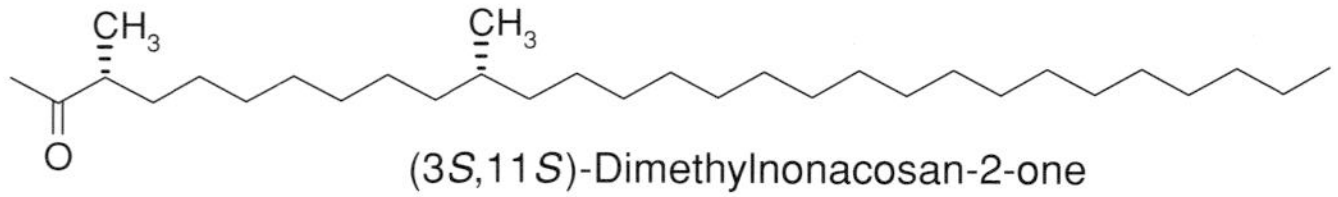 CH_3 CH_3 (3*S*,11*S*)-Dimethylnonacosan-2-one	*Blattella germanica* ♀
9	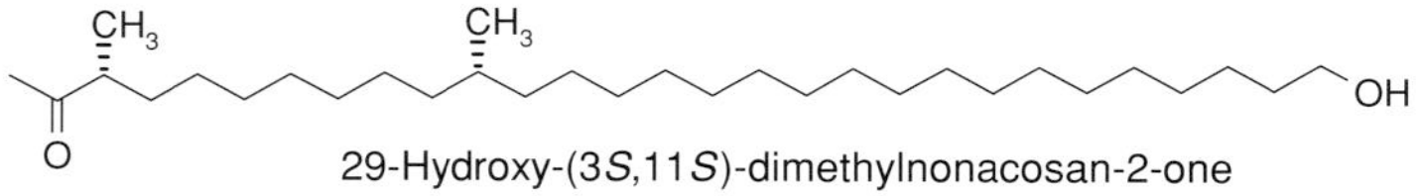 CH_3 CH_3 OH 29-Hydroxy-(3*S*,11*S*)-dimethylnonacosan-2-one	*Blattella germanica* ♀
10	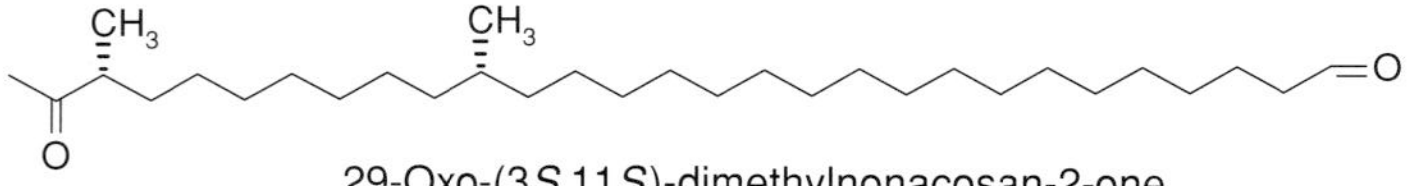 CH_3 CH_3 =O 29-Oxo-(3*S*,11*S*)-dimethylnonacosan-2-one	*Blattella germanica* ♀
11	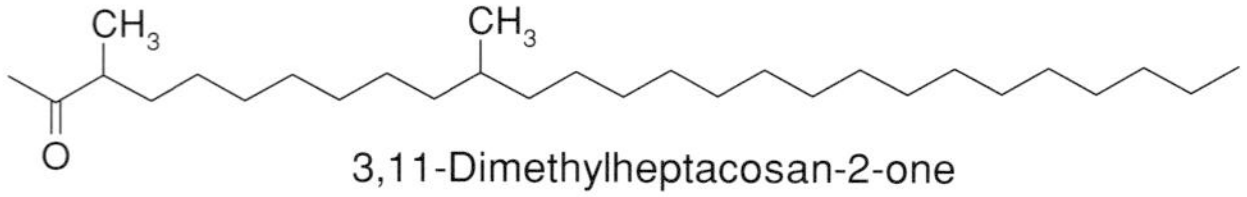 CH_3 CH_3 3,11-Dimethylheptacosan-2-one	*Blattella germanica* ♀
16	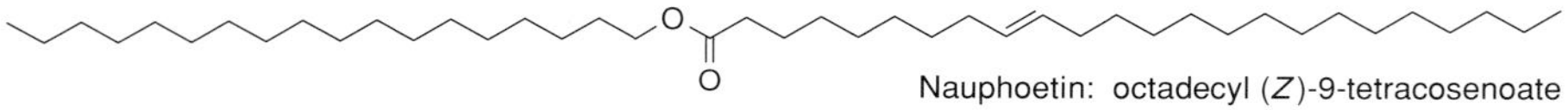 O O Nauphoetin: octadecyl (*Z*)-9-tetracosenoate	*Nauphoeta cinerea* ♂

Source of the contact pheromone and factors affecting its production Central to investigations of the biosynthetic pathway and regulation of the contact pheromone of *B. germanica* was the observation that the major cuticular hydrocarbon in this species is an isomeric mixture of 3,7-, 3,9-, and 3,11-dimethylnonacosane (Jurenka *et al.*, 1989; see reviews: Schal *et al.*, 1991, 1996, 2003; Blomquist *et al.*, 1993; Tillman *et al.*, 1999). Jurenka *et al.* (1989) suggested, based on the structural similarities between the methyl ketone pheromone component and the cuticular hydrocarbons, that production of the pheromone might result from the sex-specific oxidation of its hydrocarbon analog. Biochemical studies on the biosynthesis of methyl-branched alkanes showed that the methyl branching units are added during the early stages of chain elongation (Chase *et al.*, 1990) and that methyl-branched fatty acids are intermediates in branched alkane biosynthesis (Juarez *et al.*, 1992).

The hypothesis that the sex pheromone arises from oxidation of the preformed hydrocarbon was examined by topical application of radiolabeled synthetic 3,11-dimethylnonacosane and the putative intermediate 3,11-dimethylnonacosan-2-ol to vitellogenic *B. germanica* females. Radioactivity from the alkane was detected in 3,11-dimethylnonacosan-2-ol and 3,11-dimethylnonacosan-2-one, thus supporting the hypothesis (Chase *et al.*, 1992). Moreover, because the conversion of 3,11-dimethylnonacosan-2-ol to the corresponding methyl ketone pheromone was highly efficient in both males and females, it suggested that the sex pheromone of *B. germanica* arises via a female-specific hydroxylation of 3,11-dimethylnonacosane and a subsequent non-sex-specific oxidation. It is possible that a similar hydroxylation and subsequent oxidation at the C_{29} of 3,11-dimethylnonacosan-2-one give rise to 29-hydroxy- and 29-oxo-3,11-dimethylnonacosan-2-one. Because more radioactivity was recovered in the methyl ketone pheromone after topical application of a JH analog, this work also suggested the involvement of JH in the conversion of hydrocarbon to methyl ketone in females (Chase *et al.*, 1992).

The isolation of a fatty acid synthase and methyl-branched fatty acids from the abdominal epidermis but not from the fat body (Juarez *et al.*, 1992) suggested that epidermal tissues associated with the integument produce the precursor hydrocarbons. Localization of the site of contact pheromone biosynthesis in *B. germanica* has been accomplished with *in vitro* techniques. Gu *et al.* (1995) incubated various tissues with radiolabeled propionate and concluded that only the abdominal sternites and tergites produced methyl-branched hydrocarbons and the methyl ketone pheromone. Recently, the integument underlying the sternites was enzymatically dissociated and the cell suspension fractionated in a Percoll gradient. Only very large cells (>30 μm) produced hydrocarbons, whereas the much larger population of small cells (<15 μm) did not (Fan *et al.*, 2003). The small cells were epidermal cells, whereas the large cells had the ultrastructural features of oenocytes. It remains to be determined whether oxidation of the hydrocarbon to pheromone also occurs within the oenocytes.

Regulation of contact sex pheromone production in *B. germanica* operates at several levels, including (i) production of the 3,11-dimethylnonacosane precursor, (ii) its metabolism by female-specific oxidases to 3,11-dimethylnonacosan-2-one, and (iii) transport and distribution of the pheromone to the epicuticular surface. Early in the life of an adult female, it appears that food intake is a major regulator of hydrocarbon production. Females produce hydrocarbons only when they feed, and experimentally starved females, or gravid females that feed less, produce significantly less hydrocarbon (Schal *et al.*, 1994). Because a pool of precursor 3,11-dimethylnonacosane is required for pheromone to be made, little pheromone is produced when less hydrocarbon is available, for example in starved females.

JH plays an essential regulatory role in the metabolism of 3,11-dimethylnonacosane to 3,11-dimethylnonacosan-2-one, presumably by increasing the activity of a female-specific polysubstrate monooxygenase. Incorporation of radiolabel from [1–^{14}C]propionate into the methyl branches of the sex pheromone is low in previtellogenic and gravid females, when the JH titer is low, and high in vitellogenic females, when the JH titer is elevated (Schal *et al.*, 1994). The relationship between JH and pheromone production is also supported by experimental evidence from allatectomized females, females treated with JH analogs, and from dietary manipulations that influence hormone titer and pheromone production (Schal *et al.*, 1990b, 1991, 1994; Chase *et al.*, 1992). This subject has been reviewed in relation to pheromone biosynthesis in other insects (Schal *et al.*, 1991, 1996, 2003; Blomquist *et al.*, 1993; Tillman *et al.*, 1999).

The third, and perhaps least understood, mechanism regulating contact pheromone production involves its transport to the cuticular surface. The detection of large amounts of hydrocarbons and pheromone internally, within the hemolymph, prompted an examination of lipid transport in *B. germanica*. Gu *et al.* (1995) and Sevala *et al.* (1997) isolated and purified a high density lipoprotein, lipophorin, that carries hydrocarbons, contact pheromone, and JH within the hemolymph. The accumulated evidence supports the idea that the hydrocarbons and contact pheromone components are produced by oenocytes within the abdominal integument, carried by lipophorin, and differentially deposited in the cuticle and ovaries (Fan *et al.*, 2002). It remains to be determined whether epidermal cells are involved, and how. Also, mechanisms that regulate the uptake and deposition of pheromone by lipophorin remain unexplored. The dynamics of lipophorin-facilitated shuttling of lipids in *B. germanica* were last reviewed by Schal *et al.* (1998, 2003).

Other female pheromone components Behavioral assays have shown that males are highly responsive to isolated antennae from teneral (i.e., newly molted) females;

their response declines dramatically as the female ages, and then increases again as the female matures sexually (Nishida and Fukami, 1983). Yet, both 3,11-dimethylnonacosan-2-one and 3,11-dimethylheptacosan-2-one gradually accumulate on the female cuticle during the first gonotropic cycle (Schal *et al.*, 1990a). Schal *et al.* (1990a) speculated that this apparent discrepancy could be explained by a unique courtship-inducing substance on the newly enclosed female, which disappears or is masked over time as the contact pheromone begins to appear on the cuticle of the maturing female. Males are sexually responsive also to isolated antennae of newly molted males and even of nymphs (Roth and Willis, 1952; Nishida and Fukami, 1983), suggesting that this molt-related courtship elicitor is not sex specific. The potential adaptive benefit that newly molted cockroaches may gain from eliciting male courtship responses has not been investigated.

Male response to the contact pheromone The identification and synthesis of 3,11-dimethylnonacosan-2-one, together with a simple isolated antenna bioassay, facilitated detailed studies of male behavior (Bell *et al.*, 1978). Males exhibit several different sequences of courtship responses, but the most common overall pathway of male courtship is as follows: contact the female, erect body posture, body shake, wing-raise-turn, counterturn, and local search (Fig. 6.1). Courtship requires reciprocal acts from both sexes. During wing-raising, the male also releases a pheromone that attracts the female, and subsequent male behaviors, including a wave-like backward undulation of the abdomen and extension of the phallomere, are expressed only if his tergal gland is stimulated by the female. The male does not move back toward the female after turning, as is common in other cockroaches, such as *Periplaneta* sp. (Bell *et al.*, 1978).

The courtship response of male *B. germanica* has been the subject of detailed ethometric analyses. In encounters initiated from a head-to-head stance, the male pivots 180° through a complex rotational locomotion without appreciable translation of his body axis (Franklin *et al.*, 1981). When the male contacts the female laterally, he attempts to achieve a head-to-head position by walking sideways, but frequently he initiates a rotation through the smaller angle that will place his body axis parallel to the female (Bell and Schal, 1980). If the male does not receive mechanical stimulation from the mounting female, he counterturns and initiates a local search composed of a highly convoluted exploratory path (Schal *et al.*, 1983).

A number of chemo- and mechanoreceptors participate in the male behaviors. The female contact pheromone is detected by chemosensilla on the antennae and labial and maxillary palps (Ramaswamy and Gupta, 1981). The number of these sensilla increases dramatically during the metamorphic molt, and much more so in males than in females. Unfortunately, no electrophysiological recordings have been conducted, and the specific sensillum type that responds to the contact pheromone has not been identified. Males respond to the contact pheromone more at the beginning of the scotophase than soon after the onset of the photophase, although the differences are not dramatic (66% versus 50%) (Bell *et al.*, 1978). The response is also enhanced by isolation of males from females. Males are unresponsive after the imaginal molt and become maximally responsive 11 days later (Nishida and Fukami, 1983). Although JH appears to play a role in the development of male sexual receptivity, ablation of the CA only delays, but does not abolish, male sexual maturity (Schal and Chiang, 1995).

Male-produced pheromones

There is no evidence of long-range male-produced attractants in *B. germanica*. However, the male tergal glands produce a blend of close-range attractants, phagostimulants, and nutrients that are deployed to place the female in a precopulatory position. These glands underlie the 7th and 8th tergites and form highly specialized cuticular modifications on the tergum of male *B. germanica* (Brossut and Roth, 1977). The glands consist of transverse cuticular depressions into which numerous type 3 secretory cells empty through ducts (Sreng and Quennedey, 1976), as in the female's pygidial gland (p. 206). Sreng and Quennedey (1976) described the morphology and ontogenesis of this gland.

Nojima *et al.* (1999b) conducted a careful examination of the behavioral sequence that females express while mounting the courting male. The female first palpates and feeds upon the secretions of tergite 8 (Fig. 6.1d). The male then extends his abdomen and moves backward, displacing the female's feeding to tergite 7. At this point, the male extends his left phallomere and engages the female's subgenital plate; copulation follows. A conspicuous mechanosensory structure – the spiculum copulatus – within the male's tergal gland appears to play an important role in the final steps of courtship (Sreng, 1979a; Ramaswamy *et al.*, 1980). As the female mounts and palpates the tergal gland, signals from this structure elicit arching and extension of the male abdomen, and thrusting of the left phallomere. The spiculum copulatus develops during the metamorphic molt and its cauterization significantly interferes with courtship in *B. germanica* (Ramaswamy *et al.*, 1980).

Analytical studies of the tergal secretions of male *B. germanica* have identified a number of volatile compounds, none of which has so far been subjected to behavioral assays on females. Brossut *et al.* (1975) found *p*-hydroxybenzyl alcohol, *o*-hydroxybenzyl alcohol, di- and tri-methylnaphthalene, benzothiazole, two isomers of nonyl phenol, and myristic, palmitic, and oleic acids. The fatty acids constituted $>92\%$ of the volatile fraction; given their abundance in feces and frass, and their role as putative aggregation pheromones (Wileyto and Boush, 1983; Fuchs *et al.*, 1985; Wendler and Vlatten, 1993; Scherkenbeck *et al.*, 1999), they might serve as non-specific attractants. This is supported by the observation that males and even nymphs are attracted to the exposed tergal glands of courting males.

The non-volatile fraction of the tergal gland secretion contains lipids, proteins, and carbohydrates. The protein fraction consists of eight electrophoretic bands, including two glycoproteins (Brossut *et al.*, 1975), but their identity and role(s) in courtship behavior have not been investigated. Nojima *et al.* (1996) developed a feeding assay wherein methanolic extracts of the male tergal gland were solubilized in polyethylene glycol and dried on a glass slide; bite marks represented feeding activity of females. This innovative assay led to the identification of several male compounds that serve as phagostimulants. A mixture of oligosaccharides, including maltose, its analogs maltotriose and maltotetraose, and four analogs of α,α-trehalose with $(1\rightarrow4)$-α-glucosyl linkages were found to stimulate the feeding activity (Nojima *et al.*, 1999a). Activity of an artificial blend of these sugars is significantly enhanced by a polar lipid fraction that contains a mixture of 3-*sn*-phosphatidylcholines (lecithin). Nojima *et al.* (1999a) speculated that these secretions evolved as feeding stimulants because the sexually receptive female requires nutrients.

Other non-volatile lipids have been reported from male tergal glands, including large amounts of saturated C_{24}–C_{29} alkanes (even and odd chain lengths in equal amounts), presumably from the cuticular surface (Brossut *et al.*, 1975). However, cuticular extracts of *B. germanica* males contain mainly methyl-branched C_{27} and C_{29} alkanes (Jurenka *et al.*, 1989), suggesting that the shorter hydrocarbons in the tergal glands might play a role in sexual behavior.

Supella longipalpa *(F.) (brownbanded cockroach)*

S. longipalpa is a small cockroach, similar in size to *B. germanica*. The pale brown tegmina of the male cover his body beyond the end of the abdomen and he is an excellent flyer. The reddish-brown females, by comparison, have reduced tegmina and wings and they do not fly. Presumably of African origin (the only non-domiciliary *Supella* spp. are found there), it prefers warmer and drier areas than either *B. germanica* or *P. americana* (Cornwell, 1968).

Mating behavior

Mate finding in *S. longipalpa* is mediated by a female-produced long-range sex pheromone. Virgin females call during the scotophase (Smith and Schal, 1991); calling females are more likely to mate than non-calling females (Hales and Breed, 1983), and mated females cease to call (Smith and Schal, 1991). All this evidence suggests that calling is related to sexual receptivity and is involved in the release of sex pheromone. Calling females stand with their wings raised at a sharp angle to the abdomen's dorsum (Hales and Breed, 1983). The abdomen is flexed with the dorsal surface curved upward and the posterior end pushed toward the substrate. The legs of the female are straightened, pushing the thorax and abdomen away from the substrate. While in this position, the female pumps her abdomen and periodically exposes her genital atrium (Hales and Breed, 1983). Males orient to the volatile pheromone from some distance (Liang and Schal, 1990a), and upon contacting the female, they turn around, raise their wings, arch the abdomen with the tip pointing downward, and expose a highly specialized tergal gland on the 10th tergite (Roth, 1952). As in *B. germanica*, volatiles emitted from the tergal gland attract females to mount the male and feed on the tergal secretion. However, unlike *B. germanica*, some males initiate wing-raising without contacting the female, and this behavior may precede rotation of the body. Without contact with the female, males remain in the courtship position longer than in *B. germanica*.

Female volatile sex pheromone: source and identification

The calling behavior of *S. longipalpa*, and previous research on other cockroaches, suggested that the tergum, genital atrium, or digestive tract might be the site(s) of pheromone production. Behavioral, electrophysiological, and morphological studies have localized the sex pheromone to the 4th and 5th tergites of virgin females (Schal *et al.*, 1992). The tergal glands of *S. longipalpa* females are composed of multiple type 3 secretory units, each cell leading through a long unbranched duct to a single cuticular pore. Unlike the pygidial glands of *B. germanica*, however, the secretory units are not aggregated and there are no modifications, other than individual pores, on the cuticular surface. Moreover, cuticular pores occur on all tergites, but their density is highest on the lateral margins of the 4th and 5th tergites. As in *B. germanica* and other cockroach species, the calling female periodically exposes her genital vestibulum, but the significance of this behavior is not known.

A bioassay-driven fractionation of tergite extracts from several thousand virgin females and headspace collections from calling females identified 5-(2,4-dimethylheptyl)-3-methyl-2*H*-pyran-2-one (**12**: supellapyrone; Table 6.2) as the only behaviorally active material (Charlton *et al.*, 1993). Activity of this compound was confirmed by synthesis of racemic supellapyrone containing the four possible stereoisomers. Mori and Takeuchi (1994, 1995) speculated, based on the nuclear magnetic resonance spectra of natural and synthetic supellapyrone, that the female produced (2*R*,4*R*)-supellapyrone. Indeed, GC–EAG with a chiral phase GC column demonstrated that supellapyrone is 5-(2′*R*,4′*R*-dimethylheptyl)-3-methyl-2*H*-pyran-2-one (Leal *et al.*, 1995). Supellapyrone is not only a unique cockroach pheromone but also the first example of a new class of natural products – namely 3,5-dialkyl-substituted α-pyrones.

Factors affecting pheromone production and response

Environmental and physiological factors that regulate pheromone production and calling behavior have been studied more intensively in *S. longipalpa* than in any other cockroach. However, much of this work pre-dated the identification of supellapyrone (**12**; Table 6.2), and hence it is based on behavioral and electrophysiological assays rather than chromatographic analysis. Calling occurs discontinuously and it peaks near the middle of a 12 h dark period (Hales and Breed, 1983). Unlike *B. germanica* females, which call throughout the scotophase and photophase, *S. longipalpa* calls only during the scotophase. Calling has an endogenous circadian component, based on the observation that females exhibited a free-running calling rhythm after transfer to continuous light or dark conditions (Smith and Schal, 1991). The length of the scotophase appears to have only minor effects on calling: females under short nights extended calling into the photophase, whereas under long night conditions females terminated calling before lights-on. Therefore, the time allocated to calling by each female appears to be innately limited.

Behavioral assays using female extracts on sexually mature males showed that pheromone production starts on day 4, but calling does not commence until day 6 (Smith and Schal, 1990a). JH release by the CA *in vitro* increases in the first few days after the imaginal molt in a pattern that corresponds well with the pattern of pheromone production and calling, suggesting that JH is involved in the regulation of both (Smith *et al.*, 1989). As in *B. germanica*, this was substantiated surgically: ablation of the CA eliminated both pheromone production and calling behavior, and JH replacement with either active CA or with a JH analog restored both pheromone production and calling (Smith and Schal, 1990a). Likewise, the age of onset of pheromone production was advanced by both topical treatment with a JH analog and by severing the connections between the CA and the brain. As for *B. germanica*, it is unlikely that JH operates directly on the regulatory centers of calling and pheromone production because neither occurs in mated females that exhibit high titers of JH. However, the intervening signal molecules are not known.

After copulation, the female exhibits a shift from virgin to mated behaviors. JH production is stimulated and ovulation and oviposition are accelerated in mated females (Smith *et al.*, 1989). Mated females also become unresponsive to males, terminate pheromone production, and do not call. As in *B. germanica*, a two-stage process characterizes this switch (Smith and Schal, 1990b). First, insertion of the spermatophore causes an elevation in JH production and cessation of calling. However, the latter is transient, because females that mate with vasectomized males (no sperm), or females implanted with an artificial spermatophore (sand grain), suspend calling but then resume it after the infertile egg case is deposited. Therefore, a second, more permanent switch is required, and it has been suggested that stimulation by sperm of mechanoreceptors within the spermatheca serves this function (Smith and Schal, 1990b). This hypothesis has yet to be rigorously tested, but ventral nerve cord transection in mated females restores all the behaviors associated with the virgin state, including calling and pheromone production (Smith and Schal, 1990b).

S. longipalpa males become highly sensitive to supellapyrone 10 days after the imaginal molt (Liang and Schal, 1990a). Although they were less responsive in the photophase than in the scotophase, males responded to high pheromone concentrations during the photophase as well (Liang and Schal, 1990b). The diel periodicity of the male orientation response to the pheromone persisted for 54 h under continuous darkness, showing that this behavior too is under circadian control.

S. longipalpa offers a unique cockroach model to study enantiomeric discrimination by males. The ability of the male to discriminate among supellapyrone configurations was made possible by the recent synthesis of all four stereoisomers. Behavioral and EAG dose–response curves confirmed that (*R*,*R*), the natural stereoisomer, is the most active of the four stereoisomers (Gemeno *et al.*, 2003a). Dose–response studies revealed that 0.3 pg (*R*,*R*)-supellapyrone elicited behavioral responses in 50% of the test males, whereas 30 pg of (*S*,*R*)- and (*S*,*S*)-supellapyrone were needed to elicit similar responses. The (*R*,*S*)-isomer did not elicit responses at any concentration. Interestingly, (*R*,*R*)- and (*S*,*R*)-supellapyrone produced similar EAG peak amplitudes, revealing a poor correlation between EAG and behavior. Cross-adaptation EAG and single-cell recordings suggest that (*R*,*R*)- and (*S*,*R*)-supellapyrone stimulate the same antennal receptor neurons. The lack of correspondence between EAG and behavior probably results from isomer-specific molecular interactions within the sensillum and may be related to the much slower recovery rate of the EAG peaks in response to (*S*,*R*)-supellapyrone.

We recently sequenced complimentary DNA (cDNA) from the antennae of male *S. longipalpa* (K. O. Redding, C. Schal, and R. R. H. Anholt, unpublished data). In addition to candidate membrane receptors and signaling proteins, eight of the cDNA inserts encoded putative odorant or pheromone-binding proteins. Northern blots, *in situ* hybridization, and immunocytochemistry indicated that one odorant-binding protein was expressed almost exclusively in supporting cells throughout the male antenna, whereas female antennae showed staining in only a few cells at the same positions in successive antennal segments. This gene product is likely a pheromone-binding protein that participates in the male-specific sensillar transduction of supellapyrone signals. It will be interesting to determine whether it plays a role in enantiomeric discrimination by the peripheral nervous system.

Parcoblatta lata *(Brunner) (broad wood cockroach)*

The genus *Parcoblatta* comprises 12 forest-dwelling wood cockroach species, all endemic to North America. Recent studies have shown that more than 50% of the diet of the endangered red-cockaded woodpecker, *Picoides borealis*, consists of *Parcoblatta* spp. (Hanula *et al.*, 2000). As in *S. longipalpa*, *Parcoblatta* spp. exhibit marked sexual wing dimorphism: males have long functional wings and females are brachypterous. Based on the differential mobility of the sexes, Gemeno *et al.* (2003b) predicted that female *Parcoblatta* might release a long-range sex pheromone to which males would be attracted. This hypothesis was tested on *Parcoblatta lata*, one of the most common cockroach species in the forested areas of eastern North America.

P. lata females start calling 6 days after emergence and call mainly during the first half of the scotophase. Females in the calling posture raise the body from the substrate while performing repeated movements of the abdomen between two positions: (i) upwards and longitudinal compression, and (ii) downwards and longitudinal extension. Occasionally, the genital vestibulum is exposed during calling. Volatiles collected from calling females elicit stronger male-specific EAG responses and are preferred by males over volatiles from non-calling females. EAGs of tissue extracts showed that female-specific activity was present only in tergites 1–7; it increased with age after female emergence, and it decreased after mating. As in *S. longipalpa*, numerous cuticular pores were found on tergites, and each pore leads to a type 3 secretory cell. There is good positive correlation between the density of these glands in different tergites and their pheromone content. Taken together, this evidence indicates that virgin females of *P. lata* produce a volatile sex pheromone in tergites 1–7 and release it during calling (Gemeno *et al.*, 2003b).

This study provided necessary information toward the chemical identification of the sex pheromone of *P. lata*, which can then be used to monitor its population in relation to that of *P. borealis*, or as a means to control the cockroach itself, as it is an occasional indoor pest. We have carried out a similar study with Caudell's wood cockroach, *Parcoblatta caudelli* Hebard, and have demonstrated that in this species calling also is associated with pheromone release and that the sex pheromone is produced in tergites 1–7 (Gemeno *et al.*, 2003b).

Sex pheromones of Blaberidae

Nauphoeta cinerea *(Olivier) (lobster cockroach)*

N. cinerea is a large (25–29 mm long), ash-colored cockroach, with long wings, but neither sex can fly. It probably originated in East Africa and it now occurs primarily outdoors (Cornwell, 1968).

Mating behavior

In *N. cinerea*, as in other species in the subfamily Oxyhaloinae, females appear not to release volatile sex pheromones (Roth and Barth, 1967; Sreng, 1993). Early observations of courtship behavior suggested that *N. cinerea* males produce an attractant in the abdomen (Roth and Willis, 1954; Roth and Dateo, 1966), but there was no precise information on where it was produced or how it was released. Sreng's (1979b) detailed description of the courtship behavior of *N. cinerea* showed that males display two independent pheromone-releasing postures: a calling behavior that is independent of the presence of females, and courtship wing-raising elicited by contact with a female. The calling posture results in the release of a volatile pheromone from sternal glands that are exposed when the abdomen is curved with its tip pointing upward (Fig. 6.1a). This male-initiated behavior attracts females and is thus analogous to the female calling behaviors seen in other species. A second male pheromone is released during courtship, as the male raises his wings and curves the abdomen with its tip pointing downward, thus exposing the tergal glands, which are normally hidden between the segments (Fig. 6.1c). Volatiles released from the tergal glands stimulate the female to mount the male and feed from the tergal gland (Fig. 6.1d). This wing-raising stance is similar to that of male cockroaches of other species. If males are unsuccessful at coupling with the female they initiate stridulation (Hartman and Roth, 1967), but the role of this behavior in mate attraction and courtship is unclear.

Male volatile pheromones: sources and identification

The sternal glands have no specialized cuticular structures, are present only on sternites 3–7, and consist of glandular orifices found only on the anterior zone of each sternite (Sreng, 1979a, 1984, 1985). They are made up of four categories of cells including the type 3 glandular units seen in other species (above). The sternal glands produce a mixture of 3-hydroxy-2-butanone (**13** (acetoin); Table 6.2), 2-methylthiazolidine (**14** (2M); Table 6.2), and 4-ethyl-guaiacol (**15** (4EG); Table 6.2) (Sreng, 1990). Both 2M and 4EG are commonly known as seducin. A reconstructed mixture of acetoin, 2M, and 4EG, at a ratio of 4:4:1, respectively (total 2.7 μg), as normally occurs in the sternal gland, elicited maximal female response, whereas individual components elicited weaker responses (Sreng, 1990). In a reinvestigation of male-specific compounds in *Nauphoeta*, Sirugue *et al.* (1992) identified 23 compounds from male sternal gland extracts. Based on two-choice olfactometer assays, they concluded that 4EG and 2M attract females from a distance at low concentrations, whereas acetoin and an additional compound, 2-methyl-2-thiazoline, are attractive at much higher concentrations and probably operate at close range, keeping the female in the vicinity of the male.

Acetoin may have a secondary role in attraction (Sirugue *et al.*, 1992; Moore *et al.*, 1995).

Brossut and Roth (1977) reported sexual dimorphism in the tergal glands of *N. cinerea*, with more orifices on tergites 1 and 2 of males than in females, whereas the other tergites show no sexual dimorphism. However, the male tergal glands that attract females during courtship are distributed in tergites 2–8, averaging 7600 orifices/mm^2 on the anterior and posterior regions of each tergite. These glands lack cuticular modifications other than individual cuticular pores and are composed of five categories of cells, including type 3 glandular units. Of 23 compounds identified in sternal gland extracts, 19 were also found in the tergal glands (Sirugue *et al.*, 1992), including the three compounds described by Sreng (1990) as sternal gland sex pheromones. However, the tergal glands contained 40–147 times less material (up to 84 ng/male) than the sternites (up to 6785 ng/male) (Sirugue *et al.*, 1992), and it is not clear whether these same compounds are also involved in close-range female attraction to the male tergites.

Factors affecting pheromone production and response

Extracts from males started attracting females 4 days after the final molt, at the time calling was first observed (Sreng, 1979b). The quantities of 4EG and 2M in sternal glands increased gradually with age, reached a maximum at approximately age 15–20 days and then declined with age (Moore *et al.*, 1995). The quantity of acetoin, however, increased to age 10–15 days and then it was maintained at a relatively constant level, suggesting that this compound might have a different role from the other components (Moore *et al.*, 1995). The changes in pheromone quantity with age parallel the development of male attractiveness to females, male sexual competence, and male agonistic behavior (Moore *et al.*, 1995). Environmental quality also affected pheromone quality and quantity and male and female courtship behavior (Clark *et al.*, 1997). In poor environments (no food or water, and 35 °C for 10 days after emergence) the quantities of acetoin and 4EG were reduced and 2M remained undetectable. It is possible that these changes result from endogenous regulators of pheromone production, such as JH (see below), the titer of which may be affected by starvation, as in *B. germanica*. Moreover, various sternal gland components are biosynthesized at different times of the photophase and scotophase (Sirugue *et al.*, 1992), further confounding chemical analyses of sternal gland extracts.

The development of the tergal glands involves differentiation of four types of cell, two of which undergo apoptosis before the gland is completely formed 3–4 days after the imaginal molt. This process is orchestrated by brain-derived factors because decapitation of newly emerged males prevented the programmed death of the enveloping and ciliary cells and development of the end-apparatus (Sreng, 1998). Decapitated males also produced much less pheromone than normal males. A role for JH III was suggested because the hormone stimulated pheromone production in allatectomized males (Sreng *et al.*, 1999). The distance over which the sex pheromone is effective needs to be determined experimentally, but it has been shown that females are attracted within a radius of 10 cm in still air (Sreng, 1979b) and respond to volatile sources 50–70 cm upwind in an air flow of 3–4 cm/s (Sreng, 1990; Sirugue *et al.*, 1992).

Agonistic behavior and mate choice

Adult *N. cinerea* males establish dominant–subordinate hierarchies by means of agonistic behavior (Ewing, 1967). The dominant male establishes his position, but once established, over 90% of the encounters consist primarily of lunging by the dominant male and retreating and submissive behavior by the subordinate male (Bell and Gorton, 1978; Smith and Breed, 1982). Subordinate males experience higher mortality in agonistic encounters (Ewing, 1967), whereas dominant males enjoy greater access to females and higher reproductive success (Smith and Breed, 1982). Olfactometer tests showed that females preferred dominant males, responded earlier to them, and copulated longer with them than with subordinate males (Breed *et al.*, 1980; Moore and Moore, 1988). Because in choice tests the female preferred the dominant over the subordinate male, Moore (1988) suggested that the chemical cue used by females to discriminate between males is either the same as, or correlated with, the cue used by males in status assessment.

To test this hypothesis, Moore *et al.* (1997) altered the pheromone profile of males by gluing filter paper disks to their pronota and adding different combinations of the three sternal pheromone components. When the amount of only acetoin was elevated, males were more likely to be engaged in behavioral encounters and to react as subordinates. When either 2M or 4EG was increased, there was an increased likelihood of being dominant, an effect that was even greater when the quantities of both compounds were elevated together. The influence of acetoin was cancelled by the combined effects of 2M and 4EG, whereas increasing all three compounds simultaneously did not affect status (Moore *et al.*, 1997). Therefore, the same sex pheromone components used by males to attract females are used in male–male agonistic interactions, although the most effective blend is different in each behavioral context. The compounds have individual, additive, and contrasting effects on status. Natural and sexual selection would be expected to increase 2M and 4EG and to decrease acetoin because this would make males more competitive and increase their access to females. However, female *N. cinerea* use only one of the components of the sex pheromone to discriminate and choose among males. Moore and Moore (1999) showed that female choice is based solely on the amount of acetoin, and the amount and ratio of the other components is unimportant. Consequently, the blend that is most attractive to females differs from that which is most likely to confer higher status to males. How a male that is less successful in intrasexual interactions attracts and mates with more females remains unclear and appears to be in conflict with the observation that dominant males mate more readily than subordinates.

Females do not appear to gain any direct nutrient resources from mating with dominant males. They produce the same number of offspring whether they mate with dominant or subordinate males (Moore and Moore, 1988). Unlike other cockroach species (e.g., *B. germanica*), male *N. cinerea* do not provide tergal secretions or urates to their mate. Yet, the females appear to gain a significant indirect benefit from mating with more attractive males because the progeny from preferred males have shorter developmental times than the progeny from less-preferred males (Moore, 1994). In addition, females gain a genetic benefit from mating with dominant males by increasing the likelihood of their male progeny being dominant and their female progeny preferring dominant males (Moore, 1989, 1990a,b). In the context of a territorial male mating system, females that mate with dominant males might also gain a resting site for the long gestation period that follows oviposition (Ewing, 1973).

However, we need to take into consideration that this mating system is complicated by experiential and learning events. Schal and Bell (1983) found that the age of the male was a good predictor of status, but even here, the dominant–subordinate status of males was only transiently stable and was subject to complete reversals. A single act of copulation, or even courtship, by a subordinate male can make him dominant. This, coupled with the prevalence of the satellite-male strategy in this species (Breed *et al.*, 1980), begs the question of whether pheromones always serve as honest signals of social status and suggest that the environment–gene interaction is worthy of study in this system. Nevertheless, the work of Moore and collaborators on *N. cinerea* illustrates the potential of cockroach sex pheromone systems to address important genetic and evolutionary questions.

Male contact pheromone

N. cinerea females produce a contact pheromone that stimulates male wing-raising behavior, and males (and possibly also nymphs) produce a contact, recognition pheromone that inhibits courtship behavior in other males. The female pheromone has not been chemically identified, but the male-specific courtship depressant (nauphoetin) is octadecyl (*Z*)-9-tetracosenoate (**16**; Table 6.3) (Fukui and Takahashi, 1983). This compound elicits aggressive antennal fencing among males and is not present in extracts of females. Nauphoetin and two other isomers reduce wing-raising behavior in males. Takahashi and Fukui (1983) concluded that, regardless of the length of the carbon chain in the acid moiety, presence of a double bond is essential for the behavioral activity of nauphoetin. It is somewhat perplexing why this compound, which is used in male–male recognition, has not been tested in the context of male–male agonistic interactions.

As in *B. germanica*, newly molted males and nymphs elicit wing-raising courtship responses from sexually mature *N. cinerea* males (Fukui and Takahashi, 1983, 1999; Schal and Bell, 1983). It seems that all individuals, when newly molted, can stimulate male courtship, and that males and nymphs, but not females, produce secondary compounds as they age that reduce this effect. A compound that inhibits male courtship was identified from older nymphs as *cis*-25,26-epoxyhenpentacontadiene (Fukui and Takahashi, 1999). Whether there is an adaptive advantage for nymphs to stimulate courtship soon after the molt and later inhibit this behavior in mature males is not clear. It is possible that by mimicking females teneral cockroaches are better protected from aggressive males, but this has not been demonstrated.

Leucophaea maderae *(F.) (Madeira cockroach)*

The courtship sequence of *L. maderae* is very similar to that of *N. cinerea* and was described in detail by Roth and Barth (1967). As in *N. cinerea*, and in fact all the other Oxyhaloinae studied, females do not seem to produce a volatile sex pheromone and, instead, males call to attract females (Sreng, 1993). The male calling posture is less pronounced than in *N. cinerea*, probably because the sternal glands are more exposed than in the other species (Sreng, 1993). Tactile stimulation of the sexes results in antennal fencing, which is particularly vigorous in this species. Males may show vertical vibration movements during this period, especially if the females are not highly receptive. The male eventually raises his wings, gradually at the beginning, and he then performs a series of wing-pumping movements (wings are raised and lowered). As the male turns away from the female, his wings are raised in a continuous motion to 60–80° above the horizontal. The abdomen is extended and depressed so that the terminal sterna are pressed against the substrate. The female moves forward on the male as he moves backwards, and her mouthparts pass rapidly along the male's exposed abdominal tergites until the glandular opening on the second abdominal tergum is reached, where there are glandular exudates from which she feeds. The male then extends his phallomere and if connection occurs the pair assumes the opposed position immediately. Copulation lasts slightly less than 1 h.

Males possess glands on sternites 2–7 and tergites 2–8, similar to *N. cinerea* (Sreng, 1984). Acetoin (**13**; Table 6.2), which has attractant properties in *N. cinerea*, is also present in the sternal glands of *L. maderae* (Sreng, 1993). The number of glandular orifices varies from one segment to the next, but the anteriolateral and medial regions of tergite 2 contain aggregations of cuticular pores (Porcheron,

1975). The male tergal gland produces several male-specific proteins (Korchi *et al.*, 1998), including lipocalins and aspartic proteases, which may serve as ligand-binding proteins and carriers of pheromones or related lipids from the secretory cells to the cuticular surface (Korchi *et al.*, 1999; Cornetle *et al.*, 2001, 2002).

L. maderae has a significant place in the annals of invertebrate endocrinology, serving as an early model in studies of neurosecretion and regulation of the CA (Engelmann, 1970; Scharrer, 1987; Stay, 2000). It would be fitting to link investigations in chemical ecology to these early studies.

Blaberus craniifer *Burmeister (death's-head cockroach)*

The mating behavior of *B. craniifer* (subfamily Blaberinae) has been described by Grillou (1973) and Abed *et al.* (1993b). Females adopt a calling posture during which they release a volatile pheromone: the wings are slightly raised and contraction of the middle segments gives the tip of the abdomen a bulbous shape (Abed *et al.*, 1993b). The pygidium is lowered and its anterior part, which contains glandular tissue, is fully exposed. The male recognizes the female by touching her body with his antennae, and courtship and copulation follow, essentially as in *N. cinerea*.

Males do not discriminate between non-calling females and males (62 and 37% response, respectively), but they prefer calling females over males (87 and 12% response, respectively) (Abed *et al.*, 1993b). In tests in cages, 100% of the males visited filter papers impregnated with pentane extracts of the female pygidium whereas only 34% visited filter papers impregnated with pentane, indicating that the pygidial segment contains attractants. However, when males were given a choice between pygidia dissected from virgin females and the abdomens from which the pygidia were dissected, they were not significantly more attracted by the pygidia than by the abdomens. This suggests that the pheromone may be produced in other parts of the abdomen, besides the pygidium, or that several unrelated attractants were assayed.

Byrsotria fumigata *(Guérin) (Cuban burrowing cockroach)*

Barth (1964) described the mating behavior of *Byrsotria fumigata* and this cockroach served an important role in early studies of the neuroendocrine regulation of pheromone production. The female releases a volatile pheromone that attracts males from a distance of "up to somewhat less than a meter." Antennal contact with the female stimulates wing-pumping in the male (as in *L. maderae*). If the female is receptive, they face each other and exchange antennal fencing. The male raises his wings perpendicular to the longitudinal body axis and turns 180°. The female mounts and straddles the male's abdomen and palpates his anterior abdominal tergites, which apparently have no specialized glandular region. Upon receiving tactile information from the female, the male moves backward and copulation follows in the opposed position (Barth, 1964).

The sex pheromone of this species has not been identified. Using tissue microcauterization, Moore and Barth (1976) determined that the pheromone gland could be in the genital atrium. However, microcauterization of this region, which also sends mechanoreceptive feedback to the central nervous system from the spermatophore and ootheca, may have resulted in the inhibition of pheromone production by the females. Nevertheless, in light of more recent evidence of pheromone production by atrial glands of other cockroaches, it is likely that these results will be verified upon reevaluation. Based on bioassays with males, females start secreting pheromone within 4 days after adult emergence, and males respond to the pheromone on about the third day after emergence (Bell *et al.*, 1974). Females fail to produce pheromone if their CAs are ablated, but implantation of CA from a pheromone-producing female restores pheromone production in the operated female (Barth, 1968b). Bell and Barth (1970) proposed a dynamic model for the coordinated regulation of reproduction by JH: low titers of the hormone induce tissue competence of the fat body; higher titers stimulate vitellogenin synthesis and its uptake by oocytes; still higher levels of JH in the hemolymph cause the female to produce and emit pheromones; and, finally, the accessory glands are stimulated to make oothecal proteins. Unfortunately, this model has not been rigorously tested in any insect.

Other blaberids

Pheromone communication has been examined in several other blaberid species, but in much less detail. In addition to *N. cinerea* and *L. maderae*, males of four other Oxyhaloinae species release volatile sex pheromones that attract females (Sreng, 1993), and calling behavior has been observed in other tropical blaberids (Schal and Bell, 1985). *Pycnoscelus surinamensis* (L.) females produce a sex pheromone that attracts males from a distance. Males do not raise their wings and females do not feed on the male's tergal secretions. Male courtship consists simply of mounting the back of the female and twisting the abdomen around and under the female's to make genital connection (Roth and Barth, 1967). Allatectomy of females abolishes production of the volatile sex pheromone (Barth, 1965). Females of *Eublaberus posticus* (Erichson) mate shortly after the adult molt, while their cuticle is still soft. Several males gather around the newly molted female, but it is not clear if this is because of a volatile sex pheromone or antennal contact with the female. A few minutes later, males start to court by turning and raising their wings and tegmina, which is followed by female mounting and copulation (Roth, 1968). The mating behavior of *Eublaberus distanti* (Kirby) is similar to that of *E. posticus* (Roth, 1968).

Field and ecological observations

Cockroaches have served as important models in seminal work on insect biochemistry and physiology, particularly endocrinology. Perhaps because of the ease with which some species can be raised in the laboratory, few researchers have ventured to investigate this primitive but diverse group of insects in the field. Likewise, only a handful of studies have dealt with the chemical ecology of cockroaches under field conditions. As in agriculture, pheromones have tremendous potential in the detection, monitoring, and control of cockroaches in urban settings, and for this reason several studies have examined the efficacy of periplanone-B for attraction of *P. americana* and related species. For example, Bell *et al.* (1984) showed that addition of periplanone-B significantly enhanced the efficacy of insecticide sprays against *P. americana* in warehouses. Similarly, Liang *et al.* (1998) showed that supellapyrone was highly attractive to *S. longipalpa* males in infested apartments and university laboratories. As in studies with *P. americana*, traps with pheromone caught not only more adult males but also females and nymphs. In addition, supellapyrone could be effectively combined with food odorants to attract all life stages and, therefore, boost trap catch. Nevertheless, practical use of cockroach pheromones has been hampered by difficulties in pheromone identification and an extreme reliance on easy-to-apply pesticides.

Nocturnally active neotropical rainforest cockroaches spend the day in the leaf litter and forage at night (Schal *et al.*, 1984). Schal and Bell (1986) observed that species in the Blattellidae and Blaberidae perch on leaves, thus avoiding the intense predation on the forest floor by amphibians, spiders, and ants, and this also facilitated the search for food. Some females were also observed in calling postures, presumably emitting sex pheromones (Schal and Bell, 1985). It is particularly interesting that both flight-capable and flightless adult males tended to perch higher in the understory than their respective females (Schal and Bell, 1986). Schal (1982) suggested that female pheromones would ascend with nighttime convective currents, and, therefore, males might position themselves to intercept these odorants. The tendency of males to perch higher than females is also evident in hollow trees, caves, and even in laboratory cultures.

Summary and future directions

More than 20 years after their original isolation in 1952, the chemical structures of periplanone-A and periplanone-B from *P. americana* have been completely identified. Volatile pheromones of several other *Periplaneta* spp. have also been identified, as well as the female sex pheromone of *S. longipalpa* and the male sex pheromone blend of *N. cinerea.* A male-produced contact pheromone in *N. cinerea*, which prevents male–male courtship, and the multicomponent contact sex pheromone of *B. germanica* females have also been chemically elucidated. Identification of the volatile sex pheromone of *B. germanica* and of the contact pheromones of the other three species, when achieved, will provide material for comparative studies of the function of volatile and contact pheromones across species. Despite these advances, however, cockroach pheromone identification lags far behind other insect groups. Even extensively researched species, such as *P. americana*, produce additional pheromone components that remain to be identified. Although the few volatile sex pheromones identified to date are from species belonging to three of the five cockroach families, large gaps nonetheless remain within families. Because the known pheromones are chemically unrelated to each other, it would be of great interest to identify more sex pheromones of female Blaberidae, Blattidae, and Blattellidae. In addition, nothing is known about the sex pheromones of Polyphagidae and Cryptocercidae, the most primitive cockroach clades. Pheromone identification, therefore, remains a primary objective in cockroach chemical ecology at the present time.

Cockroaches present unique challenges in behavioral and electrophysiological assays. In contrast to many short-lived lepidopterans, in which male antennae are highly specialized for sex pheromone reception, cockroaches are long lived, and the antennae of both females and males contain numerous sensilla that respond to general odors associated with food, shelter, and the general environment. These general odorant receptors also respond to non-sex-specific odors associated with various secretions and excretions of cockroaches. Therefore, behavioral tests employed in pheromone identification must be designed so that the sex specificity of both stimulus and response are unambiguously characterized. One such approach is to collect (or extract) the candidate pheromone from the sex that produces it and to make a similar control collection from non-pheromone-producing individuals (of the same or the opposite sex, or a nymph), and then test the responses of both sexes to both stimuli. This technique was used successfully to demonstrate the production of a female sex pheromone in *B. germanica* (Liang and Schal, 1993a). An alternative procedure is to compare the responses of both sexes to a sample from the pheromone-producing sex with their responses to a general food odorant that elicits responses from both sexes. This technique was used to demonstrate the production of a volatile sex pheromone in *Parcoblatta* cockroaches (Gemeno *et al.*, 2003b). For female-produced pheromones, a male/female EAG index (defined as the mean male response divided by the mean female response to the candidate pheromone) divided by a similar response ratio to a general food odorant has been used in studies of *P. americana* pheromone components and their structural analogs (Nishino and Kimura, 1981; Nishino *et al.*, 1983). More simplified tests may be appropriate only when the response to the pheromone by the receiver is much higher than the response by the emitter (Nishino and Kimura, 1981), or when the insects do not respond to other endogenous odors that may contaminate the sample.

Pheromone production in cockroaches has been localized in the dorsal and ventral tegmina and intersegmental membranes of the abdomen in males or females, and also in the digestive tract and adjacent tissues. Although periplanone-B was the first cockroach sex pheromone identified, there is no consensus yet as to where it is produced, and even recent studies have implicated several unrelated tissues. Then again, none of these studies rigorously controlled for females that just initiated production of the pheromone, nor for contamination of various tissues with pheromone. Given the ease of tissue contamination by this pheromone, partly because of its association with the digestive tract and the prevalence of coprophagy in cockroaches, a test of *de novo* pheromone production *in vitro* will be necessary for an unambiguous determination of the site of pheromone production. This should be a feasible approach with periplanone, but its biosynthetic pathway needs to be elucidated.

Calling behavior in female cockroaches is associated with the release of volatile pheromones. However, we have only a rudimentary understanding of how the female calling posture results in the release of the pheromone and what role genital exposure plays in calling and courtship. In *B. germanica* and *S. longipalpa*, the pheromone glands are distributed on the tergites and consist of secretory cells connected to cuticular pores through long ducts. The cuticular pores are always exposed, and the secretory cells contain pheromone, but the pheromone is only emitted when the female calls. Therefore, the motor patterns that characterize calling behavior must be responsible for the release of the pheromone. Perhaps these movements facilitate the transport of the pheromone products along the ducts, but the precise mechanisms of pheromone transport and release remain unknown. In several species, the females expose their genital vestibulum during calling. Because in these species the pheromone is produced only by the tergites, the role of genital exposure is unclear and needs to be investigated. Perhaps the signals released from the genital vestibulum are not sex specific by themselves, but in the context of courtship and in combination with other signals they stimulate further male courtship.

Cockroaches are excellent models for investigations of mechanisms that regulate pheromone production. Early studies on blaberids, and more recent research with the blattellids *B. germanica* and *S. longipalpa*, have shown that almost all physiological and behavioral aspects of female reproduction are coordinately regulated and paced by JH. Food intake and mating also intervene in the regulation of pheromone production and calling behavior in females, in part by modulating production of JH. Yet, it is clear from studies of mated females that unknown regulatory elements downstream of JH must be involved. A concerted effort is needed to identify neuropeptides and other factors that activate and inactivate pheromone production, emission, and sexual receptivity.

Even less is known about the physiological regulation of male tergal and sternal secretions and maturation of male sexual receptivity. It is interesting that some species have macroscopic cuticular modifications with concentrations of tergal glands and specialized hair structures, whereas other species have individual secretory cells more or less evenly distributed throughout the tergites. Behavioral and morphological studies would be necessary to determine whether there are differences in courtship behavior between species that have specialized tergal glands and those species that do not. The tergal gland pheromones seem to act at close range, but we know of no experimental study that has tested the distance over which these pheromones are most effective under natural conditions. Although male courtship behavior (downward curving of the abdomen and wing-raising behavior) is clearly aimed at attracting females, experimental association of this behavior with pheromone release is also lacking. Further, the pheromone that attracts females to the male tergites could be a different signal from that which stimulates the females to feed on the tergal secretion.

P. americana has served as a model insect in studies of the nervous system, in particular with respect to olfaction. The response specificity of antennal sensory neurons and brain projection neurons to pheromones and food odors has been studied, as well as olfactory circuits of the brain in relation to pheromone perception (Ignell *et al.*, 2001). Identification of the sex pheromones of *P. americana* and *S. longipalpa* has stimulated research on the cellular and molecular bases of pheromone reception, and new molecular tools should facilitate this enterprise. The long cockroach antennae, biological activity of single components of the sex pheromone, and clear behavioral assays should provide an accessible, primitive, hemimetabolous model of processing of pheromone signals in specialized macroglomerular centers, as has the locust, *Locusta migratoria* L., for general odorants (Laurent *et al.*, 2001).

In comparison with volatile pheromones, little is known about chemoreception of contact pheromones. There is some indication that the physical structure, or texture, of the antenna plays a role in the response to contact pheromones, but the processing of these signals has yet to be investigated. Further, ablation experiments suggest that the palpi may be involved in pheromone reception, but electrophysiological and behavioral experimental support to demonstrate this unambiguously is lacking.

With regard to male behavior, several recent reports have delineated the time course of maturation of male sexual response, its diel periodicity and circadian regulation, and behavioral tactics employed by males in orienting to females. As in other insects, cockroaches integrate visual and olfactory cues, but the orientation mechanisms have been little studied, especially under natural conditions. In particular, it would be interesting to determine, when more pheromones are identified, whether different components play different roles in orientation, as appears to be the case in *P. americana*.

Cockroach sex pheromones may serve several concomitant functions, such as attraction of the sexes, and species and sex recognition. There is some evidence that differences in the volatile sex pheromones of *Periplaneta* spp. may be responsible for reproductive isolation among sympatric species in this genus. In addition, sex pheromones also may be involved in male recognition, agonistic behavior, and maintenance of male territories and hierarchies. As in eusocial insects, social interactions in cockroaches are overwhelmingly mediated by chemical signals. A major handicap in our understanding of cockroach chemical ecology, however, is that, with few exceptions, most studies have been carried out under oversimplified laboratory conditions with pest species for which the behavior under natural conditions is not well known. Determining the adaptive significance of specific behavioral traits in these species is difficult because factors such as operational sex ratio, previous experience and learning, and characteristics of the mating system, which are almost never considered in laboratory experiments, are variable in nature and affect mating behavior. Field studies, therefore, provide invaluable information that can be critical for the design of behavioral tests in the laboratory. The answers to general questions, such as the reasons for reversal of sex roles in pheromone signaling, the relative importance of long-range versus short-range volatile pheromones, or the involvement of male tergal secretions in mate choice will probably require a combination of field and laboratory studies. To address these questions, we need to gain a better understanding of the structure of natural populations – sex and stage composition and density in space and time – as well as of the behavior of individuals in the population, such as their foraging and mating habits.

Research on wild species in the temperate zones is limited by low species diversity and population density, but in tropical and subtropical areas cockroaches are surprisingly diverse, and valuable field observations in chemical ecology have been made in the tropics. Nevertheless, some temperate zone species, such as *Parcoblatta* in the New World and *Ectobius* in the Old, offer excellent opportunities for research under natural conditions.

Laboratory observations remain essential because they can delineate the timing of mate finding and mating in relation to other activities, and observations can provide important information about interactions between the sexes and the role of sex pheromones. The identification of sex pheromones also requires standard laboratory behavioral assays. We urge students of cockroach behavior and chemical ecology to develop more realistic behavioral assays that will facilitate the identification of sex pheromones as well as a better understanding of their role in cockroach mating systems.

The chemical ecology of cockroaches is an exciting area of research, and it has grown steadily since the 1970s. Cockroaches are particularly interesting because they rely intensely on chemical signals in their social behavior (which is highly sophisticated in some species), they represent a primitive yet diverse insect group, and because some species are important pests. The potential utility of sex pheromones in environmentally responsible pest control constitutes a major motivation for further studies. To achieve rapid advances in our understanding of the chemical ecology of this group, the identification of new sex pheromones should be given priority.

Acknowledgements

Our research on cockroach pheromones has been supported by grants from the National Science Foundation (IBN-9817075), the United States Department of Agriculture Competitive Grants Program (2002–02633), and the Blanton J. Whitmire Endowment and the W. M. Keck Center for Behavioral Biology at North Carolina State University.

References

Abed, D., Brossut, R. and Farine, J.-P. (1993a). Evidence for sex pheromones produced by males and females in *Blatta orientalis* (Dictyoptera: Blattidae). *Journal of Chemical Ecology*, **19**: 2831–2853.

Abed, D., Tokro, P., Farine, J.-P. and Brossut, R. (1993b). Pheromones in *Blattella germanica* and *Blaberus craniifer* (Blaberoidea): Glandular source, morphology and analyses of pheromonally released behaviours. *Chemoecology* **4**: 46–54.

Abed, D., Cheviet, P., Farine, J. P., Bonnard, O., Le Quéré, J. L. and Brossut, R. (1993c). Calling behaviour of female *Periplaneta americana*: behavioural analysis and identification of the pheromone source. *Journal of Insect Physiology* **39**: 709–720.

Adams, M. A., Nakanishi, K., Still, W. C., Arnold, E. V., Clardy, J. and Persoons, C. J. (1979). Sex pheromone of the American cockroach: absolute configuration of periplanone-B. *Journal of the American Chemical Society* **101**: 2495–2498.

Appel, A. G. and Rust, M. K. (1983). Temperature-mediated sex pheromone production and response of the American cockroach. *Journal of Insect Physiology* **29**: 301–305.

Barth, R. H., Jr (1964). The mating behavior of *Byrsotria fumigata* (Guérin) (Blaberidae, Blaberinae). *Behaviour* **23**: 1–30.

(1965). Insect mating behavior: endocrine control of a chemical communication system. *Science* **149**: 882–883.

(1968a). The mating behavior of *Eurycotis floridana* (Walker) (Blattaria, Blattoidea, Blattidae, Polyzosteriinae). *Psyche* **75**: 274–284.

(1968b). The comparative physiology of reproductive processes in cockroaches. Part I. Mating behaviour and its endocrine control. *Advances in Reproductive Physiology* **3**: 167–207.

Barth, R. H. (1970). The mating behavior of *Periplaneta americana* (Linnaeus) and *Blatta orientalis* Linnaeus (Blattaria, Blattinae), with notes on 3 additional species of

Periplaneta and interspecific action of female sex pheromones. *Journal of Comparative Ethology* **27**: 722–748.
Barth, R. H. and Lester, L. J. (1973). Neuro-hormonal control of sexual behavior in insects. *Annual Review of Entomology* **18**: 445–472.
Bell, W. J. (1981). Pheromones and behaviour. In *The American Cockroach*, eds. W. J. Bell and K. G. Adiyodi, pp. 371–397. New York: Chapman & Hall.
Bell, W. J. and Adiyodi, K. G. (1981). *The American Cockroach*. New York: Chapman & Hall.
Bell, W. J. and Barth R. H. (1970). Quantitative effects of juvenile hormone on reproduction in cockroach *Byrsotria fumigata*. *Journal of Insect Physiology* **16**: 2303–2313.
Bell, W. J. and Gorton, R. E., Jr. (1978). Informational analysis of agonistic behaviour and dominance hierarchy formation in a cockroach, *Nauphoeta cinerea*. *Behaviour* **67**: 217–235.
Bell, W. J. and Kramer, E. (1980). Sex pheromone-stimulated orientation of the American cockroach on a servosphere apparatus. *Journal of Chemical Ecology* **6**: 287–295.
Bell, W. J. and Schal, C. (1980). Patterns of turning in courtship orientation of the male German cockroach. *Animal Behaviour* **28**: 86–94.
Bell, W. J. and Tobin, T. R. (1981). Orientation to sex pheromone in the American cockroach: analysis of chemo-orientation mechanisms. *Journal of Insect Physiology* **27**: 501–508.
Bell, W. J., Burns, R. E. and Barth, R. H. (1974). Quantitative aspects of the male courting response in the cockroach *Byrsotria fumigata* (Guérin) (Blattaria). *Behavioral Biology* **10**: 419–433.
Bell, W. J., Vuturo, S. B. and Bennett, M. (1978). Endokinetic turning and programmed courtship acts of the male German cockroach. *Journal of Insect Physiology* **24**: 369–374.
Bell, W. J., Fromm, J., Quisumbing, A. R. and Kydonieus, A. F. (1984). Attraction of American cockroaches (Orthoptera: Blattidae) to traps containing periplanone B and to insecticide–periplanone B mixtures. *Environmental Entomology* **13**: 448–450.
Bellés, X., Cassier, P., Cerdá, X. *et al.* (1993). Induction of choriogenesis by 20-hydroxyecdysone in the German-cockroach. *Tissue and Cell* **25**: 195–204.
Biendl, M., Hauptmann, H. and Sass, H. (1989). Periplanone D_1 and periplanone D_2: two new biologically active germacranoid sesquiterpenes from *Periplaneta americana*. *Tetrahedron Letters* **30**: 2367–2368.
Blomquist, G. J., Tillman-Wall, J. A., Guo, L., Quilici, D. R., Gu, P. and Schal, C. (1993). Hydrocarbon and hydrocarbon-derived sex pheromones in insects: biochemistry and endocrine regulation. In *Insect Lipids: Chemistry, Biochemistry and Biology*, eds. D. W. Stanley-Samuelson and D. R. Nelson, pp. 317–351. Lincoln, NE: University of Nebraska Press.
Bodenstein, W. G. (1970). Distribution of female sex pheromone in the gut of *Periplaneta americana* (Orthoptera: Blattidae). *Annals of the Entomological Society of America* **63**: 336–337.
Boeckh, J. and Ernst, K. D. (1987). Contribution of single unit analysis in insects to an understanding of olfactory function. *Journal of Comparative Physiology A* **161**: 549–565.
Boeckh, J. and Selsam, P. (1984). Quantitative investigation of the odour specificity of central olfactory neurons in the American cockroach. *Chemical Senses* **9**: 369–380.
Boeckh, J., Priesner, E., Schneider, D. and Jacobson, M. (1963). Olfactory receptor response to the cockroach sexual attractant. *Science* **141**: 716–717.

Boeckh, J., Sass, H. and Wharton, D. R. A. (1970). Antennal receptors: reactions to female sex attractant in *Periplaneta americana*. *Science* **168**: 589.

Boeckh, J., Ernst, K. D., Sass, H. and Waldow, U. (1984). Anatomical and physiological characteristics of individual neurones in the central antennal pathway of insects. *Journal of Insect Physiology* **30**: 15–26.

Bowers, W. S. and Bodenstein, W. G. (1971). Sex pheromone mimics of the American cockroach. *Nature* **232**: 259–261.

Bray, S. and Amrein, H. A. (2003). Putative *Drosophila* pheromone receptor expressed in male-specific taste neurons is required for efficient courtship. *Neuron* **39**: 1019–1029.

Breed, M. D., Smith, S. K. and Gall, B. G. (1980). Systems of mate selection in a cockroach species with male dominance hierarchies. *Animal Behaviour* **28**: 130–134.

Brenner, R. J. (1995). Medical and economic significance. In *Understanding and Controlling the German Cockroach*, eds. M. K. Rust, J. M. Owens and D. A. Reierson, pp. 77–92. New York: Oxford University Press.

Brossut, R. and Roth, L. M. (1977). Tergal modifications associated with abdominal glandular cells in the Blattaria. *Journal of Morphology* **151**: 259–297.

Brossut, R., Dubois, P., Rigaud, J. and Sreng, L. (1975). Biochemical study of the secretion of the tergal glands of the Blattaria. *Insect Biochemistry* **5**: 719–732.

Burrows, M., Boeckh, J. and Esslen, J. (1982). Physiological and morphological properties of interneurones in the deuterocerebrum of male cockroaches which respond to female pheromone. *Journal of Comparative Physiology A* **145**: 447–457.

Bykhovskaia, M. B. and Zhorov, B. S. (1996). Atomic model of the recognition site of the American cockroach pheromone receptor. *Journal of Chemical Ecology* **22**: 869–883.

Cardé, R. T. and Baker, T. C. (1984). Sexual communication with pheromones. In *Chemical Ecology of Insects*, eds. W. J. Bell and R. T. Cardé, pp. 355–383. London: Chapman & Hall.

Charlton, R. E., Webster, F. X., Zhang, A. *et al.* (1993). Sex pheromone for the brownbanded cockroach is an unusual dialkyl-substituted α-pyrone. *Proceedings of the National Academy of Sciences, USA* **90**: 10202–10205.

Chase, J., Jurenka, R. A., Schal, C., Halarnkar, P. P. and Blomquist, G. J. (1990). Biosynthesis of methyl branched hydrocarbons of the German cockroach *Blattella germanica* (L.) (Orthoptera, Blattellidae). *Insect Biochemistry* **20**: 149–156.

Chase, J. Touhara, K., Prestwich, G. D., Schal, C. and Blomquist, G. J. (1992). Biosynthesis and endocrine control of the production of the German cockroach sex pheromone, 3,11-dimethylnonacosan-2-one. *Proceedings of the National Academy of Sciences, USA* **89**: 6050–6054.

Chow, Y.-S., Lin, Y.-M., Lee, M.-Y., Wang, Y.-T. and Wang, C.-S. (1976). Sex pheromone of the American cockroach, *Periplaneta americana* (L.). I. Isolation techniques and attraction test for the pheromone in a heavily infested room. *Bulletin of the Institute of Zoology Academia Sinica* **15**: 39–45.

Clark, D. C., DeBano, S. J. and Moore, A. J. (1997). The influence of environmental quality on sexual selection in *Nauphoeta cinerea* (Dictyoptera: Blaberidae). *Behavioral Ecology* **8**: 46–53.

Cornette, R., Farine, J-P., Quennedey, B. and Brossut, R. (2001). Molecular characterization of a new adult male putative calycin specific to tergal aphrodisiac secretion in the cockroach *Leucophaea maderae*. *FEBS Letters* **507**: 313–317.

(2002). Molecular characterization of Lma-p54, a new epicuticular surface protein in the cockroach *Leucophaea maderae* (Dictyoptera, Oxyhaloinae). *Insect Biochemistry and Molecular Biology* **32**: 1635–1642.

Cornwell, P. B. (1968). *The Cockroach: A Laboratory Insect and an Industrial Pest*, vol. 1. London: Rentokil Library.

Day, A. C. and Whiting, M. C. (1964). The structure of the sex-attractant of the American cockroach. *Proceedings of the Chemical Society of London* **1964**: 368.
(1966). On the structure of the sex attractant of the American cockroach. *Journal of the Chemical Society C, Organic* **4**: 464–467.
de Bruyne, M., Foster, K. and Carlson, J. R. (2001). Odor coding in the *Drosophila* antenna. *Neuron* **30**: 537–552.
Dusham, E. H. (1918). The dorsal pygidial glands of the female cockroach, *Blattella germanica. Canadian Entomologist* **50**: 278–280.
Engelmann, F. (1970). *The Physiology of Insect Reproduction*. Oxford: Pergamon Press.
Evans, L. D. and Stay, B. (1995). Regulation of competence for milk production in *Diploptera punctata*: interaction between mating, ovaries and the corpus allatum. *Invertebrate Reproduction and Development* **28**: 161–170.
Ewing, L. S. (1967). Fighting and death from stress in a cockroach. *Science* **155**: 1035–1036.
(1973). Territoriality and the influence of females on the spacing of males in the cockroach *Nauphoeta cinerea. Behaviour* **45**: 282–303.
Fan, Y., Chase, J., Sevala, V. and Schal, C. (2002). Lipophorin-facilitated hydrocarbon uptake by oocytes in the German cockroach, *Blattella germanica* (L.). *Journal of Experimental Biology* **205**: 781–790.
Fan, Y., Zurek, L., Dykstra, M. J. and Schal, C. (2003). Hydrocarbon synthesis by enzymatically dissociated oenocytes of the abdominal integument of the German Cockroach, *Blattella germanica. Naturwissenschaften* **90**: 121–126.
Farine, J.-P., Le Quere, J.-L., Duffy, J., Semon, E. and Brossut, R. (1993). 4-hydroxy-5-methyl-3(2*H*)-furanone and 4-hydroxy-2,5-dimethyl-3(2*H*)-furanone, two components of the male sex pheromone of *Eurycotis floridana* (Walker) (Insecta, Blattidae, Polyzosteriinae). *Bioscience Biotechnology and Biochemistry* **57**: 2026–2030.
Farine, J.-P., Le Quere, J.-L., Duffy, J., Everaerts, C. and Brossut, R. (1994). Male sex pheromone of cockroach *Eurycotis floridana* (Walker) (Blattidae, Polyzosteriinae), role and composition of tergites 2 and 8 secretions. *Journal of Chemical Ecology* **20**: 2291–2306.
Farine, J.-P., Everaerts, C., Abed, D., Ntari, M. and Brossut, R. (1996). Pheromonal emission during the mating behavior of *Eurycotis floridana* (Walker) (Dictyoptera: Blattidae). *Journal of Insect Behavior* **9**: 197–213.
Franklin, R., Bell, W. J. and Jander, R. (1981). Rotational locomotion by the cockroach *Blattella germanica. Journal of Insect Physiology* **27**: 249–255.
Fuchs, von M. E. A., Franke, S. and Francke, W. (1985). Carboxylic acids in the feces of *Blattella germanica* (L.) and their possible role as part of the aggregation pheromone. *Journal of Applied Entomology* **99**: 499–503.
Fukui, M. and Takahashi, S. (1983). Studies on the mating behavior of the cockroach, *Nauphoeta cinerea* (Olivier) (Dictyoptera: Blaberidae) III. Isolation and identification of intermale recognition pheromone. *Applied Entomology and Zoology* **18**: 351–356.
(1999). Characterization of the nymph recognition pheromone of a cockroach, *Nauphoeta cinerea* (Olivier) (Dictyoptera: Blaberidae), that depresses wing-raising activity in adult males. *Applied Entomology and Zoology* **34**: 39–47.
Gadot, M., Burns, E. and Schal, C. (1989). Juvenile hormone biosynthesis and oocyte development in adult female *Blattella germanica*: effects of grouping and mating. *Archives of Insect Biochemistry and Physiology* **11**: 189–200.
Gautier, J. Y., Deleporte, P. and Rivault, C. (1988). Relationships between ecology and social behavior in cockroaches. In *The Ecology of Social Behavior*, ed. C. N. Slobodchikoff, pp. 335–351. New York: Academic Press.

Gemeno, C., Leal, W. S., Mori, K. and Schal, C. (2003a). Behavioral and electrophysiological responses of the brownbanded cockroach, *Supella longipalpa*, to stereoisomers of its sex pheromone, supellapyrone. *Journal of Chemical Ecology* **29**: 1169–1783.

Gemeno, C., Snook, K., Benda, N. and Schal, C. (2003b). Behavioral and electrophysiological evidence for volatile sex pheromones in *Parcoblatta* wood cockroaches. *Journal of Chemical Ecology* **29**: 37–54.

Grandcolas, P. (1996). The phylogeny of cockroach families: a cladistic appraisal of morpho-anatomical data. *Canadian Journal of Zoology* **74**: 508–527.

Grillou, H. (1973). A study of sexual receptivity in *Blabera craniifer* Burm. (Blattaria). *Journal of Insect Physiology* **19**: 173–193.

Gu, X., Quilici, D., Juarez, P., Blomquist, G. J. and Schal, C. (1995). Biosynthesis of hydrocarbons and contact sex pheromone and their transport by lipophorin in females of the German cockroach (*Blattella germanica*). *Journal of Insect Physiology* **41**: 257–267.

Guthrie, D. M. and Tindall, A. R. (1968). *The Biology of the Cockroach*. New York: St Martin's Press.

Hales, R. A. and Breed, M. D. (1983). Female calling and reproductive behavior in the brown banded cockroach, *Supella longipalpa* (F.) (Orthoptera: Blattellidae). *Annals of the Entomological Society of America* **76**: 239–241.

Hanula, J. L., Lipscomb, D., Franzreb, K. E. and Loeb, S. C. (2000). Diet of nestling red-cockaded woodpeckers at three locations. *Journal of Field Ornithology* **71**: 126–134.

Hartman, H. B. and Roth, L. M. (1967). Stridulation by a cockroach during courtship behavior. *Nature* **25**: 1243–1244.

Hauptmann, H., Mühlbauer, G. and Sass, H. (1986). Identifizierung und synthese von periplanon A. *Tetrahedron Letters* **27**: 6189–6192.

Hawkins, W. A. (1978). Effects of sex pheromone on locomotion in the male American cockroach, *Periplaneta americana*. *Journal of Chemical Ecology* **4**: 149–160.

Hawkins, W. A. and Gorton, R. E., Jr (1982). Sex pheromone-induced chemolocation in the male American cockroach, *Periplaneta americana*. *Journal of Chemical Ecology* **8**: 219–231.

Hawkins, W. A. and Rust, M. K. (1977). Factors influencing male sexual response in the American cockroach *Periplaneta americana*. *Journal of Chemical Ecology* **3**: 85–99.

Ho, H.-Y., Yang, H.-T., Kou, R. and Chow, Y.-S. (1992). Sex pheromone of the brown cockroach, *Periplaneta brunnea* Burmeister. I. Isolation. *Bulletin of the Institute of Zoology Academia Sinica* **31**: 225–230.

Hösl, M. (1990). Pheromone-sensitive neurons in the deutocerebrum of *Periplaneta americana*: receptive fields on the antenna. *Journal of Comparative Physiology A* **167**: 321–327.

Ignell, R., Anton. S. and Hansson, B. S. (2001). The antennal lobe of orthoptera: anatomy and evolution. *Brain Behavior and Evolution* **57**: 1–17.

Ishii, S. (1972). Sex discrimination by males of the German cockroach, *Blattella germanica* (L.) (Orthoptera: Blattidae). *Applied Entomology and Zoology* **7**: 226–233.

Jacobson, M. and Beroza, M. (1965). American cockroach sex attractant. *Science* **147**: 748–749.

Jacobson, M., Beroza, M. and Yamamoto, R. T. (1963). Isolation and identification of the sex attractant of the American cockroach. *Science* **139**: 48–49.

Juarez, P., Chase, J. and Blomquist, G. J. (1992). A microsomal fatty acid synthetase from the integument of *Blattella germanica* synthesizes methyl-branched fatty acids, precursors to hydrocarbon and contact sex pheromone. *Archives of Biochemistry and Biophysics* **293**: 333–341.

Jurenka, R. A., Schal, C., Burns, E., Chase, J. and Blomquist, G. J. (1989). Structural correlation between cuticular hydrocarbons and female contact sex pheromone of German cockroach *Blattella germanica* (L.). *Journal of Chemical Ecology* **15**: 939–949.

Kaissling, K.-E. and Priesner, E. (1970). Die Riechschwelle des seidenspinners. *Naturwissenschaften* **57**: 23–28.

Kambhampati, S. (1995). A phylogeny of cockroaches and related insects based on DNA-sequence of mitochondrial ribosomal RNA genes. *Proceedings of the National Academy of Sciences, USA* **92**: 2017–2020.

(1996). Phylogenetic relationship among cockroach families inferred from mitochondrial 12S rRNA gene sequence. *Systematic Entomology* **21**: 89–98.

Kitahara, T., Mori, M., Koseki, K. and Mori, K. (1986). Total synthesis of (−)-periplanone-B, the sex pheromone of the American cockroach. *Tetrahedron Letters* **27**: 1343–1346.

Kitamura, C. and Takahashi, S. (1973). The mating behavior and evidence for a sex stimulant of the Japanese cockroach, *Periplaneta japonica* Karny (Orthoptera: Blattidae). *Kontyû* **41**: 383–388.

(1976). Isolation procedure of the sex pheromone of the American cockroach, *Periplaneta americana* L. *Applied Entomology and Zoology* **11**: 373–375.

Kitamura, C., Takahashi, S., Tahara, S. and Mizutani, J. (1976). A sex stimulant to the male American cockroach in plants. *Agricultural Biology and Chemistry* **40**: 1965–1969.

Klass, K.-D. (1997). The external male genitalia and the phylogeny of Blattaria and Mantoidea. *Bonner Zoologische Monographien* **42**: 1–341.

(1998). The ovipositor of Dictyoptera (Insecta): homology and ground-plan of the main elements. *Zoologischer Anzeiger* **236**: 69–101.

Korchi, A., Farine, J.-P. and Brossut, R. (1998). Characterization of two male-specific polypeptides in the tergal glands secretions of the cockroach *Leucophaea maderae* (Dictyoptera, Blaberidae). *Insect Biochemistry and Molecular Biology* **28**: 113–120.

Korchi, A., Brossut, R., Bouhin, H. and Delachambre, J. (1999). cDNA cloning of an adult male putative lipocalin specific to tergal gland aphrodisiac secretion in an insect (*Leucophaea maderae*). *FEBS Letters* **449**: 125–128.

Kuwahara, S. and Mori, K. (1990). Synthesis of both the enantiomers of Hauptmann's periplanone-A and clarification of the structure of Persoon's periplanone-A. *Tetrahedron* **46**: 8083–8092.

Laurent, G., Stopfer, M., Friedrich, R. W., Rabinovich, M. I., Volkovskii, A. and Abarbanel, H. D. I. (2001). Odor encoding as an active, dynamical process: experiments, computation, and theory. *Annual Reviews of Neuroscience* **24**: 263–297.

Leal, W. S., Shi, X., Liang, D., Schal, C. and Meinwald, J. (1995). Application of chiral gas chromatography with electroantennographic detection to the determination of the stereochemistry of a cockroach sex pheromone. *Proceedings of the National Academy of Sciences, USA* **92**: 1033–1037.

Liang, D. and Schal, C. (1990a). Circadian rhythmicity and development of the behavioural response to sex pheromone in male brown-banded cockroaches, *Supella longipalpa*. *Physiological Entomology* **15**: 355–361.

(1990b). Effects of pheromone concentration and photoperiod on the behavioral response sequence to sex pheromone in the male brown-banded cockroach, *Supella longipalpa*. *Journal of Insect Behavior* **3**: 211–223.

(1993a). Volatile sex pheromone in the female German cockroach. *Experientia* **49**: 324–328.

(1993b). Calling behavior of the female German cockroach, *Blattella germanica* (Dictyoptera: Blattellidae). *Journal of Insect Behavior* **6**: 603–614.

(1993c). Ultrastructure and maturation of a sex pheromone gland in the female German cockroach, *Blattella germanica*. *Tissue and Cell* **25**: 763–776.

(1994). Neural and hormonal regulation of calling behavior in *Blattella germanica* females. *Journal of Insect Physiology* **40**: 251–258.

Liang, D., Zhang, A., Kopanic, R. J., Jr, Roelofs, W. L. and Schal, C. S. (1998). Field and laboratory evaluation of female sex pheromone for the detection, monitoring, and management of brownbanded cockroaches (Dictyoptera: Blattellidae). *Journal of Economic Entomology* **91**: 480–485.

Macdonald, T. L., Delahunty, C. M. and Sawyer, J. S. (1987). Synthesis of periplanone-A: a sex pheromone of *Periplaneta americana*. *Heterocycles* **25**: 305–313.

Manabe, S. and Nishino, C. (1983). Sex pheromonal activity of (+)-bornyl acetate and related compounds to the American cockroach. *Journal of Chemical Ecology* **9**: 433–448.

Manabe, S., Takayanagi, H. and Nishino, C. (1983). Structural significance of the geminal-dimethyl group of (+)-*trans*-verbenyl acetate, sex pheromone mimic of the American cockroach. *Journal of Chemical Ecology* **9**: 533–549.

Manabe, S., Nishino, C. and Matsushita, K. (1985). Studies on relationship between activity and electron density on carbonyl oxygen in sex pheromone mimics of the American cockroach, part XI. *Journal of Chemical Ecology* **11**: 1275–1287.

McCluskey, R., Wright, C. G. and Yamamoto, R. T. (1969). Effect of starvation on the responses of male American cockroaches to sex and food stimuli. *Journal of Economic Entomology* **62**: 1465–1468.

McKittrick, F. A. (1964). Evolutionary studies of cockroaches. *Cornell University Agricultural Experimental Station Memoirs* **389**: 1–197.

Moore, A. J. (1988). Female preferences, male social status, and sexual selection in *Nauphoeta cinerea*. *Animal Behaviour* **36**: 303–305.

(1989). Sexual selection in *Nauphoeta cinerea*: inherited mating preference? *Behavior Genetics* **19**: 717–724.

(1990a). The inheritance of social dominance, mating behaviour and attractiveness to mates in male *Nauphoeta cinerea*. *Animal Behaviour* **39**: 388–397.

(1990b). Sexual selection and the genetics of pheromonally mediated social behavior in *Nauphoeta cinerea* (Dictyoptera: Blaberidae). *Entomologia Generalis* **15**: 133–147.

(1994). Genetic evidence for the "good genes" process of sexual selection. *Behavioral Ecology and Sociobiology* **35**: 235–241.

Moore, A. J. and Moore, P. J. (1988). Female strategy during mate choice: threshold assessment. *Evolution* **42**: 387–391.

(1999). Balancing sexual selection through opposing mate choice and male competition. *Proceedings of the Royal Society of London, Series B* **266**: 711–716.

Moore, A. J., Reagan, N. L. and Haynes, K. F. (1995). Conditional signaling strategies: effects of ontogeny, social experience and social status on the pheromonal signal of male cockroaches. *Animal Behaviour* **50**:, 191–202.

Moore, J. K. and Barth, R. H. (1976). Studies on the site of sex pheromone production in the cockroach, *Byrsotria fumigata*. *Annals of the Entomological Society of America* **69**: 911–916.

Moore, P. J., Reagan-Wallin, N. L., Haynes, K. F. and Moore, A. J. (1997). Odour conveys status on cockroaches. *Nature* **389**: 25.
Mori, K. and Takeuchi, Y. (1994). Synthesis of (2*R*,4*R*)-supellapyrone, the sex pheromone of the brownbanded cockroach. *Proceedings of the Japan Academy, Series B, Physical and Biological Sciences* **70**: 143–145.
(1995). Synthesis of (2*R**,4*R**)-supellapyrone, the sex pheromone of the brownbanded cockroach. *Natural Product Letters* **5**: 275–280.
Mori, K., Suguro, T. and Masuda, S. (1978). Stereocontrolled synthesis of all of the four possible stereoisomers of 3,11-dimethyl-2-nonacosanone, the female sex pheromone of the German cockroach. *Tetrahedron Letters* **37**: 3447–3450.
Mukha, D. V., Wiegmann, B. M. and Schal, C. (2002). Evolution and phylogenetic information content of the ribosomal DNA repeat unit in the Blattodea (Insecta). *Insect Biochemistry and Molecular Biology* **32**: 951–960.
Nalepa, C. and Bandi, C. (1999). Phylogenetic status, distribution, and biogeography of *Cryptocercus* (Dictyoptera : Cryptocercidae). *Annals of the Entomological Society of America* **92**: 292–302.
Nicolaou, K. C. and Sorensen, E. J. (1996). *Classics in Total Synthesis: Targets, Strategies, Methods*. New York: Weinheim.
Nishida, R. and Fukami, H. (1983). Female sex pheromone of the German cockroach, *Blattella germanica. Memoirs of the College of Agriculture of the Kyoto University* **122**: 1–24.
Nishida, R., Fukami, H. and Ishii, S. (1974). Sex pheromone of the German cockroach (*Blattella germanica* L.) responsible for male wing-raising: 3,11-dimethyl-2-nonacosanone. *Experientia* **30**: 978–979.
(1975). Female sex pheromone of the German cockroach, *Blattella germanica* (L.) (Orthoptera, Blattellidae), responsible for male wing-raising (I). *Applied Entomology and Zoology* **10**: 10–18.
Nishida, R., Sato, T., Kuwahara, Y., Fukami, H. and Ishii, S. (1976). Female sex pheromone of the German cockroach, *Blattella germanica* (L.) (Orthoptera: Blatellidae), responsible for male wing-raising. II. 29-Hydroxy-3,11-dimethyl-2-nonacosanone. *Journal of Chemical Ecology* **2**: 449–455.
Nishida, R., Kuwahara, Y., Fukami, H. and Ishii, S. (1979). Female sex pheromone of the German cockroach, *Blattella germanica* (L.) (Orthoptera: Blattellidae), responsible for male wing-raising: IV. The absolute configuration of the pheromone, 3,11-dimethyl-2-nonacosanone. *Journal of Chemical Ecology* **5**: 289–297.
Nishii, Y., Watanabe, K., Yoshida, T., Okayama, T., Takahashi, S. and Tanabe, Y. (1997). Total synthesis of (−)-periplanones C and D. Their pheromonal activities against three *Periplaneta* species. *Tetrahedron* **53**: 7209–7218.
Nishino, C. and Kimura, R. (1981). Isolation of sex pheromone mimic of the American cockroach by monitoring with male/female ratio in electroantennogram. *Journal of Insect Physiology* **27**: 305–311.
Nishino, C. and Kuwahara, K. (1983). Threshold dose values for sex pheromones of the American cockroach in electroantennogram and behavioral responses. *Comparative Biochemistry and Physiology A* **74**: 909–914.
Nishino, C. and Manabe, S. (1983). Olfactory receptor systems for sex pheromone mimics in the American cockroach, *Periplaneta americana* L. *Experientia* **39**: 1340–1342.
(1985a). Behavioral and electroantennogram responses of male *Periplaneta australasiae* to sex pheromones of conspecies and *Periplaneta americana*. *Journal of Pesticide Science* **10**: 721–726.

(1985b). Application of the differential saturation electroantennogram method for characterizing sex pheromone of *Periplaneta brunnea* (the brown cockroach). *Comparative Biochemistry and Physiology A* **82**: 775–780.

Nishino, C. and Usui, K. (1985). Olfactory sensitivity of *Periplaneta* cockroaches to functional group and molecular size of general odors. *Comparative Biochemistry and Physiology A* **81**: 43–47.

Nishino, C. and Washio, H. (1976). Electroantennograms of the American cockroach (Orthoptera: Blattidae) to odorous straight chain compounds. *Applied Entomology and Zoology* **11**: 222–228.

Nishino, C., Tobin, T. R. and Bowers, W. S. (1977a). Sex pheromone mimics of the American cockroach (Orthoptera: Blattidae) in monoterpenoids. *Applied Entomology and Zoology* **12**: 287–290.

(1977b). Electroantennogram responses of the American cockroach to germacrene D sex pheromone mimic. *Journal of Insect Physiology* **23**: 415–419.

Nishino, C., Washio, H., Tsuzuki, K., Bowers, W. S. and Tobin, T. R. (1977c). Electroantennogram responses to a stimulant, T-cadinol, in the American cockroach. *Agricultural and Biological Chemistry* **41**: 405–406.

Nishino, C., Kimura, R. and Takayanagi, H. (1980). External appearance of *Periplaneta brunnea* antennae and their electroantennogram responses to odorous compounds. *Agricultural and Biological Chemistry* **44**: 1461–1467.

Nishino, C., Takayanagi, H. and Manabe, S. (1982). Comparison of sex pheromonal activity on the American cockroach between acetates and propionates of verbenyl type alcohols. *Agricultural and Biological Chemistry* **46**: 2781–2785.

Nishino, C., Manabe, S., Kuwabara, K., Kimura, R. and Takayanagi, H. (1983). Isolation of sex pheromones of the American cockroach by monitoring with electroantennogram responses. *Insect Biochemistry* **13**: 65–70.

Nishino, C., Kobayashi, K., Fukushima, M., Imanari, M., Nojima, K. and Kohno, S. (1988). Structure and receptor participation of periplanone A, the sex pheromone of the American cockroach. *Chemical Letters* **3**: 517–520.

Noirot, C. and Quennedey, A. (1974). Fine structure of insect epidermal glands. *Annual Review of Entomology* **19**: 61–80.

Nojima, S., Sakuma, M. and Kuwahara, Y. (1996). Polyethylene glycol film method: a test for feeding stimulants of the German cockroach, *Blattella germanica* (L.) (Dictyoptera: Blattellidae). *Applied Entomology and Zoology* **31**: 537–546.

Nojima, S., Nishida, R., Kuwahara, Y. and Sakuma, M. (1999a). Nuptial feeding stimulants: a male courtship pheromone of the German cockroach, *Blattella germanica* (L.) (Dictyoptera: Blattellidae). *Naturwissenschaften* **86**: 193–196.

Nojima, S., Sakuma, M., Nishida, R. and Kuwahara, Y. (1999b). A glandular gift in the German cockroach, *Blattella germanica* (L.) (Dictyoptera: Blattellidae): the courtship feeding of a female on secretions from male tergal glands. *Journal of Insect Behavior* **12**: 627–640.

Okada, K., Mori, M., Kuwahara, S. *et al.* (1990a). Behavioral and electroantennogram responses of male American cockroaches to periplanones and their analogs. *Agricultural and Biological Chemistry* **54**: 575–576.

Okada, K., Mori, M., Shimazaki, K. and Chuman, T. (1990b). Behavioral responses of male *Periplaneta americana* L. to female sex pheromone components, periplanone-A and periplanone-B. *Journal of Chemical Ecology* **16**: 2605–2614.

Persoons, C. J., Ritter, F. J. and Lichtendonk, W. J. (1974). Sex pheromones of the American cockroach, *Periplaneta americana*. Isolation and partial identification of two excitants. *Proceedings of the Koninklijke Nederlandse Akademie van Weteschappen, Series C, Biological and Medical Sciences* **77**: 201–204.

Persoons, C. J., Verwiel, P. E. J., Ritter, F. J., Talman, E., Nooijen, P. J. F. and Nooijen, W. J. (1976). Sex pheromones of the American cockroach, *Periplaneta americana*: a tentative structure of periplanone-B. *Tetrahedron Letters* **24**: 2055–2058.

Persoons, C. J., Verwiel, P. E. J., Talman, E. and Ritter, F. J. (1979). Sex pheromone of the American cockroach, *Periplaneta americana*: isolation and structure elucidation of periplanone-B. *Journal of Chemical Ecology* **5**: 221–236.

Persoons, C. J., Verwiel, P. E. J., Ritter, F. J. and Nooyen, W. J. (1982). Studies on sex pheromone of American cockroach, with emphasis on structure elucidation of periplanone-A. *Journal of Chemical Ecology* **8**: 439–451.

Persoons, C. J., Ritter, F. J., Verwiel, P. E. J., Hauptmann, H. and Mori, K. (1990). Nomenclature of American cockroach sex pheromones. *Tetrahedron Letters* **31**: 1747–1750.

Picimbon, J. F. and Leal, W. S. (1999). Olfactory soluble proteins of cockroaches. *Insect Biochemistry and Molecular Biology* **29**: 973–978.

Porcheron, P. (1975). Histological and cytological study of the tergal gland of *Leucophaea maderae* and of its development during sexual maturation of the male. *Archives d'Anatomie Microscopique et de Morphologie Experimentale* **64**: 157–181.

Prillinger, L. (1981). Postembryonic development of the antennal lobes in *Periplaneta americana* L. *Cell and Tissue Research* **215**: 563–575.

Ramaswamy, S. B. and Gupta, A. P. (1981). Sensilla of the antennae and the labial and maxillary palps of *Blattella germanica* (L.) (Dictyoptera: Blattellidae): their classification and distribution. *Journal of Morphology* **168**: 269–279.

Ramaswamy, S. B., Gupta, A. P. and Fowler, H. G. (1980). External ultrastructure and function of the 'spiculum copulatus' (SC) of the German cockroach, *Blattella germanica* (L.) (Dictyoptera: Blattellidae). *Journal of Experimental Zoology* **214**: 287–292.

Roth, L. M. (1952). The tergal gland of the male cockroach, *Supella supellectilium*. *Journal of Morphology* **91**: 469–477.

(1968). Reproduction in some poorly known species of Blattaria. *Annals of the Entomological Society of America* **61**: 571–579.

(1969). The evolution of male tergal glands in the Blattaria. *Annals of the Entomological Society of America* **62**: 176–208.

(1970). Evolution and taxonomic significance of reproduction in Blattaria. *Annual Review of Entomology* **15**: 75–96.

Roth, L. M. and Barth, R. H., Jr. (1967). The sense organs employed by cockroaches in mating behavior. *Behaviour* **28**: 58–94.

Roth, L. M. and Dateo, G. P. (1966). A sex pheromone produced by males of the cockroach *Nauphoeta cinerea*. *Journal of Insect Physiology* **12**: 255–265.

Roth, L. M. and Stay, B. (1962). Oocyte development in *Blattella germanica* and *Blattella vaga* (Blattaria). *Annals of the Entomological Society of America* **55**: 633–642.

Roth, L. M. and Willis, E. R. (1952). A study of cockroach behavior. *American Midland Naturalist* **47**: 66–129.

(1954). The reproduction of cockroaches. *Smithsonian Miscellaneous Collections* **122**: 1–49.

(1960). The biotic associations of cockroaches. *Smithsonian Miscellaneous Collections* **141**: 1–439.

Rust, M. K. (1976). Quantitative analysis of male responses released by female sex pheromone in *Periplaneta americana*. *Animal Behaviour* **24**: 681–685.

Rust, M. K. and Bell, W. J. (1976). Chemo-anemotaxis: a behavioral response to sex pheromone in nonflying insects. *Proceedings of the National Academy of Sciences, USA* **73**: 2524–2526.

Rust, M. K., Owens, J. M. and Reierson, D. A. (1995). *Understanding and Controlling the German Cockroach*. New York: Oxford University Press.
Salecker, I. and Boeckh, J. (1995). Embryonic development of the antennal lobes of a hemimetabolous insect, the cockroach *Periplaneta americana*: light and electron microscopic observations. *Journal of Comparative Neurology* **352**: 33–54.
(1996). Influence of receptor axons on the formation of olfactory glomeruli in a hemimetabolous insect, the cockroach *Periplaneta americana*. *Journal of Comparative Neurology* **370**: 262–279.
Sass, H. (1976). Sensory encoding of odour stimuli in *Periplaneta americana*. *Journal of Comparative Physiology A* **107**: 49–65.
(1978). Olfactory receptors on the antenna of *Periplaneta*: response constellations that encode food odors. *Journal of Comparative Physiology A* **128**: 227–233.
(1983). Production, release and effectiveness of two female sex pheromone components of *Periplaneta americana*. *Journal of Comparative Physiology A* **152**: 309–317.
Sato, T., Nishida, R., Kuwahara, Y., Fukami, H. and Ishii, S. (1976). Synthesis of female sex pheromone analogues of the German cockroach and their biological activity. *Agricultural and Biological Chemistry* **40**: 1407–1410.
Schafer, R. (1977a). The nature and development of sex attractant specificity in cockroaches of the genus *Periplaneta*. III. Normal intra- and interspecific behavioral responses and responses of insects with juvenile hormone-altered antennae. *Journal of Experimental Zoology* **199**: 73–84.
(1977b). The nature and development of sex attractant specificity in cockroaches of the genus *Periplaneta*. IV. Electrophysiological study of attractant specificity and its determination by juvenile hormone. *Journal of Experimental Zoology* **199**: 189–208.
Schafer, R. and Sanchez, T. V. (1976). The nature and development of sex attractant specificity in cockroaches of the genus *Periplaneta*. II. Juvenile hormone regulates sexual dimorphism in the distribution of antennal olfactory receptors. *Journal of Experimental Zoology* **198**: 323–336.
Schal, C. (1982). Intraspecific vertical stratification as a mate-finding mechanism in tropical cockroaches. *Science* **215**: 1405–1407.
Schal, C. and Bell, W. J. (1982). Ecological correlates of paternal investment of urates in a tropical cockroach. *Science* **218**: 170–173.
(1983). Determinants of dominant–subordinate interactions in males of the cockroach *Nauphoeta cinerea*. *Biology and Behavior* **8**: 117–139.
(1985). Calling behavior in female cockroaches (Dictyoptera, Blattaria). *Journal of the Kansas Entomological Society* **58**: 261–268.
(1986). Vertical community structure and resource utilization in neotropical forest cockroaches. *Ecological Entomology* **11**: 411–423.
Schal, C. and Chiang, A.-S. (1995). Hormonal control of sexual receptivity in cockroaches. *Experientia* **51**: 994–998.
Schal, C. and Smith, A. F. (1990). Neuroendocrine regulation of pheromone production in cockroaches. In *Cockroaches as Models for Neurobiology: Applications in Biomedical Research*, eds. I. Huber, E. P. Masler and B. R. Rao, pp. 179–200. Boca Raton, FL: CRC Press.
Schal, C., Fan, Y. and Blomquist, G. J. (2003). Regulation of pheromone biosynthesis, transport, and emission in cockroaches. In *Insect Pheromones: Biochemistry and Molecular Biology*, eds. G. J. Blomquist and R. Vogt, pp. 283–322. New York: Academic Press.

Schal, C., Tobin, T. R., Surber, J. L. *et al.* (1983). Search strategy of sex pheromone-stimulated male German cockroaches. *Journal of Insect Physiology* **29**: 575–579.
Schal, C., Gautier, J.-Y. and Bell, W. J. (1984). Behavioural ecology of cockroaches. *Biological Reviews* **59**: 209–254.
Schal, C., Burns, E. L., Jurenka, R. A. and Blomquist, G. J. (1990a). A new component of the female sex pheromone of *Blattella germanica* (L.) (Dictyoptera: Blattellidae) and interaction with other pheromone components. *Journal of Chemical Ecology* **16**: 1997–2008.
Schal, C., Burns, E. L. and Blomquist, G. J. (1990b). Endocrine regulation of female contact sex pheromone production in the German cockroach, *Blattella germanica. Physiological Entomology* **15**: 81–91.
Schal, C., Burns, E. L., Gadot, M., Chase, J. and Blomquist, G. J. (1991). Biochemistry and regulation of pheromone production in *Blattella germanica* (L.) (Dictyoptera, Blattellidae). *Insect Biochemistry* **21**: 73–79.
Schal, C., Liang, D., Hazarika, L. K., Charlton, R. E. and Roelofs, W. L. (1992). Site of pheromone production in female *Supella longipalpa* (Dictyoptera: Blattellidae): behavioral, electrophysiological, and morphological evidence. *Annals of the Entomological Society of America* **85**: 605–611.
Schal, C., Gu, X., Burns, E. L. and Blomquist, G. J. (1994). Patterns of biosynthesis and accumulation of hydrocarbons and contact sex pheromone in the female German cockroach, *Blattella germanica. Archives of Insect Biochemistry and Physiology* **25**: 375–391.
Schal, C., Liang, D. and Blomquist, G. J. (1996). Neural and endocrine control of pheromone production and release in cockroaches. In *Insect Pheromone Research: New Directions*, eds. R. T. Cardé and A. K. Minks, pp. 3–20. New York: Chapman & Hall.
Schal, C., Holbrook, G. L., Bachmann, J. A. S. and Sevala, V. L. (1997). Reproductive biology of the German cockroach, *Blattella germanica*: juvenile hormone as a pleiotropic master regulator. *Archives of Insect Biochemistry and Physiology* **35**: 405–426.
Schal, C., Sevala, V. L., Young, H. P. and Bachmann, J. A. S. (1998). Sites of synthesis and transport pathways of insect hydrocarbons: cuticle and ovary as target tissues. *American Zoologist* **38**: 382–393.
Schaller, D. (1978). Antennal sensory system of *Periplaneta americana* L. Distribution and frequency of morphologic types of sensilla and their sex-specific changes during postembryonic development. *Cell and Tissue Research* **191**: 121–139.
Schaller-Selzer, L. (1984). Physiology and morphology of the larval sexual pheromone-sensitive neurones in the olfactory lobe of the cockroach, *Periplaneta americana. Journal of Insect Physiology* **30**: 537–546.
Scharrer, B. (1987). Insects as models in neuroendocrine research. *Annual Review of Entomology* **32**: 1–16.
Scherkenbeck, J., Nentwig, G., Justus, K. *et al.* (1999). Aggregation agents in German cockroach *Blattella germanica*: examination of efficacy. *Journal of Chemical Ecology* **25**: 1105–1119.
Schneiderman, A. M. and Hildebrand, J. G. (1985). Sexually dimorphic development of the insect olfactory pathway. *Trends in Neurosciences* **8**: 494–499.
Seelinger, G. (1984). Sex-specific activity patterns in *Periplaneta americana* and their relation to mate-finding. *Journal of Comparative Ethology* **65**: 309–326.
(1985a). Behavioural responses to female sex pheromone components in *Periplaneta americana. Animal Behaviour* **33**: 591–598.

(1985b). Interspecific attractivity of female sex pheromone components of *Periplaneta americana*. *Journal of Chemical Ecology* **11**: 137–148.
Seelinger, G. and Gagel, S. (1985). On the function of sex pheromone components in *Periplaneta americana*: improved odour source localization with periplanone-A. *Physiological Entomology* **10**: 221–234.
Seelinger, G. and Schuderer, B. (1985). Release of male courtship display in *Periplaneta americana*: evidence for female contact sex pheromone. *Animal Behaviour* **33**: 599–607.
Selzer, R. (1981). The processing of a complex food odor by antennal olfactory receptors of *Periplaneta americana*. *Journal of Comparative Physiology A* **144**: 509–519.
(1984). On the specificities of antennal olfactory receptor cells of *Periplaneta americana*. *Chemical Senses* **8**: 375–395.
Sevala, V. L., Bachmann, J. A. S. and Schal, C. (1997). Lipophorin: a hemolymph juvenile hormone binding protein in the German cockroach, *Blattella germanica*. *Insect Biochemistry and Molecular Biology* **27**: 663–670.
Sevala, V., Shu, S., Ramaswamy, S. B. and Schal, C. (1999). Lipophorin of female *Blattella germanica* (L.): characterization and relation to hemolymph titers of juvenile hormone and hydrocarbons. *Journal of Insect Physiology* **45**: 431–441.
Seybold, S. J. (1993). Role of chirality in olfactory-directed behavior: aggregation of pine engraver beetles in the genus *Ips* (Coleoptera: Scolytidae). *Journal of Chemical Ecology* **19**: 1809–1831.
Shizuri, Y., Yamaguchi, S., Terada, Y. and Yamamura, S. (1987a). Biomimetic reaction of germacrene-D epoxides in connection with periplanone A. *Tetrahedron Letters* **28**: 1791–1794.
Shizuri, Y., Yamaguchi, S., Yamamura, S. *et al.* (1987b). The synthesis of a tricyclic hydroazulenone from exo-epoxygermacrene-D in connection with periplanone A. *Tetrahedron Letters* **28**: 3831–3834.
Shizuri, Y., Yamaguchi, S., Terada, Y. and Yamamura, S. (1987c). What is the correct structure for periplanone A? *Tetrahedron Letters* **28**: 1795–1798.
Silverman, J. M. (1977). Patterns of response to sex pheromone by young and mature adult male cockroaches, *Periplaneta americana*. *Journal of Insect Physiology* **23**: 1015–1019.
Simon, D. and Barth, R. H. (1977a). Sexual behavior in the cockroach genera *Periplaneta* and *Blatta*. II. Sex pheromones and behavioral responses. *Journal of Comparative Ethology* **44**: 162–177.
(1977b). Sexual behavior in the cockroach genera *Periplaneta* and *Blatta*. I. Descriptive aspects. *Journal of Comparative Ethology* **44**: 80–107.
(1977c). Sexual behavior in the cockroach genera *Periplaneta* and *Blatta*. III. Aggression and sexual behavior. *Journal of Comparative Ethology* **44**: 305–322.
(1977d). Sexual behavior in the cockroach genera *Periplaneta* and *Blatta*. IV. Interspecific interactions. *Journal of Comparative Ethology* **45**: 85–103.
Sirugue, D., Bonnard, O., Le Quere, J. L., Farine, J.-P. and Brossut, R. (1992). 2-Methylthiazolidine and 4-ethylguaiacol, male sex pheromone components of the cockroach *Nauphoeta cinerea* (Dictyoptera, Blaberidae): a reinvestigation. *Journal of Chemical Ecology* **18**: 2261–2276.
Smith, A. F. and Schal, C. (1990a). Corpus allatum control of sex pheromone production and calling in the female brown-banded cockroach, *Supella longipalpa* (F.) (Dictyoptera: Blattellidae). *Journal of Insect Physiology* **36**: 251–257.
(1990b). The physiological basis for the termination of pheromone-releasing behaviour in the female brown-banded cockroach, *Supella longipalpa* (F.) (Dictyoptera: Blattellidae). *Journal of Insect Physiology* **36**: 369–373.

(1991). Circadian calling behavior of the adult female brown-banded cockroach, *Supella longipalpa* (F.) (Dictyoptera: Blattellidae). *Journal of Insect Behavior* **4**: 1–14.

Smith, A. F., Yagi, K., Tobe, S. S. and Schal, C. (1989). *In vitro* juvenile hormone biosynthesis in adult virgin and mated female brown-banded cockroaches, *Supella longipalpa*. *Journal of Insect Physiology* **35**: 781–785.

Smith, S. K. and Breed, M. D. (1982). Olfactory cues in discrimination among individuals in dominance hierarchies in the cockroach, *Nauphoeta cinerea*. *Physiological Entomology* **7**: 337–341.

Sreng, L. (1979a). Ultrastructure and chemistry of the tergal gland secretion of the male of *Blattella germanica* (L.) (Dictyoptera: Blattelidae). *International Journal of Insect Morphology and Embryology* **8**: 213–227.

(1979b). Pheromones and sexual behaviour in *Nauphoeta cinerea* (Olivier) (Insecta, Dictyoptera). *Comptes Rendus Hebdomadaires des Seances de L'Academie des Sciences, Serie D* **289**: 687–690.

(1984). Morphology of the sternal and tergal glands producing the sexual pheromones and the aphrodisiacs among the cockroaches of the subfamily Oxyhaloinae. *Journal of Morphology* **182**: 279–294.

(1985). Ultrastructure of the glands producing sex pheromones of the male *Nauphoeta cinerea* (Insecta, Dictyoptera). *Zoomorphology* **105**: 133–142.

(1990). Seducin, male sex pheromone of the cockroach *Nauphoeta cinerea*: isolation, identification, and bioassay. *Journal of Chemical Ecology* **16**: 2899–2912.

(1993). Cockroach mating behaviors, sex pheromones, and abdominal glands (Dictyoptera: Blaberidae). *Journal of Insect Behavior* **6**: 715–735.

(1998). Apoptosis-inducing brain factors in maturation of an insect sex pheromone gland during differentiation. *Differentiation* **63**: 53–58.

Sreng, L. and Quennedey, A. (1976). Role of a temporary ciliary structure in the morphogenesis of insect glands. An electron microscope study of the tergal glands of male *Blattella germanica* L. (Dictyoptera, Blattellidae). *Journal of Ultrastructural Research* **56**: 78–95.

Sreng, L., Leoncini, I. and Clement, J.-L. (1999). Regulation of sex pheromone production in the male *Nauphoeta cinerea* cockroach: role of brain extracts, corpora allata (CA), and juvenile hormone (JH). *Archives of Insect Biochemistry and Physiology* **40**: 165–172.

Stay, B. (2000). A review of the role of neurosecretion in the control of juvenile hormone synthesis: a tribute to Berta Scharrer. *Insect Biochemistry and Molecular Biology* **30**: 653–662.

Still, W. C. (1979). (±)-Periplanone-B. Total synthesis and structure of the sex excitant pheromone of the American cockroach. *Journal of the American Chemical Society* **101**: 2493–2495.

Stinson, S. C. (1979). Scientists synthesize roach sex attractant. *Chemical and Engineering News* **30**: 24–26.

Stürckow, B. and Bodenstein, W. G. (1966). Location of the sex pheromone in the American cockroach, *Periplaneta americana* (L.). *Experientia* **22**: 851–853.

Tahara, S., Yoshida, M., Mizutani, J., Kitamura, C. and Takahashi, S. (1975). A sex stimulant to the male American cockroach in the composite plants. *Agricultural and Biological Chemistry* **39**: 1517–1518.

Takahashi, S. and Fukui, M. (1983). Studies on the mating behavior of the cockroach, *Nauphoeta cinerea* (Olivier) (Dictyoptera: Blaberidae) IV. Synthesis and biological activity of nauphoetin and related compounds. *Applied Entomology and Zoology* **18**: 357–360.

Takahashi, S. and Kitamura, C. (1972). Bioassay procedure of the sex stimulant of the American cockroach, *Periplaneta americana* (L.) (Orthoptera: Blattidae). *Applied Entomology and Zoology* **7**: 133–141.

Takahashi, S., Kitamura, C. and Waku, Y. (1976). Site of the sex pheromone production in the American cockroach, *Periplaneta americana* L. *Applied Entomology and Zoology* **11**: 215–221.

Takahashi, S., Takegawa, H., Takabayashi, J., Abdullah, M., Fatimah, A. S. and Mohamed, M. (1988a). Sex pheromone activity of synthetic periplanone-B in male cockroaches of genera *Periplaneta* and *Blatta*. *Journal of Pesticide Science* **13**: 125–127.

Takahashi, S., Takegawa, H., Takahashi, T. and Doi, T. (1988b). Sex pheromone activity of synthetic (±)-periplanone-A and (±)-epiperiplanone-A to males of *Periplaneta* and *Blatta*. *Journal of Pesticide Science* **13**: 501–503.

Takahashi, S., Watanabe, K., Saito, S. and Nomura, Y. (1995). Isolation and biological activity of the sex pheromone of the smoky brown cockroach, *Periplaneta fuliginosa* Serville (Dictyoptera: Blattidae). *Applied Entomology and Zoology* **30**: 357–360.

Takegawa, H. and Takahashi, S. (1989). Sex pheromone of the Japanese cockroach, *Periplaneta japonica* Karny (Dictyoptera: Blattidae). *Applied Entomology and Zoology* **24**: 435–440.

Talman, E., Verwiel, P. E. J., Ritter, F. J. and Persoons, C. J. (1978). Sex pheromones of the American cockroach, *Periplaneta americana*. *Israel Journal of Chemistry* **17**: 227–235.

Tillman, J. A., Seybold, S. J., Jurenka, R. A. and Blomquist, G. J. (1999). Insect pheromones: an overview of biosynthesis and endocrine regulation. *Insect Biochemistry and Molecular Biology* **29**: 481–514.

Tobin, T. R. (1981). Pheromone orientation: role of internal control mechanisms. *Science* **214**: 1147–1149.

Tobin, T. R., Seelinger, G. and Bell, W. J. (1981). Behavioral responses of male *Periplaneta americana* to periplanone B, a synthetic component of the female sex pheromone. *Journal of Chemical Ecology* **7**: 969–979.

Tokro, P. G., Brossut, R. and Sreng, L. (1993). Studies on the sex pheromone of female *Blattella germanica* L. *Insect Science and its Application* **14**: 115–126.

Tsai, C.-W. and Lee, H.-J. (1997). Volatile pheromone detection and calling behavior exhibition: secondary mate-finding strategy of the German cockroach, *Blattella germanica* (L.). *Zoological Studies* **36**: 325–332.

Vilaplana, L., Maestro, J. L., Piulachs, M. D. and Bellés, X. (1999). Determination of allatostatin levels in relation to the gonadotropic cycle in the female of *Blattella germanica* (L.) (Dictyoptera, Blattellidae). *Physiological Entomology* **24**: 213–219.

Volkov, Y. P., Poleshchuk, V. D., Zharov, V. G. and Vashkov, V. I. (1967). Investigation of the sexual attracting substance of female *Blattella germanica* L. *Medical Parazitology i Parazitology Bolezni* **36**: 45–48.

Waldow, U. and Sass, H. (1984). The attractivity of the female sex pheromone of *Periplaneta americana* and its components for conspecific males and males of *Periplaneta australasiae* in the field. *Journal of Chemical Ecology* **10**: 997–1006.

Warthen, J. D., Jr, Uebel, E. C., Lusby, W. R. and Adler, V. E. (1983). Investigation of a sex pheromone for the oriental cockroach, *Blatta orientalis*. *Journal of Insect Physiology* **29**: 605–609.

Washio, H. and Nishino, C. (1976). Electroantennogram responses to the sex pheromone and other odours in the American cockroach. *Journal of Insect Physiology* **22**: 735–741.

Washio, H., Nishino, C. and Bowers, W. S. (1976). Antennal receptor response to sex pheromone mimics in the American cockroach. *Nature* **262**: 487–489.

Wendler, G. and Vlatten, R. (1993). The influence of aggregation pheromone on walking behaviour of cockroach males (*Blattella germanica* L.). *Journal of Insect Physiology* **39**: 1041–1050.

Wharton, D. R. A., Miller, G. L. and Wharton, M. L. (1954). The odorous attractant of the American cockroach, *Periplaneta americana* (L.). I. Quantitative aspects of the response to the attractant. *Journal of General Physiology* **37**: 461–469.

Wharton, D. R. A., Black, E. D., Merritt, C., Jr, Wharton, M. L., Bazinet, M. and Walsh, J. T. (1962). Isolation of the sex attractant of the American cockroach. *Science* **28**: 1062–1063.

Wharton, D. R. A., Black, E. D. and Merritt, C., Jr (1963). Sex attractant of the American cockroach. *Science* **142**: 1257–1258.

Wharton, M. L. and Wharton, D. R. A. (1957). The production of sex attractant substance and of oöthecae by the normal and irradiated American cockroach, *Periplaneta americana* L. *Journal of Insect Physiology* **1**: 229–239.

Wileyto, E. P. and Boush, G. M. (1983). Attraction of the German cockroach, *Blattella germanica* (Orthoptera: Blatellidae), to some volatile food components. *Journal of Economic Entomology* **76**: 752–756.

Yamamoto, R. (1963). Collection of the sex attractant from female American cockroaches. *Journal of Economic Entomology* **56**: 119–120.

Yang, H.-T., Wang, C. H., Kou, R. and Chow, Y. S. (1992). Electroantennogram responses of synthetic periplanone-A and periplanone-B in the American cockroach. *Journal of Chemical Ecology* **18**: 371–378.

Yang, H.-T., Chow, Y.-S., Peng, W.-K. and Hsu, E.-L. (1998). Evidence for the site of female sex pheromone production in *Periplaneta americana*. *Journal of Chemical Ecology* **24**: 1831–1843.

Zhukovskaya, M. I. (1995). Circadian rhythm of sex pheromone perception in the male American cockroach, *Periplaneta americana* L. *Journal of Insect Physiology* **41**: 941–946.

7

A quest for alkaloids: the curious relationship between tiger moths and plants containing pyrrolizidine alkaloids

William E. Conner

Department of Biology, Wake Forest University Winston-Salem, USA

Susan J. Weller

J. F. Bell Museum of Natural History, University of Minnesota St Paul, USA

Introduction

A curious relationship exists between a group of plants, the pyrrolizidine alkaloids they contain, and tiger moths of the family Arctiidae. Tiger moths possess an impressive array of chemicals, either produced *de novo* or sequestered from plants, that protect them to a greater or lesser degree from predators and parasites. These chemicals include cyossin (Teas *et al.*, 1966), biogenic amines (Bisset *et al.*, 1959, 1960; Rothschild and Aplin, 1971), pyrazines (Rothschild *et al.*, 1984), polyphenolics (Hesbacher *et al.*, 1995), iridoid glycosides (Bowers and Stamp, 1997), and cardenolides (Rothschild *et al.*, 1970, 1973; Wink and von Nickisch-Rosenegk, 1997); however, no group of compounds, it seems, has influenced the natural history and behavior of tiger moths as the pyrrolizidine alkaloids (PAs) have done (Weller *et al.*, 2000a). Several excellent reviews have been written about these compounds from the perspective of their chemistry, the plants that produce them (Bull *et al.*; 1968; Mattocks, 1986; Hartmann and Witte, 1995), and the insects that utilize them (Schneider, 1986; Boppré, 1990; Hartmann and Ober, 2000), but none has focussed exclusively, on their intimate relationships with tiger moths.

The members of the family Arctiidae, which numbers over 11 000 species, are often brilliantly colored (Watson and Goodger, 1986; Holloway, 1988; Weller *et al.*, 2000a). In addition to standard aposematic red, yellow, or black patterns, adults and larvae may have iridescent blue and green, or even pearly white, coloration. White can be considered aposematic when individuals rest conspicuously on green vegetation. Numerous species are involved in mimicry rings with other distasteful species. Many adults are superb Müllerian mimics of lycid beetles, bees, wasps,

This chapter is dedicated to Professor Thomas Eisner, a remarkable scientist and a pioneer in the field of chemical ecology.

Advances in Insect Chemical Ecology, ed. R. T. Cardé and J. G. Millar. Published by Cambridge University Press.

hemipterans, and even colorful cockroaches (Rothschild, 1961, 1963, 1972a,b,c; Blest, 1964; Rothschild *et al.*, 1973; Simmons and Weller, 2002; Waller *et al.*, 2000b). The mimetic coloration is often supported by a behavior repertoire that mimics their Müllerian models (Blest, 1964; Adams, 1987). Such behaviors include antennal and wing waving behavior reminiscent of wasps and even false stinging behavior. Some tiger moth adults produce a colorful and sometimes sticky froth from dorsal openings at a base of the patagia, located in front of the forewings (Dethier, 1939). The froth provides a memorable visual display and a sample of hemolymph – a distinct gustatory warning – to potential predators. The combination of all of these characteristics is quite effective: arctiids are usually discriminated against in palatability tests with visual predators such as birds (Jones, 1932, 1934; Sargent, 1995).

In addition to visual signals, many adults also emit high-frequency sounds (ultrasound) when they are handled roughly or when they hear the echolocation cries of approaching bats (Blest, 1964; Roeder, 1974; Fullard *et al.*, 1994). Tiger moths use thoracic tymbals (Fullard and Heller, 1990) to produce short trains of intense high-frequency clicks, and bats do not attack clicking moths (Dunning and Roeder, 1965). Some researchers have postulated that these clicks startle the bats or jam the bats' sonar (Fullard *et al.*, 1979, 1994; Stoneman and Fenton, 1988; Miller, 1991; Tougaard *et al.*, 1998), but others have suggested that the clicks are additional aposematic signals (Dunning and Roeder, 1965; Dunning, 1968; Acharya and Fenton, 1992; Dunning *et al.*, 1992; Dunning and Krüger, 1995). Tymbals are a synapomorphy, or shared derived character, for the arctiids (Jacobson and Weller, 2001), indicating that an acoustic component of defense was important in the early evolution of the group. Some tiger moths are also secondarily diurnal, and presumably it is their potent chemical defenses that allow them to exploit the daylight hours and repel the visual predators that hunt diurnally (Rothschild, 1985). Many of the brightest and noisiest arctiids contain high levels of plant-derived compounds including PAs.

PAs are found in many species of the plant families Asteraceae, Boraginaceae, Fabaceae, and Orchidaceae (Bull *et al.*, 1968; Mattocks, 1986), and sporadically in the Apocynaceae, Celastracease, Convolvulaceae, Ranuculaceae, Rhizophoraceae, Santalaceae, and Sapotaceae (Hartmann and Ober, 2000). Pyrrolizidine alkaloids are bitter-tasting, hepatotoxic, and carcinogenic (Bull *et al.*, 1968; Mattocks, 1986). They are presumed to have evolved as defenses against herbivory, although evidence in this regard is surprisingly sparse (Ritchey and McKee, 1941; Bentley *et al.*, 1984; Dreyer *et al.*, 1985; de Boer, 1999; Hägele and Rowell-Rahier, 2000). Each PA (and they number in the hundreds) is an ester composed of a necine amino alcohol derived from a hydroxylated 1-methyl-pyrrolizine and one or two esterifying necic acids attached at C_1 and C_7. Two of the necic acids may be combined to form a macrocyclic ring. Based on their taxonomic distribution and biosynthetic origin, PAs are classified into five structural types: the senecionine, triangularine, monocrotaline,

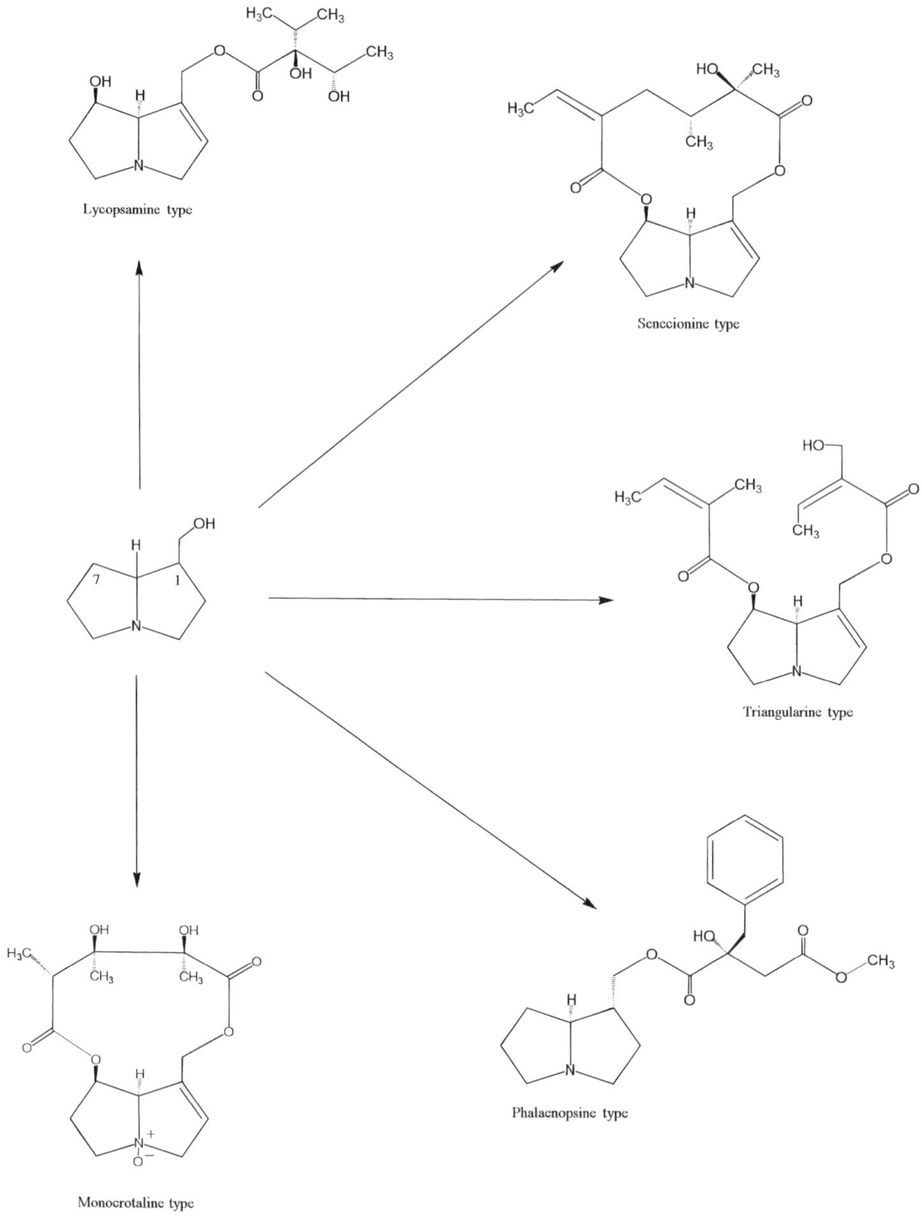

Lycopsamine type

Senecionine type

Triangularine type

Phalaenopsine type

Monocrotaline type

Fig. 7.1. Pyrrolizidine alkaloid diversity. (After Hartmann and Ober, 2000.)

lycopsamine, and phalaenopsine types (Hartmann and Ober, 2000; Fig. 7.1). Most may exist in either a reduced free-base form or in a form in which the nitrogen in the pyrrolizidine ring is oxidized, the *N*-oxide form. The free-base forms of the alkaloids are lipid soluble; the *N*-oxide forms are polar and thus more water soluble. PAs are most often stored and transported in plants and insects as relatively innocuous *N*-oxides. The free-base forms, in contrast, are sometimes called pretoxic compounds (Hartmann and Ober, 2000) because they are frequently bioactivated in vertebrates and insects by the normally detoxifying mixed function oxidases to form highly reactive pyrroles. The pyrroles attack biological nucleophiles, resulting in the cytotoxicity for which the PAs are infamous (Mattock, 1986).

Most PAs are ingested and sequestered intact from the host plants of larval arctiids (larval PA feeders). Others are ingested by the adults from excrescences on the surface of alkaloid-containing plants. These adult PA feeders may or may not sequester PAs as larvae. Last, other PAs are produced only by the insects from PA-related precursors ingested from plants. Herein we concentrate on arctiids that sequester PAs as larvae and adults.

Larval pyrrolizidine alkaloid feeders

The most thoroughly documented larval PA feeder is *Utetheisa ornatrix* (Plate 7.1, p. 252). This callimorphine arctiid has been studied intensively by Professors Thomas Eisner and Jerrold Meinwald and their colleagues. *U. ornatrix* is the quintessential example of a specialist that sequesters its PAs from its larval food. At the Archbold Biological Station in south-central Florida (where much of the work on *U. ornatrix* has been carried out) *U. ornatrix* feeds on the leaves and seeds of *Crotalaria spectabilis* and *Crotalaria mucronata* (Fabaceae), introduced species that inhabit disturbed sandy habitats. *U. ornatrix* once survived on the much smaller native *Crotalaria* spp. (Tietz, 1972). All of these *Crotalaria* spp. are rich in macrocyclic monocrotaline-type PAs (Fig. 7.1). In *C. spectabilis*, the specific compounds are monocrotaline and monocrotaline *N*-oxide; in *C. mucronata* the main alkaloids are usaramine and usaramine *N*-oxide. Larval *U. ornatrix* that are fed either plant species sequester the alkaloids and retain them through the pupal stage and into adulthood. These PAs have been shown to protect *U. ornatrix* from a variety of invertebrate predators (Eisner, 1980; Eisner and Meinwald 1987, 1995a). Larval and adult *U. ornatrix* were found to be unpalatable in feeding bioassays using the ground-dwelling lycosid spider *Lycosa ceratiola* (Eisner and Eisner, 1991). Since *U. ornatrix* frequently wander off their host plants and pupate under leaf litter, ground-dwelling spiders are relevant predators. The adult is also rejected by the golden orb-weaving spider, *Nephila clavipes* (Araneidae), a normally voracious predator, which nonetheless cuts *U. ornatrix* from its web (Eisner, 1980; Eisner and Meinwald, 1987). The discrimating feeding habits of *N. clavipes* are so reliable that they have become a standard and extremely sensitive bioassay for PA-containing insects (Brown, 1984; Vasoconcellos-Neto and Lewinsohn, 1984). In all cases, the unpalatability of *U. ornatrix* can be traced to PAs. Invertebrate predators readily eat *U. ornatrix* rendered PA-free by raising them on semisynthetic pinto bean-based diets, whereas the same predators reject individuals raised on semisynthetic diets with PAs added (Eisner, 1980;

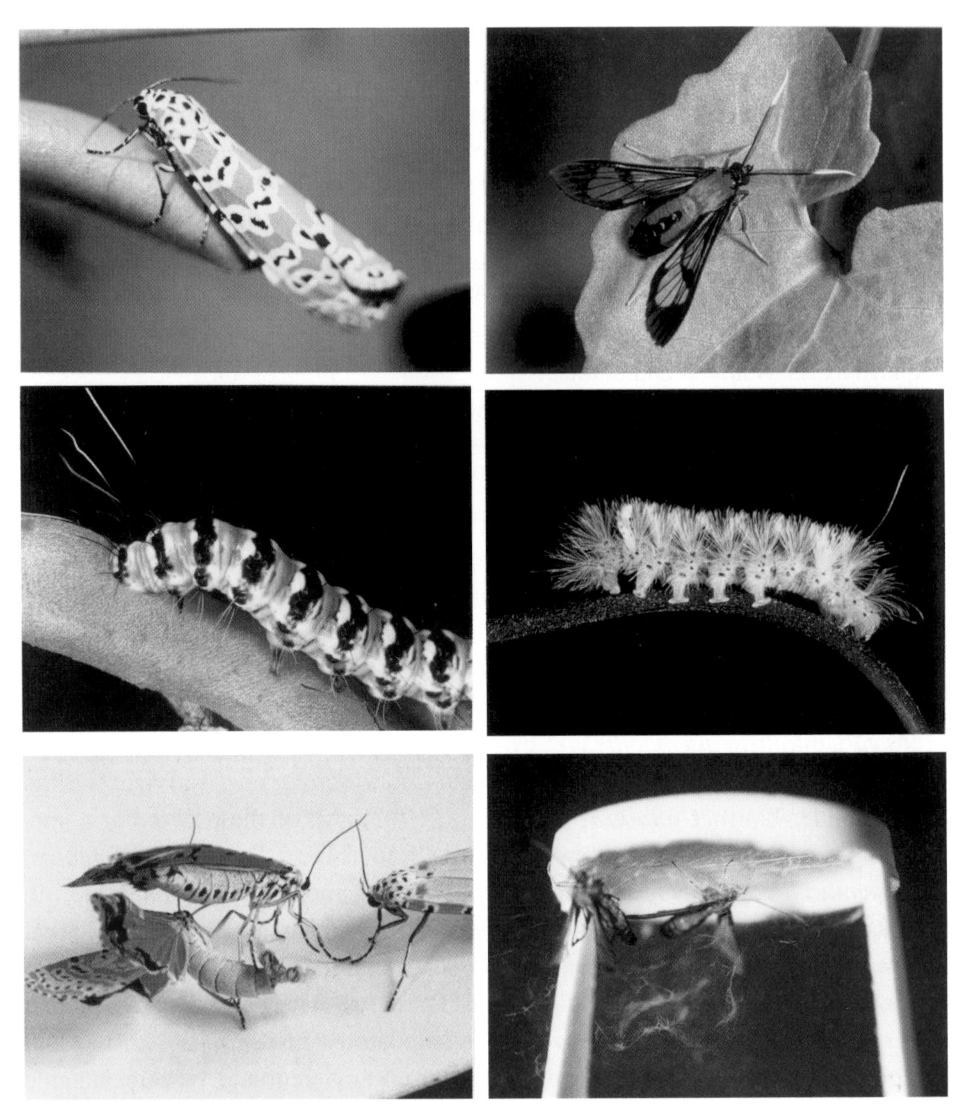

Plate 7.1. Natural history of *Utetheisa ornatrix* (left panel). Adult male (top); larva feeding on seedpod of *Crotalaria mucronata* (middle); adult male courting a female (bottom). The yellow genitalic coremata are partially inflated and thrust toward the female. Natural history of *Cosmosoma myrodora* (right panel). Adult male resting on leaf of the larval host plant *Mikania scandens* (top); larva (middle); male courting female (bottom). The flocculent has been released and forms a cloud around the pair.

Dussourd *et al.*, 1988; Eisner and Eisner, 1991). They also reject food items that have been topically treated with PAs (Eisner, 1980). PAs are effective deterrents against pileated finches *Coryphospingus pileatus* (Cardoso, 1997) and big brown bats *Eptesicus fuscus* (N. Hristov and W. E. Conner, unpublished), presumably because of the chemicals' bitter taste. The larvae of *U. ornatrix* are classically aposematic with black, yellow, and white markings readily visible, and the larvae feed openly on the seedpods of *Crotalaria* spp. (Plate 7.1, p. 252). The warningly colored pink, black, and white adults fly during the day with a nonchalance typical of a chemically protected insect. Interestingly, PAs do not render the moths universally distasteful to predators. Loggerhead shrikes (*Lanius ludovicianus*) accept small numbers of adult *U. ornatrix* as food items and show no ill effects (Yosef *et al.*, 1996). PAs are also not effective taste deterrents to mice (Glendinning *et al.*, 1990).

Similarly, the PAs in *U. ornatrix* do not appear to protect against larval parasitoids. Parasitoid loads can reach levels as high as 20% in field-collected animals. Parasitoids reported include flies (Tachinidae) and wasps (Brachonidae, Chalcididae and Ichneumonidae). Such parasitoids presumably derive protection from developing within a generally unpalatable host. It is not clear whether the parasitoids also benefit by incorporating PAs from within *U. ornatrix*. Two specimens of the tachinid fly, *Archytas aterrimus*, a parasitoid that ecloses from the pupa of *U. ornatrix*, did not contain significant levels of PAs (Iyengar *et al.*, 1999; Rossini *et al.*, 2000). Another tachinid (*Lespesia aletiae*) and an ichneumid (*Corsonus* sp.) contained trace amounts of PAs (Rossini *et al.*, 2000). The effects of PAs on larval arctiid parasitoids have been investigated in two other species. Benn *et al.* (1979) found detectable levels of PAs in the braconid parasitoid *Microplitis* sp., which emerges from the late larval instars of the pericopine *Nyctemera annulata*. Recent studies by Singer (2000) have shown that the extremely polyphagous arctiid *Grammia geneura* can derive protection from generalist larval parasitoids when larvae feed sequentially on host plants with different chemical makeups. The host plants of *G. geneura* include several PA-containing members of the Boraginaceae. Perhaps the PAs can influence levels of parasitoids in this manner.

Throughout the world, *Utetheisa* spp. are specialists that feed on *Crotalaria*, *Heliotropium*, and *Tournefortia* spp., all of which contain PAs. PA defense coupled with aposematic coloration appears to be a common theme for the genus. One set of exceptions occurs in the Galapágos Islands, where the genus has undergone a small radiation. Here, *Utetheisa devriesi, Utetheisa perryi*, and *Utetheisa galapagenesis* have all lost their aposematic coloration in both the adult and larval stages even though their larvae feed on a *Tournefortia* sp. (Roque-Albelo, 2000), which contains significant levels of PAs (Roque-Albelo *et al.*, 2002). Why the Galapágos *Utetheisa* spp. have lost their aposematic coloration is an open question.

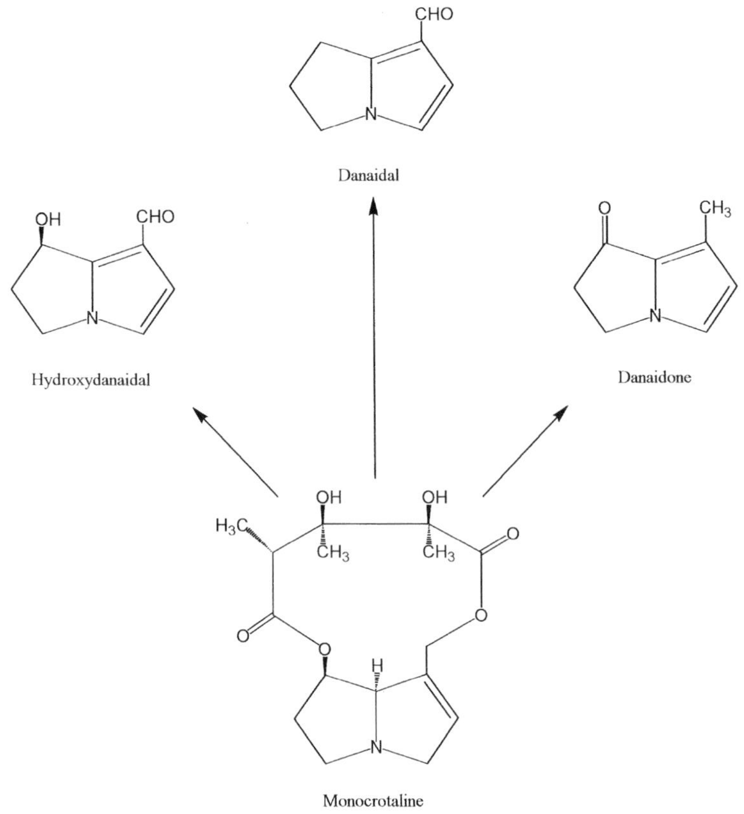

Fig. 7.2. Pheromonal compounds derived from pyrrolizidine alkaloids.

Utetheisa spp. rely on PAs for far more than defense. In 1972, Culvenor and Edgar showed that the male inflatable scent brushes, or coremata, of the Australian species *Utetheisa puchelloides* and *Utetheisa lotrix* contained apparent derivatives of PAs called dihydropyrrolizines (Fig. 7.2). Later the coremata of *U. ornatrix* were found to contain the PA-derived dihydropyrrolizine hydroxydanaidal (Conner *et al.*, 1981). The coremata are exposed briefly and repeatedly during courtship. They are thrust toward the head and antennae of the female. Males without coremata or males deficient in hydroxydanaidal (larvae raised on a PA-free diet) have lower mating success because they cannot elicit genital exposure behavior in females. Hydroxydanaidal is a diet-dependent male courtship pheromone in *U. ornatrix*. This courtship system is a classic example of chemically mediated female choice (Conner *et al.*, 1981; Iyengar *et al.*, 2001).

Why use a volatile form of a defensive compound as a pheromone? It was first suggested that the female would benefit from choosing hydroxydanaidal-laden males because her offspring would also have the genes necessary for sequestering these important defensive materials (Conner *et al.*, 1981): a "good genes" argument. The benefit to the female turns out to be much more interesting and complex than that. When males mate with females they transfer a spermatophore containing sperm and seminal fluid enriched with nutrients and defensive alkaloids. Dussourd *et al.* (1991) found that hydroxydanaidal is an accurate predictor of the amount of alkaloid in the nuptial gift. Male diet can vary considerably in nature, and PA titers vary among individuals (Conner *et al.*, 1990). Hydroxydanaidal acts as a chemical indicator of the male's larval diet and potential alkaloid content of his spermatophore (Dussourd *et al.*, 1991).

The nuptial gift of alkaloid is the key to the system. The alkaloids obtained by the discriminating female contribute to her protection from predators (González *et al.*, 1999), and a sizable portion finds its way quickly and precisely from the spermatophore to her reproductive system. The alkaloids are added to the eggs and supplement the maternal contribution of alkaloids for defense against egg predators such as coccinelid beetles, lacewings, and ants (Dussourd *et al.*, 1988; Hare and Eisner, 1993; Eisner *et al.*, 2000). Although the female contributes most of an egg's defenses, the male's contribution ($\sim 30\%$) proved significant in predation assays. The alkaloid, in other words, is a chemical form of male parental investment, and further investigations in other species have proven the link between chemical defenses and sex to be a potent and common one (Eisner and Meinwald, 1995b).

In addition to his PA status, hydroxydanaidal titers also advertise male size. Since PAs are potent phagostimulants for larval *U. ornatrix* (Bogner and Eisner, 1991), the PA content of a male is proportional to his body mass (Conner *et al.*, 1990). This allows hydroxydanaidal content to be an indicator of PA load and simultaneously male size. Large males provide large, nutritious spermatophores. Females who choose males with a high titer of hydroxydanaidal get the added bonus of a large spermatophores (Dussourd *et al.*, 1991) and the fecundity benefits that accompany them (LaMunyon, 1997). This association raised an interesting question. Do females really choose males based on the strength of their hydroxydanaidal signal or are they really just choosing by measuring the size of the male? Subsequent experiments support both alternatives. When forced to choose between males in which the hydroxydanaidal level and body size have been artificially disassociated, females clearly chose mates with the higher hydroxydanaidal levels (Iyengar *et al.*, 2001). However, females can also exercise choice after copulation. When given a choice between the sperm of two males that differed in size (and spermatophore size), they preferentially utilized the sperm of the male with the larger spermatophore regardless of the relative defensive value of the nuptial gift (LaMunyon and Eisner, 1993, 1994). Female *U. ornatrix* thus discriminate at two levels and the determining factor is different on each level. Precopulatory choice is based on the male's hydroxydanaidal titer and postcopulatory sperm choice is based on spermatophore size. This two-tiered system has an interesting, strategic advantage: the female gets a second postcopulatory chance to verify her precopulatory choice. She can detect potential cheaters and adjust her choice of sperm accordingly.

In addition to the immediate phenotypic benefits of a large PA-laden nuptial gift, females derive genetic benefits as well. Body mass of male *U. ornatrix* has been shown to be heritable. Females that choose large males assure that their male offspring will also be larger and thereby more likely to be chosen as mates (Iyengar and Eisner, 1999a), thus gaining a Fisherian genetic advantage. Offspring of both sexes will also be larger and more fecund (Iyengar and Eisner, 1999b): a "good genes" advantage. The story of *U. ornatrix* has now come full circle. It began with the supposition of a genetic advantage for female choice (genes to acquire PAs) and it has ended with the illustration of a genetic advantage, albeit not the one originally suggested.

The primary selective forces in the evolution and maintenance of male courtship pheromones in the subfamily Arctiinae, as exemplified by *U. ornatrix*, appear to be somewhat unusual when compared with the most important factors in other families of moths. It has been shown that androconia (elongate scent scales emitting pheromones in males) of male moths have most often evolved through Fisherian runaway sexual selection in the context of reproductive isolation (Phelan and Baker, 1987; Birch *et al.*, 1990; Phelan, 1992). Androconia and species-specific pheromone blends arise repeatedly in species pairs that share common host plants (Phelan and Baker, 1987). In Lepidoptera, host plants are often used as mating sites and precise species-specific signaling prevents interbreeding. These differentiated mating signals can then become exaggerated and spread through runaway sexual selection (Phelan and Baker, 1987; Birch *et al.*, 1990; Phelan, 1992). Although Fisherian runaway sexual selection occurs in *U. ornatrix* (Iyengar and Eisner, 1999a), the context is not just species recognition but also includes the additional component of male nuptial gifts. Female preference for PA volatiles is reinforced by the direct defensive benefits conferred by the PA-laden spermatophores. Nuptial PA gifts are probably common in the Arctiinae, based on the widespread use of PA-derived pheromone components in several arctiine lineages (Table 7.1). This argues strongly in favor of the importance of the direct defensive benefits of female mate choice throughout the subfamily.

Larval PA sequestration, similar to that in *U. ornatrix*, is common within the Arctiidae (Table 7.1). Arctiid larvae, as exemplified by *Creatonotus transiens* and *Tyria jacobaeae*, feed on plants with PAs predominantly in the *N*-oxide form. The *N*-oxides are reduced to the lipophilic free bases in the midgut and are passively absorbed (Hartmann and Ober, 2000; but see Wink and Schneider, 1988). The free bases are then reoxidized by a soluble NADPH-dependent mixed function

Table 7.1. *Arctiids with a known association with pyrrolizidine alkaloids (PAs)*[a]

Tribe and species in the subfamily Arctiinae	PA source or feeding behavior	Sequestering life stage	PA-derived pheromone[b]	Pheromone disseminating structure[c]	References
Arctiini (s.l.)					
Amphicallia bellatrix Dalman	*Crotalaria*	Larva	–	–	Rothschild and Aplin, 1971; Rothschild, 1972b,c; Rothschild *et al.*, 1979
Arctia caja L.	Polyphagous	Larva	–	Hair brushes	Bisset *et al.*, 1959, 1960; Rothschild and Aplin, 1971; Rothschild, 1972b,c; Aplin and Rothschild, 1972; Rothschild *et al.*, 1979; von Nickisch-Rosenegk and Wink, 1993
Creatonotus gangis (L.)	Polyphagous	Larva	Hodal	Abdominal (4) coremata	Schneider *et al.*, 1982
Creatonotus transiens (Wlk.)	Polyphagous	Larva	Hodal	Abdominal (4) coremata	Schneider *et al.*, 1982; Bell *et al.*, 1984; Boppré and Schneider, 1985, 1989; Bell and Meinwald, 1986; Wunderer *et al.*, 1986; Wink and Schneider, 1988; Wink *et al.*, 1988; Schmitz *et al.*, 1989; Egalhaaf *et al.*, 1990; Schultz *et al.*, 1993; von Nickisch-Rosenegk and Wink, 1994
Diaphora mendica Clerk	*Senecio*	Larva	–	–	Rothschild, 1972b; Rothschild *et al.*, 1979
Diacrisia sannio (L.)	Polyphagous	Larva	–	–	von Nickisch-Rosenegk and Wink, 1993
Estigmene acrea (Drury)	Polyphagous	Larva	Hodal	Abdominal (2) coremata	Krasnoff and Roelofs, 1989; Davenport and Conner, 2003
Grammia geneura (Strecker)	Polyphagous	Larva	–	–	Singer, 2000

(*cont.*)

Table 7.1. (*cont.*)

Tribe and species in the subfamily Arctiinae	PA source or feeding behavior	Sequestering life stage	PA-derived pheromone[b]	Pheromone disseminating structure[c]	References
Phragmatobia fuliginosa (L.)	Polyphagous	Larva	Danaidal, hodal	Abdominal (4) coremata	Krasnoff *et al.*, 1987; Krasnoff and Roelofs, 1989, 1990; von Nickisch-Rosenegk and Wink, 1993
Pyrrharctia isabella (JE Smith)	Polyphagous	Larva	Hodal	Abdominal (4) coremata	Krasnoff *et al.*, 1987; Krasnoff and Yager, 1988; Krasnoff and Roelofs, 1989, 1990
Scearctia figulina	*Heliotropium*	Larva	–	–	Trigo *et al.*, 1996
Spilosoma lubricipeda (L.)	Polyphagous	Larva	–	Abdominal coremata	Rothschild, 1963; Rothschild, *et al.*,1979; von Nickisch-Rosenegk and Wink, 1993
Spilosoma luteum Hufnagel	Polyphagous	Larva	–	–	Rothschild, 1963; Rothschild *et al.*, 1979
Spilosoma virginica (F.)	Polyphagous	Larva	–	None	Tietz, 1972; Covell, 1984
Callimorphini (s.l.)					
Amerila phaedra	Pharmacophagous	Adult	–	–	Boppré, 1981
14 *Amerila* species	Pharmacophagous	Adult	–	Various	Pinhey, 1975; Holloway, 1988; Boppré, 1990
35 *Amerila* species	Pharmacophagous	Adult	–	Various	Häuser and Boppré, 1997
Argina cribaria Clerk	*Crotalaria*	Larva	–	–	Rothschild *et al.*, 1979
Callimorpha (= *Panaxia*) *dominula*	Polyphagous	Larva	–	–	von Nickisch-Rosenegk and Wink, 1993
Callimorpha (= *Panaxia*) *quadripunctaria* Poda	Polyphagous	Larva	Hodal, danaidal ethyl esters	Abdominal (2)	Schneider *et al.*, 1998

Haploa clymene (Brown)	*Eupatorium* (early), polyphagous (late)	Larva	Hodal	Abdominal (2) coremata	Davidson *et al.*, 1997
Haploa confusa (Lyman)	*Eupatorium* (early), polyphagous (late)	Larva	Hodal	Abdominal (2) coremata	Davidson *et al.*, 1997
Nyctemera annulata Boisduval	*Senecio*	Larva	–	–	Benn *et al.*, 1979
Nyctemera coleta Cramer	Polyphagous Pharmacophagous	Larva Adult	–	–	Hartmann and Witte, 1995
Nyctemera restricta	–	Larva	–	–	Boppré, 1990
Various species	–	Larva	–	–	Holloway, 1988
Pareuchaetes insulata Wlk.	*Ageratum*	–	–	Abdominal (2) coremata	Covell, 1984
Pareuchaetes pseudoinsulata Rego Barros	*Chromolaena Ageratum*	Larva	Hodal	Hair brush	Cock, 1982; Schneider *et al.*, 1992
Tyria jacobaeae (L.)	*Senecio*	Larva	–	Hair brush	Aplin *et al.*, 1968; Aplin and Rothschild, 1972; Rothschild *et al.*, 1979; Ehmke *et al.*, 1990; von Nickisch-Rosenegk and Wink, 1993

(*cont.*)

Table 7.1. (*cont.*)

Tribe and species in the subfamily Arctiinae	PA source or feeding behavior	Sequestering life stage	PA-derived pheromone[b]	Pheromone disseminating structure[c]	References
Utetheisa bella (L.)	*Crotalaria*	Larva	Hodal	Genital coremata	Aplin *et al.*, 1968
Utetheisa galapagensis (Wallengren)	*Tournefortia*	Larva	–	Genital coremata	Roque-Albelo *et al.*, 2002
Utetheisa lotrix Cramer	*Crotalaria*	Larva	Hodal, danaidal	Genital coremata	Culvenor and Edgar, 1972; Holloway, 1988
Utetheisa ornatrix (L.)	*Crotalaria*	Larva	Hodal	Genital coremata	Eisner, 1980; Conner *et al.*, 1981; Eisner and Meinwald, 1987, 1995a,b; Dussourd *et al.*, 1988, 1991; Grant *et al.*, 1989; Conner *et al.*, 1990; Eisner and Eisner, 1991; Hare and Eisner, 1993; LaMunyon and Eisner, 1993, 1995; González *et al.*, 1999; Iyengar and Eisner, 1999a,b; Eisner *et al.*, 2000; Iyengar *et al.*, 2001
Utetheisa pulchella L.	*Crotalaria*	Larva	–	Genital coremata	Holloway, 1988
Utetheisa pulchelloides Hampson	*Heliotropium*	Larva	Hodal	Genital coremata	Rothschild and Aplin, 1971; Culvenor and Edgar, 1972; Rothschild, 1972b

Pericopini					
Dysschema s.l. various species	–	–	–	–	Trigo *et al.*, 1996
Gnophaela latipennis Bois.	*Hackelia*	Larva	–	–	Stretch, 1882; L'Empereur *et al.*, 1989
Gnophaela vermiculata (Grt and Robin)	–	–	–	–	Bruce, 1888; Cockerell, 1889
Hyalurga syma Wlk.	*Heliotropium*	Larva	–	–	Trigo *et al.*, 1993
Hypocrita (= *Eucyane*) *uranicola* Walker	Pharmacophagous	Adult	–	–	Pliske, 1975a
Hypocrita (= *Eucyane*) *excellens* Walker	Pharmacophagous	Adult	–	–	Pliske, 1975a
Hypocrita (= *Eucyane*) *diana* Walker	Pharmacophagous	Adult	–	–	Pliske, 1975a
Phaegopterini					
Halysidota harrisii Walsh	Pharmacophagous	Adult	–	–	Dussourd, 1986
Halysidota tesselaris (Smith)	Pharmacophagous	Adult	–	–	Goss, 1979; Krasnoff and Dussourd, 1989
Leucanopsis longa (Grote)	Pharmacophagous	Adult	–	–	Goss, 1979
Leucanopsis sp.	Pharmacophagous	Adult	–	–	Goss, 1979
36 *Leucanopsis* neotropical spp. Genera include *Baritus, Pachydota, Elysius, Opharus, Calidota, Ischnocampa, Pelochyta, Pseudapistrodea, Halysidota, Thalesa, Agoraea, Psychophasma, Scotura, Pitane, Amastus*	Pharmacophagous	Adult	–	–	Pliske, 1975a,b

(*cont.*)

Table 7.1. (*cont.*)

Tribe and species in the subfamily Arctiinae	PA source or feeding behavior	Sequestering life stage	PA-derived pheromone[b]	Pheromone disseminating structure[c]	References
Ctenuchini					
Artichloris eriphea F.	Pharmacophagous	Adult	–	–	Pliske, 1975b
Belemnia sp.	Pharmacophagous	Adult	–	–	Boppré, 1995
Cisseps fulvicollis (Hbn.)	Pharmacophagous	Adult	Hodal	Abdominal (2) coremata	Beebe and Kenedy, 1957; Pliske, 1975a; Dussourd, 1986; Krasnoff and Dussourd, 1989
Corematura sp.	Pharmacophagous	Adult	–	–	Boppré, 1995
Ctenucha virginica (Esper.)	Pharmacophagous	Adult	Hodal	Abdominal (2) coremata	Beebe and Kenedy, 1957; Krasnoff and Dussourd, 1989
Eucereon carolina Hy – Edwards	Pharmacophagous	Adult	–	Abdominal (2) coremata	Goss, 1979; Covell, 1984
Several *Eucereon* species	Pharmacophagous	Adult	–	Various	Hampson, 1898; Beebe and Kenedy, 1957; Pliske, 1975a; Goss, 1979
Lymire edwardsii (Grt.)	Pharmacophagous	Adult	–	Abdominal (2) coremata	Goss, 1979
95 neotropical species Genera include *Euceron, Episcepsis, Aclytia, Delphyre, Heliura, Trichodesma, Hypocladia, Euagra, Agyrta, Antichloris, Cyanopepla. Teucer, Ceropimorpha, Correbia, Hyaleucerea, Atyphopsis, Napata, Aethria, Corematura, Trichura, Dinia, Argyroeides, Chrysostola*	Pharmacophagous	Adult	–	Various	Pliske, 1975a,b

Euchromiini					
Cosmosoma auge (L.)	Pharmacophagous	Adult	–	–	Pliske, 1975a
Cosmosoma myrodora (Dyar)	Pharmacophagous	Adult	–	Subabdominal pouch	Goss, 1979; Boada, 1997; Conner *et al.*, 2000
Didasys belae Grt.	Pharmacophagous	Adult	–	Subabdominal pouch	M. Conner, personal communication
Euchromia amoena (Moschler)	Pharmacophagous	Adult	–	–	Boppré, 1981
Euchromia formosa (L.)	Pharmacophagous	Adult	–	–	Boppré, 1981
Euchromia sp.	Pharmacophagous	Adult	–	–	Boppré, 1990
44 neotropical *Euchromia* species	Pharmacophagous	Adult	–	Various	Pliske, 1975a,b
Genera include *Cosmosoma, Sarosa, Ichoria, Isanthrene, Sphecosoma, Pseudophex, Myrmecopsis, Syntrichura, Eumenogaster, Saurita, Leucotmemis, Dixophlebia, Loxophlebia, Mistrocneme, Nyridela, Ixylasia, Methysia, Hypocharis, Pompiliodes, Eupyra, Macrocneme, Poliopastea*					

[a]Classification follows Jacobson and Weller (2001).
[b]Hydroxydanaidal is abbreviated hodal.
[c]Number in parentheses indicates number of branches in coremata.

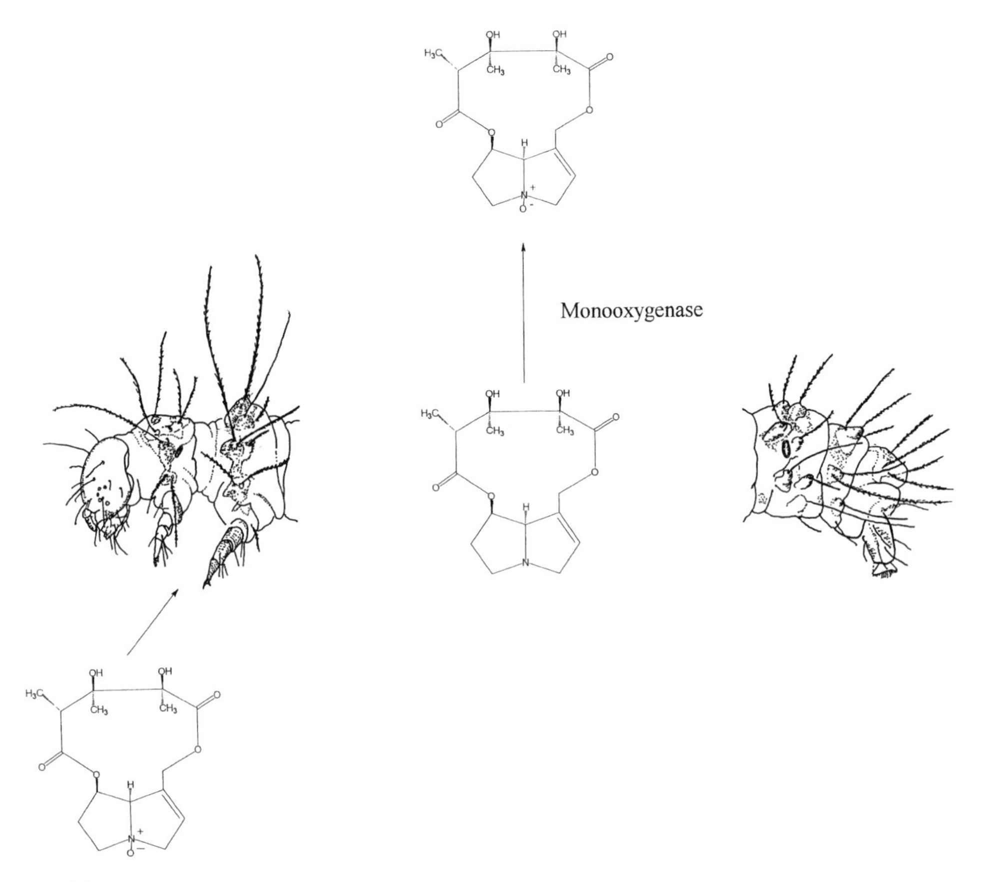

Fig. 7.3. Metabolism of pyrrolizidine alkaloids in arctiid larvae.

monooxygenase (Fig. 7.3). The *N*-oxides are distributed through the caterpillar, with the highest accumulations in the integument (Hartmann and Ober, 2000). Members of several genera, including *Amphicallia*, *Argina*, *Gnophaela*, *Hyalurga*, *Nyctemera*, and *Tyria*, feed exclusively on PA-containing plants, and most species of these genera investigated have been shown to sequester PAs. Others, including *Haploa* spp., feed on PA-containing plants during their early larval life and retain PA derivatives throughout their later life stages. Yet others, such as *Arctia*, *Creatonotus*, *Diacrisia*, *Estigmene*, *Grammia*, *Phragmatobia*, *Pyrrharctia*, and *Spilosoma*, include PA plants in their generalist diets and some sequester the compounds when they are available (Table 7.1). The larvae of at least some of these species have galeal chemoreceptors that respond very strongly to PAs and presumable allow them to identify plants with high levels of the alkaloids (Bernays *et al.*, 2002). When the ability of arctiid larvae to feed on PA plants and to sequester PAs is mapped on a phylogeny of the Arctiidae, it is clear that this ability evolved once within the family at the ancestral node of the subfamily Arctiinae (node 1, Fig. 7.4).

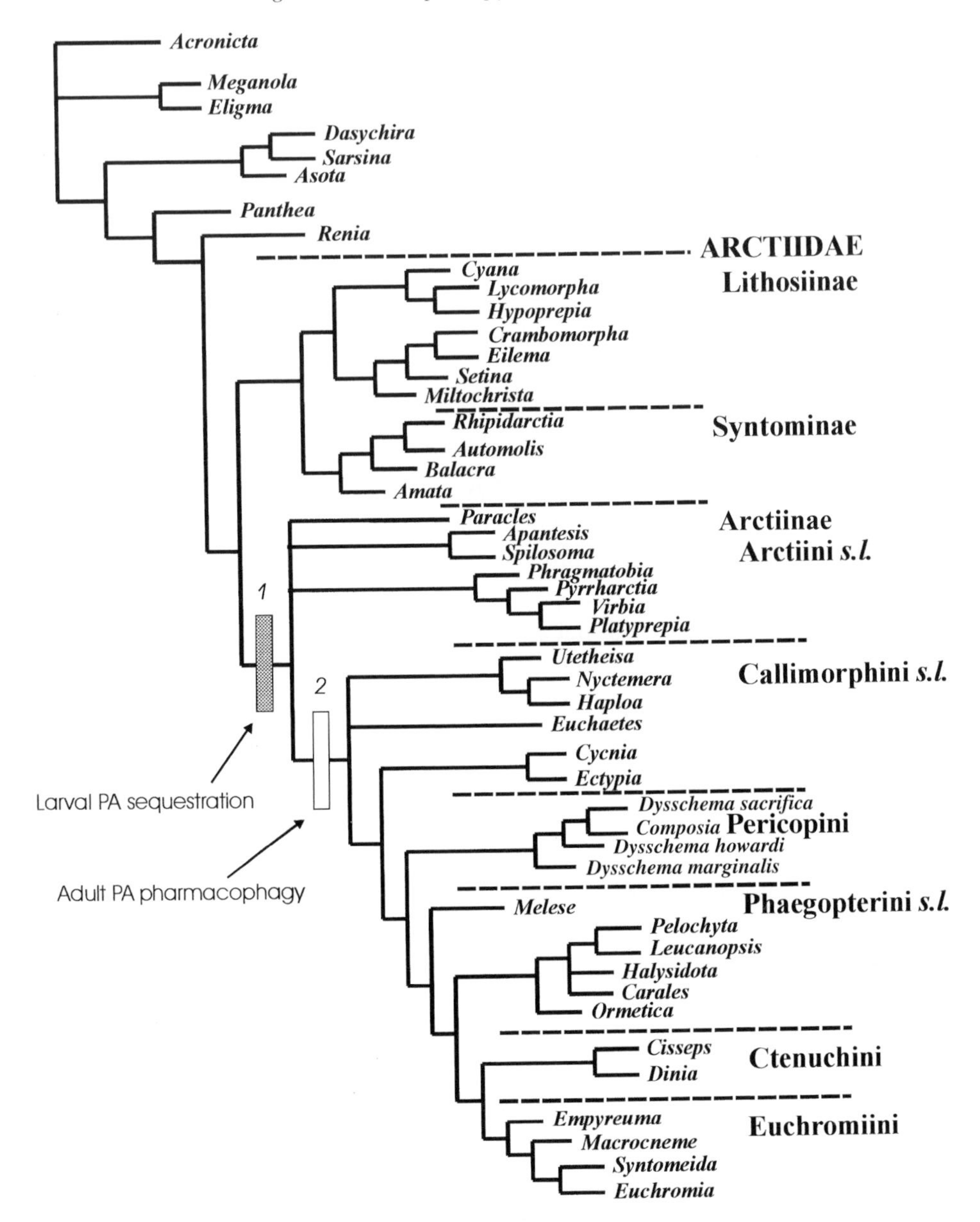

Fig. 7.4. Phylogeny of the Arctiidae (after Jacobson and Weller, 2001) showing the evolutionary interrelationship of larval pyrrolizidine alkaloid (PA) sequestration and adult pharmacophagous behavior. Node 1, larval PA detoxification/sequestration abilities evolve; node 2, adult pharmacophagy originates.

A fundamental change occurred in the biochemistry of the ancestor that allowed ingested PAs to be detoxified, retained, and transported safely throughout the larval body. The acquisition of the PA-specific monooxygenase may have been the critical evolutionary innovation for the Arctiinae that sealed their intimate relationship with PA-containing plants. This specific enzyme allows them simultaneously to detoxify the PAs and render them mobile. Once sequestered, PAs could deter predators.

On numerous occasions arctiines appear to have switched from one group of PA-containing hosts to another or to have secondarily lost their reliance on PA plants in favor of more generalist feeding habits. The latter is most apparent in the tribe Arctiini (Krasnoff and Roelofs, 1990), where many species (e.g., *C. transiens*, *Pyrrharctia isabella*, *Spilosoma lubricipeda*, and *Diacrisia sannio*) feed on a variety of herbaceous plants that sometimes includes PA-containing species. Interestingly, they usually retain the ability to sequester the alkaloids (Wink and Schneider, 1988). A parallel example of larval detoxification abilities being retained through evolutionary time is found in swallowtail butterflies (Miller and Feeney, 1989).

Adult males of many of the species that feed on PA plants have elaborate scent-disseminating structures that are involved in courtship (Birch *et al.*, 1990; Weller *et al.*, 2000a; Jacobson and Weller, 2001; Simmons, 2001). PA-containing host plants, PA-derived defense, PA-derived pheromones, and extraordinary male scent-disseminating structures appear to be inextricably entwined. The scent-disseminating structures, which include hair brushes, abdominal coremata, genital coremata, subabdominal pouches, and dorsal abdominal glands, are clearly not homologous and have evolved repeatedly (Table 7.1; Birch, 1969, 1979; Jacobson, 1994; Häuser and Boppré, 1997). Conversely, the scent-disseminating structures or their behavioral functions are frequently lost in arctiine lineages that abandon PA feeding (Krasnoff and Roelofs, 1989, 1990; Jacobson and Weller, 2001; Simmons, 2001).

The male scent-disseminating structures of arctiids are used in a variety of ways. In most species studied (including *Haploa*, *Cisseps*, *Phragmatobia*, *Pyrrharctia*, and *Amerila* spp.), the structures are displayed briefly in the moments just prior to mating as in *U. ornatrix* (Meyer, 1984; Krasnoff and Roelofs, 1990; Davidson *et al.*, 1997; Häuser and Boppré, 1997). The structures and their PA-derived pheromones have sometimes been shown to mediate female choice (Conner *et al.*, 1981; Davidson *et al.*, 1997), but in other cases, removal of the structures or their PA-derived scents by removing the PAs from the larval diet had no effect on pre-copulatory female choice (Krasnoff and Yager, 1989; Krasnoff and Roelofs, 1990). However, experiments involving the deletion of a single cue are difficult to interpret. Multiple redundant courtship cues may exist, rendering the loss of a single cue inconsequential (Conner, 1987). The discovery of cryptic female choice (Eberhard, 1996) through sperm choice in *U. ornatrix* (LaMunyon and Eisner, 1993, 1994) also complicates the issue and requires that many systems be revisited to check for its relevance.

In contrast to brief male displays, the spectacular coremata of the Asian *Creatonotus* spp. are deployed very differently. In the early evening, males initiate courtship by gathering in "leks" and slowly inflating their quadrifid coremata (Pagden, 1957). Females are attracted to groups of displaying males and mate with one of them. Curiously, the females do not appear to choose among the males, but mate with the first male encountered. Again the emphasis has been placed on precopulatory choice (Wunderer *et al.*, 1986); postcopulatory choice mechanisms remain to be explored. The active principle of the coremata of *C. transiens* and *Creatonotus gangis* is the dihydropyrrolizine hydroxydanaidal (Bell and Meinwald, 1986; Wunderer *et al.*, 1986), which is derived from dietary alkaloids with one spectacular twist. The size of the coremata in the adult is dependent on the amount of PAs ingested by the larva (Schneider *et al.*, 1982; Boppré and Schneider, 1985, 1989; Schmitz *et al.*, 1989). This morphogenetic effect again underlines the importance of PAs in the evolution of courtship in arctiids.

The attraction of females to males in *Creatonotus* spp. may be considered an example of sex role reversal. The attraction of the female to the males is a logical prediction according to the tenets of the sex allocation theory (Andersson, 1994). If the contributions of male *Creatonotus* to the offspring exceed that of the female, then the males should be the choosy sex and the females compete for males. Sex role reversal is the logical evolutionary endpoint of a system in which males compete for females using larger and larger spermatophores, eventually tipping the parental investment scale. A testable prediction would be that the spermatophores of *Creatonotus* spp. would be the largest relative to their body weight for arctiids. A similar mating system was noted in the salt marsh caterpillar moth, *Estigmene acrea*, but this observation has been difficult to replicate. In this case, the "lekking" behavior appears to be facultative and depends on the density of individuals (Willis and Birch, 1982). A facultative system is also a prediction of the sex allocation theory if the ability of males to offer very large spermatophore is dependent on environment conditions (Gwynne, 1993). As in *Creatonotus* spp., the coremata of *E. acrea* express a pronounced PA-dependent polyphenism (Davenport and Conner, 2003), underscoring the importance of PAs to the reproductive biology of this species.

Adult pyrrolizidine alkaloid feeding

Some arctiid larvae do not feed on plants containing PAs. Individuals of these species obtain PAs in a different way. They seek out PA-rich plants as adults and extract the alkaloids from deposits on the plant's surface (Plate 7.2, p. 268). This

Plate 7.2. Pharmacophagous adults attracted to the roots of *Eupatorium capillifolium*: adult male *Cosmosoma myrodora* (top left); female *Halisidota tessellaris* (top right): male *Cisseps fulvicollis* (bottom left); male queen butterfly *Danaus gilippus bernice* (bottom right). This last photograph was taken during mid afternoon. The ring flash blackened the background. All other photos were taken at night.

behavior is well known to lepidopterists. Butterfly and moth collectors in the tropics (Beebe, 1955; Beebe and Kenedy, 1957) have long known about and exploited the remarkable attractive power of plants containing PAs. *Heliotropium indicum*, commonly called "fedegoso," is usually the plant of choice. It is uprooted, sometimes damaged or moistened, and then hung as bait. The foliage and especially the roots of the plant attract Lepidoptera from several families including danaine and ithomiine nymphalid butterflies (Pliske 1975a,b; Brown, 1984; Boppré, 1990). Individuals approach the baits from downwind, land, extend their proboscis, and regurgitate on the plants. The regurgitant dissolves alkaloids on the plant's surface and the extract is then reimbibed. The process can continue for hours, with the insect sometimes becoming engorged and even lethargic (Boppré, 1990). Michael Boppré has defined this extraordinary behavior as a form of pharmacophagy: "Insects are pharmacophagous if they search for certain secondary plant substances directly, take them up, and utilize them for specific purpose other than primary metabolism or (merely) host plant recognition" (Boppré, 1984a, p. 1152).

Arctiids frequently appear on lists of pharmacophagous species (Pliske, 1975a,b; Goss, 1979; Boppré, 1981, 1986, 1990, 1995). The arctiids known to exhibit pharmacophagy are listed in Table 7.1. They include arctiines in the tribes Callimorphini, Percopini, Phaegopterini, Ctenuchini, and Euchromiini. The lithosiine and syntomiine arctiids are conspicuously absent, as are members of the tribe Arctiini of the Arctiinae (Weller *et al.*, 2000a; Table 7.1). Adult PA feeding is practiced primarily by males, although in some species the activity is carried out exclusively by females (e.g., *Halisidota tessellaris*) or more equally by both sexes (e.g., *Eucereon carolina*; Pliske, 1975a; Goss, 1979). The distribution of this behavior among arctiid lineages indicates that larval feeding on PAs (node 1) precedes the collection of PAs by adult Callimorphini, Pericopini, Phaegopterini, Ctenuchini, and Euchromiini (node 2, Fig. 7.4; Weller *et al.*, 2000a). Antennal receptors for the PA derivatives (Bogner and Boppré, 1989; Grant *et al.*, 1989) emitted by larval host plants and used in host plant location may have set the stage for PA collection by adults, as suggested for danaine butterflies (Edgar *et al.*, 1974). Alternatively, nectar feeding could also have led to PA pharmacophagy (Boppré, 1979). There are also multiple larval host plant switches away from PA-containing plants in groups beyond node 2. In many cases, the switch is to plants containing cardenolides, another group of effective defensive chemicals (Weller *et al.*, 2000a).

The actual attractants released from PA-containing plants appear not be the PAs themselves. Monoester and diester PAs have remarkably little potency as arctiid attractants. In contrast, small, volatile PA derivatives containing the bicyclic pyrrolizidine nucleus are potent attractants (Krasnoff and Dussourd, 1989). These include hydroxydanaidal, heliotridine, retronecine, and, perhaps, trachelanthamidine. Krasnoff and Dussourd (1989) have suggested that PA-containing plants release these compounds or similar derivatives in low concentrations and it is these chemicals that attract the moths. This suggestion was supported by the discovery of hydroxydanaidal receptors on the antennae of male *Amerila* (*Rhodagastria*), a highly pharmacophagous genus (Bogner and Boppré, 1989). It is also clear from the trapping experiments of Krasnoff and Dussourd (1989) that the specific chemical used as the attractant can influence the sex ratio of the moths attracted. Paradoxically, they also found that the most attractive compound in their tests was (*S*)-(+)-hydroxydanaidal, an enantiomer not to be expected from the alkaloidal biosynthetic pathways of most PA-containing plants (Hartmann and Ober, 2000). This finding is still unexplained.

The function of adult PA feeding in arctiids is most clearly illustrated by a member of the Euchromini, the scarlet-bodied wasp moth *Cosmosoma myrodora*.

In south central Florida, *C. myrodora* males are common visitors to baits composed of dried *Eupatorium capillifolium* roots (Goss, 1979). They are also attracted to the stems, leaves, and blossoms of *E. capillifolium* and its congener *Eupatorium moryii* (Conner *et al.*, 2000). Most individuals attracted to baits are pristine and likely just emerged. Older individuals, as evidenced by their loss of scales and faded coloration, visit occasionally. The males sequester two small enantiomeric PAs from *E. capillifolium* – lycopsamine and intermedine – and their respective *N*-oxides. The alkaloids imbibed by the male have been shown to be effective feeding deterrents against the orb-weaving spider *N. clavipes* (Boada, 1997; Conner *et al.*, 2000). Prior to imbibing PAs, adult males raised on their normal PA-free host plant, *Mikania scandens* (Compositae), are quite palatable to the spider as are adult (virgin) females, which do not normally visit *E. capillifolium* (Conner *et al.*, 2000).

Males transport a portion of the ingested alkaloid to two ventral pouches at the base of the abdomen. The subabdominal pouches (the ventral valves of Hampson, 1898) are packed with brilliant white cuticular fibers or flocculent. Each individual fiber is a modified scale, which arises from a socket in the lining of the pouch. The flocculent is sticky yet breaks away easily and is readily carried on air currents. The flocculent is a rich source of alkaloids, containing concentrations up to 20 times the average found in other body tissues (Conner *et al.*, 2000). During close-range courtship, flocculent is released in bursts, enveloping and adhering to the female. The embellished female is instantaneously endowed with a protective covering of alkaloids. In feeding bioassays, such females are cut from the web by *N. clavipes*, and the spiders immediately groom any flocculent from their mouthparts. The alkaloidal shield provided by the male is a unique form of paternal investment. Unlike female *U. ornatrix*, which must wait for the PAs in the spermatophore to be transferred during the long mating process (up to 12 h), *C. myrodora* females are immediately protected. The copulating male then supplements the female's defense by transferring an alkaloid – and nutrient-laden spermatophore over the next several hours. After mating, the female transfers a portion of the alkaloid to her eggs and reserves the remainder for personal defense (Conner *et al.*, 2000). The PA levels transferred are sufficient to protect the male's offspring from predators (Hare and Eisner, 1993). Surprisingly, preliminary tests (using monocrotaline *N*-oxide as the ingested PA) indicate that *C. myrodora* females show no precopulatory preference for males that have visited PA sources (Boada, 1997). Perhaps males without PA can provide significant benefits to the female in non-alkaloidal spermatophores and a choice based on PA is not warranted. Alternatively, the female may assess the quality of successive mates and exercise postcopulatory sperm choice. This hypothesis remains to be tested.

Male structures to disperse flocculent appear to be limited to the tribes Ctenuchini and Euchromiine; however, the Arctiidae is poorly surveyed for androconia. Non-homologous flocculent pouches may occur on mesothoracic tibial leg segments (e.g., *Orcynia calcarata*), between the forecoxae (*Eumanogaster* spp.), between the 9th tergite and the tegumen (*Pompiliopsis tarsalis*), between the 2nd and 3rd abdominal sternites (single abdominal pouch, *Pseudosphex fassli*), or between the 2nd, 3rd, and 4th, abdominal sternites (double abdominal pouch, e.g., *C. myrodora*) (Weller *et al.*, 2000b; Simmons, 2001). The double and single pouches appear to occur only in the Euchromiini; however, the tribe is doubtfully monophyletic (Simmons and Weller, 2002) and further study is warranted. Flocculent has been presumed to have only a defensive role in some species. Blest (1964) reported that male *Homeocera stricta* (Ctenuchini) release a yellow and pungent smelling flocculent when handled roughly. Adams (1987) described similar defensive behavior in several species. The rough congruence of a list of species with flocculent and a list of pharmacophagous species (Weller *et al.*, 2000a,b) suggests that many of these species will be shown to sequester PAs and transmit them during copulation.

There are also examples of males lacking flocculent that are pharmacophagous for PA as well as female visitors at PA sources. In south-central Florida *E. capillifolium* attracts predominantly males of *Lymire edwardsii, Cisseps fullvicollis, Didasys belae*, predominantly females of *H. tessellaris* and *E. carolina*, and both sexes of *Leucanopsis longa*. In the case of the ctenuchine *L. edwardsii*, Goss (1979) has argued that pharmacophagously acquired PAs are primarily nutritional and function as a nitrogen source. Goss noted that females mated with PA-laden males produce a greater number of eggs. In caged experiments, females did not exercise precopulatory choice between groups of males with and without access to a source of PAs. From these results, Goss (1979) concluded that the PAs are not relevant to the sexual behavior of this species. His arguments were made, however, before it was known that alkaloids are retained essentially intact within most pharmacophagous species (i.e., they are not metabolized as a nitrogen source: Dussourd, 1986; Conner *et al.*, 2000). Further, his study predates the cryptic female choice literature (Eberhard, 1996), and subsequent experiments designed to test for presence of postcopulatory choice in arctiids (LaMunyon and Eisner, 1993, 1994). The fact that PAs are collected by the females or by both sexes in some species indicates that females are not always dependent on males for their defensive compounds.

The most extraordinary aspect of PA visitation is that it has evolved repeatedly in the class Insecta. PA-containing plants are attractive to select butterflies (Nymphalidae), beetles (Chrysomelidae), flies (Chloropidae), and even locusts (Boppré, 1990, 1995). The similarity of the behavior and morphology of arctiids and nymphalid (dainaine and ithomiine) butterflies is uncanny. For example, male queen butterflies, *Danaus gilippus bernice*, visit PA-containing plants including *E. capillifolium* (Pliske and Eisner, 1969). From the imbided alkaloids, they produce the PA-derived courtship pheromone danaidone. This pheromone is released from hairpencils that are exposed during aerial courtship. Danaidone-endowed hairpencils ensure mating success. The hairpencils even have breakaway particles that are transferred to the female during courtship, in a fashion similar to the flocculent transfer in *C. myrodora*. PAs transferred from the male to the female queen butterfly find their way into the eggs in substantial quantities (Dussourd *et al.*, 1989). This transfer is not an isolated example. Studies by Brown (1984) indicated the presence of PAs in the male reproductive tract of several ithomiine species, and Boppré (1984b) has documented a diversity of breakaway particles in several danaines. The intimate relationships between the insects, their PAs, courtship behavior, and defense are clearly the result of convergence on an effective method of antipredatory defense (Wink and von Nickisch-Rosenegk, 1997).

Why pyrrolizidine alkaloids?

What is it about PAs that makes them the center of so much insect attention? One key to the importance of the PAs appears to be what Hartmann and Ober (2000) have termed the "two faces" of the PAs: the facility with which they can be converted from the pretoxic free-base form to the non-toxic *N*-oxide form. In the larval arctiid gut, plant PAs, which occur predominantly in the *N*-oxide form, are reduced to the free base, which is then absorbed through the wall of the midgut and into the hemolymph. The critical enzyme for arctiids that sequester PAs is the mixed function monooxygenase, which reconverts the alkaloid to its non-toxic *N*-oxide form. It can then be conjugated and transported throughout the insect's body with impunity (Fig. 7.3). Non-sequestering insects are incapable of this crucial transformation, and the pretoxic free base is instead bioactivated by microsomal P450 enzymes; the resultant toxic pyrroles are highly reactive with biological nucleophiles and result in cytotoxic affects. The possession of the monooxygenase apparently launched the arctiines on their current evolutionary trajectory, with the PAs playing critical roles in their defense against predators.

Bowers (1992) lists six attributes required for the evolution of unpalatability through sequestration of plant secondary metabolites. They include feeding on a toxin-laden plant, the loss of normal detoxification mechanisms, transfer of the toxins through the wall of the midgut, protection of tissues and organs from the toxins, concentration of defensive compounds in the hemolymph, and retention of the compounds through the pupal and adult stages. The "two faces" of the PAs appear to satisfy several of them. The ability to switch between the forms provides a mechanism for traversing the midgut and circumvents the normal "detoxification/bioactivation" problem, simultaneously producing a form that protects the tissues and organs of the insect from damage. The transfer of PA use from the larval to the adult stage is a key evolutionary event. During pupal metamorphosis, polyploid larval tissues are broken down and replaced by the growing diploid imaginal (adult) cells. Many lepidopteran species are chemically defended as larvae but not as adults. Bowers (1993) has suggested that sequestered larval toxins could poison the adult tissues when liberated during the pupal stage unless specialized adaptations have evolved. The monooxygenase appears to be just such an adaptation. Many species have thus solved the problem of autotoxicity, allowing metamorphosis and retention of the pretoxic *N*-oxides in the adult stage.

A second factor in the importance of PAs in the natural history of insects is that the compounds can be readily broken down into two small, relatively volatile components. In fact, this may happen spontaneously on the surface of some plants, hence their attractiveness in nature (Pliske, 1975a; Krasnoff and Dussourd, 1989). These volatile components probably facilitate the location of potential hosts or nectar sources by adult females. If so, it was then a simple step to involve PAs in courtship. As described above, the necine base portion of the molecules has routinely been transformed into dihydropyrrolizine-based male courtship signals in Arctiidae and Danaiine butterflies. In some butterflies (Danaiinae and Ithomiinae), the necic acid portion of the molecule has been utilized to form male pheromonal lactones (Edgar *et al.*, 1976; Nishida *et al.*, 1996). The suggestion that the pheromonal communication in arctiids evolved through the exploitation of preexisting receptors for plant volatiles is supported by the morphology of the courtship pheromone receptors. The hydroxydanaidal-sensitive neurons on the antennae of female *U. ornatrix* are housed in sensilla basiconica (Grant *et al.*, 1989). These sensilla are ordinarily reserved for the receptors of plant volatiles (Cuperus, 1985). Therefore, the localization of the hydroxydanaidal receptors on these sensilla indicates their original function.

The third factor is that PAs are extremely effective broad-spectrum feeding deterrents. Numerous species spanning 11 plant families have made use of this attribute (Hartmann and Ober, 2000). Arctiids and other PA-pharmocophagous insects have converted antiherbivore defenses to antipredator defenses. We know little about the mechanisms by which PAs affect their unpalatability. It seems unlikely that the long-term cytotoxic and genotoxic effects of PAs are relevant to their fast-acting deterrency. Recent work has indicated that some PAs bind to acetylcholine receptors (Schmeller *et al.*, 1997); however, further study is required to understand the mode of action of the PAs.

The PAs and the insects that seek them have fascinated scientists for more than 40 years. More importantly, PAs and arctiids have brought attention to the interesting and complex relationships between insect defenses and sex. The degree to which animal defenses have played a role in the evolution of courtship signals has probably been underestimated. Identifying mates that can defend themselves, their mates, and their offspring against predators, parasites, and disease has proven supremely useful to some moths, butterflies, beetles, and flies (Eisner and Meinwald, 1995a,b; Eisner *et al.*, 2002). Future studies will undoubtedly expand this list.

Acknowledgements

We thank Nickolay Hristov and Thomas Eisner for the use of photographs (Plate 1, lower left by T. E.; Plate 1, lower right by N. H.) and Rebecca Simmons for comments on the manuscript.

References

Acharya, L. and Fenton, M. B. (1992). Echolocation behavior of vespertillionid bats (*Lasiurus cinereus* and *Lasiurus borealis*) attacking aerial targets including arctiid moths. *Canadian Journal of Zoology* **70**: 1292–1298.

Adams, J. K. (1987). The Defenses of Adult Tiger Moths (Lepidoptera: Arctiidae): Phylogenetic and Ecological Factors Influencing the Array of Defenses in Individual Species. Ph.D. Thesis. University of Kansas, Lawrence.

Andersson, M. (1994). *Monographs in Behavior and Ecology*: *Sexual Selection*. Princeton, NJ: Princeton University Press.

Aplin, R. T. and Rothschild, M. (1972). Poisonous alkaloids in the body tissues of the garden tiger moth (*Arctia caja* L.) and the cinnabar moth (*Tyria* (= *Callimorpha*) *jacobaeae* L.) (Lepidoptera). In *Toxins of Animal and Plant Origin*, eds. A. de Vries and K. Kochva. pp. 579–595. London: Gordon and Breach.

Aplin, R. T., Benn, M. H. and Rothschild, M. (1968). Poisonous alkaloids in the body tissues of the cinnabar moth (*Callimorpha jacobaeae* L.). *Nature* **219**: 747–748.

Beebe, W. (1955). Two little-known selective insect attractants. *Zoologica* (*New York Zoological Society*) **40**: 27–36.

Beebe, W. and Kenedy, R. (1957). Habits, palatability, and mimicry in thirteen ctenuchid moth species from Trinidad, B. W. I. *Zoologica* (*New York Zoological Society*) **42**: 147–158.

Bell, T. W. and Meinwald, J. (1986). Pheromones of two arctiid moths (*Creatonotos transiens* and *C. gangis*): chiral components from both sexes and achiral female components. *Journal of Chemical Ecology* **12**: 385–409.

Bell, T. W., Boppré, M., Schneider, D. and Meinwald, J. (1984). Stereochemical course of pheromone biosynthesis in the arctiid moth, *Creatonotos transiens*. *Experientia* **40**: 713–714.

Benn, M., DeGrave, J., Gnanasunderam, C. and Hutchins, R. (1979). Host-plant pyrrolizidine alkaloids in *Nyctemera annulata* Boisduval: their persistence through the life cycle and transfer to a parasite. *Experientia* **35**: 731–732.

Bentley, M. D., Leonard, D. E., Stoddard, W. F. and Zalkow, L. H. (1984). Pyrrolizidine alkaloids as larval feeding deterrents for spruce budworm, *Choristoneura fumiferana* (Lepidoptera: Tortricidae). *Annals of the Entomological Society of America* **77**: 393–397.

Bernays, E. A., Chapman, R. F. and Hartmann, T. (2002). A highly sensitive taste receptor cell for pyrrolizidine alkaloids in the lateral galeal sensillum of a polyphagous caterpillar, *Estigmene acrea*. *Journal of Comparative Physiology A* **188**: 715–723.
Birch, M. C. (1969). Scent Organs in Male Lepidoptera. D.Phil. Thesis. Oxford University, Oxford.
(1979). Eversible structures. In *Moths and Butterflies of Great Britain and Ireland*, vol. 9, eds. J. Heath and A. M. Emmet, pp. 9–18. London: Curwen.
Birch, M. C., Poppy, G. M. and Baker, T. C. (1990). Scents and eversible scent structures of male moths. *Annual Review of Entomology* **35**: 25–58.
Bisset, G. W., Grazer, J. F. D., Rothschild, M. and Schachter, M. (1959). A choline ester and other substances in the garden tiger moth, *Arctia caja* (L.). *Journal of Physiology* **146**: 38–39.
(1960). A pharmacologically active choline ester and other substances in the garden tiger moth, *Arctia caja* (L.). *Proceedings of the Royal Society of London B* **152**: 225–262.
Blest, A. D. (1964). Protective display and sound production in some new world arctiid and ctenuchid moths. *Zoologica* **49**: 161–181.
Boada, R. (1997). Courtship and Defense of the Scarlet-bodied Wasp Moth *Cosmosoma myrodora* Dyar (Lepidoptera: Arctiidae) with Notes on Related Euchromiines. M.Sc. Thesis. Wake Forest University, Winston-Salem.
Bogner, F. and Boppré, M. (1989). Single cell recordings reveal hydroxydanaidal as the volatile compound attracting insect to pyrrolizidine alkaloids. *Entomologia Experimentalis et Applicata* **50**: 171–184.
Bogner, F. and Eisner, T. (1991). Chemical basis of egg cannibalism in a caterpillar (*Utetheisa ornatrix*). *Journal of Chemical Ecology* **17**: 2063–2075.
Boppré, M. (1979). Chemical communication, plant relationships, and mimicry in the evolution of danaid butterflies. *Entomologia Experimentalis et Applicata* **24**: 264–277.
(1981). Adult Lepidoptera "feeding" at withered *Heliotropium* plants (Boraginaceae) in East Africa. *Ecological Entomology* **6**: 449–452.
(1984a). Redefining "pharmacophagy." *Journal of Chemical Ecology* **10**: 1151–1154.
(1984b). Chemically mediated interactions between butterflies. In *The Biology of Butterflies, Symposium of the Royal Entomological Society of London*, No. 11, eds. R. I. Vane-Wright and P. R. Ackery, pp. 259–275. London: Academic Press.
(1986). Insects pharmacophageously utilizing defensive plant chemicals (pyrrolizidine alkaloids). *Naturwissenschaften* **73**: 17–26.
(1990). Lepidoptera and pyrrolizidine alkaloids: exemplification of complexity in chemical ecology. *Journal of Chemical Ecology* **16**: 165–185.
(1995). Pharmakophagie: Drogen, Sex und Schmetterlinge. *Biologie in unserer Zeit* **25**: 8–17.
Boppré, M. and Schneider, D. (1985). Pyrrolizidine alkaloids quantitatively regulate both scent organ morphogenesis and pheromone biosynthesis in male *Creatonotos* moths (Lepidoptera: Arctiidae). *Journal of Comparative Physiology A* **157**: 569–577.
(1989). The biology of *Creatonotos* (Lepidoptera: Arctiidae) with special reference to the androconial system. *Zoological Journal of the Linnaean Society* **96**: 339–356.
Bowers, D. (1992). The evolution of unpalatability and the cost of chemical defense in insects. In *Insect Chemical Ecology: An Evolutionary Approach*, eds. B. D. Roitberg and M. B. Isman, pp. 216–244. London: Chapman & Hall.
(1993). Aposematic caterpillars: life styles of the warningly colored and unpalatable. In *Caterpillars: Ecological and Evolutionary Constraints on Foraging*, eds. N. E. Stamp and T. M. Casey, pp. 331–371. New York: Chapman & Hall.

Bowers, M. D. and Stamp, N. E. (1997). Fate of host–plant iridoid glycosides in lepidopteran larvae of nymphalidae and arctiidae. *Journal of Chemical Ecology* **23**: 2955–2965.

Brown, K. S., Jr (1984). Adult-obtained pyrrolizidine alkaloids defend ithomiine butterflies against a spider predator. *Nature* **309**: 707–709.

Bruce, D. (1888). Description of mature larva of *Gnophaela vermiculata* G&R. *Entomologia Americana* **4**: 24.

Bull, L. B., Culvenor, C. C. J. and Dick, A. T. (1968). *The Pyrrolizidine Alkaloids*. Amsterdam: North-Holland Biomedical Press.

Cardoso, M. Z. O. (1997). Testing chemical defense based on pyrrolizidine alkaloids. *Animal Behaviour* **54**: 985–991.

Cock, M. J. W. (1982). The history of, and prospects for, the biological control of *Chromolaena odorata* (Compositae) by *Pareuchaetes pseudoinsulata* Rego Barros and allies (Lepidoptera: Arctiidae). *Bulletin of Entomological Research* **72**: 193–205.

Cockerell, T. D. A. (1889). The larva of *Gnophaela vermiculata* G&R. *Entomologia Americana* **5**: 57–58.

Conner, W. E. (1987). Ultrasound: its role in the courtship of the arctiid moth, *Cycnia tenera*. *Experientia* **43**: 1029–1031.

Conner, W. E., Eisner T., Vander Meer, R. K., Guerrero, A. and Meinwald, J. (1981). Precopulatory sexual interactions in an arctiid moth (*Utetheisa ornatrix*): role of pheromone derived from alkaloids. *Behavioral Ecology and Sociobiology* **9**: 227–235.

Conner, W. E. Roach, B., Benedict, E., Meinwald, J. and Eisner, T. (1990). Courtship pheromone production and body size as correlates of larval diet in males of the arctiid moth, *Utetheisa ornatrix*. *Journal of Chemical Ecology* **16**: 543–551.

Conner, W. E., Boada, R., Schroeder, F. and Eisner, T. (2000). Chemical defense: bestowal of a nuptial alkaloidal garment by a male moth on its mate. *Proceedings of the National Academy of Sciences, USA* **97**: 14406–14411.

Covell, C. V. (1984). *A Field Guide to the Moths of Eastern North America*. Boston: Houghton Mifflin.

Culvenor, C. C. J. and Edgar, J. A. (1972). Dihydropyrrolizidine secretions associated with coremata of *Utetheisa* moths (family Arctiidae). *Experientia* **28**: 627–628.

Cuperus, P. L. (1985). Inventory of pores in antennal sensilla of *Yponomeuta* spp. (Lepidoptera: Yponomeutidae) and *Adoxophyes orana* F.v. R. (Lepidoptera: Tortricidae). *International Journal of Insect Morphology Embryology* **14**: 347–359.

Davenport, J. W. and Conner, W. E. (2003). Dietary alkaloids and the development of androconial organs in *Estigmene acrea*. *Journal of Insect Science* **3**: 3.

Davidson, R. B., Baker, C., McElveen, M. and Conner, W. E. (1997). Hydroxydanaidal and the courtship of *Haploa* (Arctiidae). *Journal of the Lepidopterists' Society* **51**: 288–294.

de Boer, N. J. (1999). Pyrrolizidine alkaloid distribution in *Senecio jacobaea* rosettes minimizes losses to generalist feeding. *Entomologia Experimentalis et Applicata* **91**: 169–173.

Dethier, V. G. (1939). Prothoracic glands of adult lepidoptera. *Journal of the New York Entomological Society* **47**: 131–144.

Dreyer, D. L., Jones, K. C. and Molyneux, R. L. (1985). Feeding deterrency of some pyrrolizidine, indolizidine, and quinolizidine alkaloids towards pea aphid (*Acryrthosiphon pisum*) and evidence of phloem transport of indolizidine alkaloid swainsonine. *Journal of Chemical Ecology* **11**: 1045–1051.

Dunning, D. C. (1968). Warning sounds of moths. *Zeitschrift für Tierpsychologie* **25**: 129–138.

Dunning, D. C. and Krüger, M. (1995). Aposematic sounds in African moths. *Biotropica* **27**: 227–231.
Dunning, D. C. and Roeder, K. D. (1965). Moth sounds and the insect catching behavior of bats. *Science* **147**: 173–174.
Dunning, D. C., Acharya, L., Merriman, C. B. and Ferro, L. D. (1992). Interactions between bats and arctiid moths. *Canadian Journal of Zoology* **70**: 2218–2223.
Dussourd, D. E. (1986). Adaptations of Insect Herbivores to Plant Defenses. Ph.D. Thesis. Cornell University, Ithaca.
Dussourd, D. E., Ubik, K., Harvis, C., Resch, J., Meinwald, J. and Eisner, T. (1988). Biparental defensive endowment of eggs with acquired plant alkaloid in the moth *Utetheisa ornatrix*. *Proceedings of the National Academy of Sciences, USA* **85**: 5992–5996.
Dussourd, D. E., Harvis, C. A. and Eisner, T. (1989). Paternal allocation of sequestered plant pyrrolizidine alkaloid to eggs in the danaine butterfly, *Danaus gilippus*. *Experientia* **45**: 896–898.
Dussourd, D. E., Harvis, C. A., Meinwald, J. and Eisner, T. (1991). Pheromonal advertisement of a nuptial gift by a male moth (*Utetheisa ornatrix*). *Proceedings of the National Academy of Sciences, USA* **88**: 9224–9227.
Eberhard, W. G. (1996). *Female Control: Sexual Selection by Cryptic Female Choice*. Princeton, NJ: Princeton University Press.
Edgar, J. A., Culvenor, C. C. J. and Pliske, T. E. (1974). Coevolution of danaid butterflies and their host plants. *Nature* **250**: 646–648.
(1976). Isolation of a lactone, structurally related to the esterifying acids of pyrrolizidine alkaloids, from the costal fringes of male Ithomiinae. *Journal of Chemical Ecology* **2**: 263–270.
Egalhaaf, A., Coelln, K., Schmitz, B., Buck, M., Wink, M. and Schneider, D. (1990). Organ specific storage of dietary pyrrolizidine alkaloids in the arctiid moth *Creatonotus transiens*. *Zeitschrift für Naturforschung* **45c**: 172–177.
Ehmke, A., Witte, L., Biller, A. and Hartmann, T. (1990). Sequestration, *N*-oxidation, and transformation of plant pyrrolizidine alkaloids by the arctiid moth *Tyria jacobaeae* L. *Zeitschrift für Naturforschung* **45c**: 1185–1192.
Eisner, T. (1980). Chemistry, defense, and survival: case studies and selected topics. In *Insect Biology in the Future*, ed. M. Locke and D. S. Smith, pp. 847–878. New York: Academic Press.
Eisner, T. and Eisner, M. (1991). Unpalatability of the pyrrolizidine alkaloid-containing moth *Utetheisa ornatrix*, and its larva, to wolf spiders. *Psyche* **98**: 111–118.
Eisner, T. and Meinwald, J. (1987). Alkaloid-derived pheromones and sexual selection in Lepidoptera. In *Pheromone Biochemistry*, eds. G. D. Prestwich and G. J. Blomquist, pp. 251–269. Orlando, FL: Academic Press.
(1995a). The chemistry of phyletic dominance. *Proceedings of the National Academy of Sciences, USA* **92**: 14–18.
(1995b). The chemistry of sexual selection. *Proceedings of the National Academy of Sciences, USA* **92**: 50–55.
Eisner, T., Eisner, M., Rossini, C. *et al.* (2000). Chemical defense against predation in an insect egg. *Proceedings of the National Academy of Sciences, USA* **97**: 1634–1639.
Eisner, T., Rossini, C., Gonzales, A., Iyengar, V. K., Siegler, M. V. S. and Smedley, S. R. (2002). Parental investment in egg defense. In *Chemoecology of Insect Eggs and Egg Deposition*, eds. M. Hilker and T. Meiners, pp. 91–116. Berlin: Blackwell.
Fullard, J. H. and Heller, B. (1990). Functional organization of the arctiid moth tymbal (Insecta, Lepidoptera). *Journal of Morphology* **204**: 57–65.

Fullard, J. H., Fenton, M. B. and Simmons, J. A. (1979). Jamming bat echolocation: the clicks of arctiid moths. *Canadian Journal of Zoology* **57**: 647–649.
Fullard, J. H., Simmons, J. A. and Saillant, P. A. (1994). Jamming bat echolocation: the dogbane tiger moth *Cycnia tenera* times its clicks to the terminal attack calls of the big brown bat *Eptesicus fuscus*. *Journal of Experimental Biology* **194**: 285–298.
Glendinning, J. I., Brower, L. P. and Montgomery, C. A. (1990). Responses of three mouse species to deterrent chemicals in the monarch butterfly. I. Taste and toxicity tests using artificial diets laced with digitoxin or monocrotaline. *Chemoecology* **1**: 114–123.
González, A., Rossini, C., Eisner, M. and Eisner, T. (1999). Sexually transmitted chemical defense in a moth (*Utetheisa ornatrix*). *Proceedings of the National Academy of Sciences, USA* **96**: 5570–5574.
Goss, G. J. (1979). The interaction between moths and plants containing pyrrolizidine alkaloids. *Environmental Entomology* **8**: 487–493.
Grant, A. J., O'Connell, R. J. and Eisner, T. (1989). Pheromone-mediated sexual selection in the moth *Utetheisa ornatrix*: olfactory receptor neurons responsive to a male-produced pheromone. *Journal of Insect Behavior* **2**: 371–385.
Gwynne, D. (1993). Food quality controls sexual selection in Mormon crickets by altering male mating investment. *Ecology* **74**: 1406–1413.
Hägele, B. F. and Rowell-Rahier, M. (2000). Choice, performance, and heritability of performance of specialist and generalist insect herbivores towards cacalol and seneciphylline, two allelochemics of *Adenostyles alpina* (Asteraceae). *Journal of Evolutionary Biology* **13**: 131–142.
Hampson, G. F. (1898). *Catalogue of the Lepidoptera Phalaenae*, vol. 1. London: British Museum (Natural History).
Hare, J. F. and Eisner, T. (1993). Pyrrolizidine alkaloid deters ant predators of *Utetheisa ornatrix* eggs: effects of alkaloid concentration, oxidation state, and prior exposure of ants to alkaloid-laden prey. *Oecologia* **96**: 9–18.
Hartmann, T. and Ober, D. (2000). Biosynthesis and metabolism of pyrrolizidine alkaloids in plants and specialized insect herbivores. *Topics in Current Chemistry* **209**: 207–243.
Hartmann, T. and Witte, L. (1995). Chemistry, biology, and chemoecology of the pyrrolizidine alkaloids. In *Alkaloids: Chemical and Biological Perspectives*, vol. 9, ed. S. W. Pelletier, pp. 155–233. Oxford: Pergamon Press.
Häuser, C. L. and Boppré, M. (1997). A revision of the Afrotropical taxa of the genus *Amerila* Walker (Lepidoptera, Arctiidae). *Systematic Entomology* **22**: 1–44.
Hesbacher, S., Giez, I., Embacher, G. *et al.* (1995). Sequestration of lichen compounds by lichen feeding members of the Arctiidae (Lepidoptera). *Journal of Chemical Ecology* **21**: 2079–2089.
Holloway, J. D. (1988). *The Moths of Borneo*, Part 6: *Arctiidae, Syntominae, Euchromiinae, Arctiinae, Aganainae (to Noctuidae)*. Malaysia: Southdene Sdn. Bhd.
Iyengar, V. K. and Eisner, T. (1999a). Heritability of body mass, a sexually selected trait, in an arctiid moth (*Utetheisa ornatrix*). *Proceedings of the National Academy of Sciences, USA* **96**: 9169–9171.
(1999b). Female choice increases offspring fitness in an arctiid moth (*Utetheisa ornatrix*). *Proceedings of the National Academy of Sciences, USA* **96**: 15013–15016.
Iyengar, V. K., Rossini, C., Hoebeke, E. R., Conner, W. E. and Eisner, T. (1999). First records of the parasitoid *Archytas aterrimus* (Diptera: Tachinidae) for *Utetheisa ornatrix* (Lepidoptera: Arctiidae). *Entomological News* **110**: 144–146.
Iyengar, V. K., Rossini, C. and Eisner, T. (2001). Precopulatory assessment of male quality in an arctiid moth (*Utetheisa ornatrix*): hyroxydanaidal is the only criterion of choice. *Behavioral Ecology and Sociobiology* **49**: 283–288.

Jacobson, N. (1994). Cladistic Studies of the Arctiidae (Lepidoptera) and the Genus *Agylla* (Arctiidae: Lithosiinae) Using Characters of Adults and Larvae. Ph.D. Thesis, Cornell University, Ithaca.

Jacobson, N. and Weller, S. J. (2001). *A Cladistic Study of the Tiger Moth Family Arctiidae (Noctuoidea) Based on Larval and Adult Morphology*. Lanham, MD: Thomas Say Publications, Entomological Society of America.

Jones, F. M. (1932). Insect coloration and the relative acceptability of insects to birds. *Transactions of the Royal Entomological Society of London* **80**: 345–385.

(1934). Further experiments on coloration and relative acceptability of insects to birds. *Transactions of the Royal Entomological Society of London* **82**: 443–453.

Krasnoff, S. B. and Dussourd, D. E. (1989). Dihydropyrrolizidine attractants for arctiid moths that visit plants containing pyrrolizidine alkaloids. *Journal of Chemical Ecology* **15**: 47–60.

Krasnoff, S. B. and Roelofs, W. L. (1989). Quantitative and qualitative effects of larval diet on male scent secretions of *Estigmene acrea, Phragmatobia fuliginosa*, and *Pyrrharctia isabella* (Lepidoptera: Arctiidae). *Journal of Chemical Ecology* **15**: 1077–1093.

(1990). Evolutionary trends in the male pheromone systems of arctiid moths: evidence from studies of courtship in *Phragmatobia fuliginosa* and *Pyrrharctia isabella* (Lepidoptera: Arctiidae). *Zoological Journal of the Linnaean Society* **99**: 319–338.

Krasnoff, S. B. and Yager, D. D. (1988). Acoustic response to a pheromonal cue in the arctiid moth *Pyrrharctia isabella*. *Physiological Entomology* **13**: 433–440.

Krasnoff, S. B., Bjostad, L. B. and Roelofs, W. L. (1987). Quantitative and qualitative variation in male pheromones of *Phragmatobia fuliginosa* and *Pyrrarctia isabella* (Lepidoptera, Arctiidae). *Journal of Chemical Ecology* **13**: 807–822.

LaMunyon, C. W. (1997). Increased fecundity, as a function of mutiple mating, in an arctiid moth, *Utetheisa ornatrix*. *Ecological Entomology* **22**: 69–73.

LaMunyon, C. W. and Eisner, T. (1993). Postcopulatory sexual selection in an arctiid moth (*Utetheisa ornatrix*). *Proceedings of the National Academy of Sciences, USA* **90**: 4689–4692.

(1994). Spermatophore size as determinant of paternity in an arctiid moth (*Utetheisa ornatrix*). *Proceedings of the National Academy of Sciences, USA* **91**: 7081–7084.

L'Empereur, K. M., Li, Y. and Stermitz, F. R. (1989). Pyrrolizidine alkaloids from *Hackelia californica* and *Gnophaela latipennis*, an *H. californica*-hosted arctiid moth. *Journal of Natural Products* **52**: 360–366.

Mattocks, A. R. (1986). *Chemistry and Toxicology of Pyrrolizidine Alkaloids*. London: Academic Press.

Meyer, W. (1984). Sex pheromone chemistry and biology of some arctiid moths (Lepidoptera: Arctiidae): enantiomeric differences in pheromone perception. M.Sc. Thesis. Cornell University, Ithaca.

Miller, J. S. and Feeney, P. P. (1989). Interspecific differences among swallowtail larvae (Lepidoptera: Papilionidae) in susceptibility to aristolochic acids and berberine. *Ecological Entomology* **14**: 287–296.

Miller, L. (1991). Arctiid moth clicks can degrade the accuracy of range difference discrimination in echolocating big brown bats, *Eptesicus fuscus*. *Journal of Comparative Physiology A* **168**: 571–579.

Nishida, R., Schulz, S., Kim, C. S. *et al.* (1996). Male sex pheromone of a giant danaine butterfly, *Idea leuconoe*. *Journal of Chemical Ecology* **22**: 949–972.

Pagden, H. T. (1957). The presence of coremata in *Creatonotus gangis* (L.) (Lepidoptera: Arctiidae). *Proceedings of the Royal Entomological Society of London A* **32**: 90–94.

Phelan, P. L. (1992). Evolution of sex pheromones and the role of asymmetric tracking. In *Insect Chemical Ecology: An Evolutionary Approach*, eds. B. D. Roitberg and M. B. Isman, pp. 265–314. New York: Chapman & Hall.

Phelan, P. L. and Baker T. C. (1987). Evolution of male pheromones in moths: reproductive isolation through sexual selection? *Science* **235**: 205–207.

Pinhey, E. C. G. (1975). *Moths of Southern Africa*. Capetown: Tafeble.

Pliske, T. E. (1975a). Attraction of Lepidoptera to plants containing pyrrolizidine alkaloids. *Environmental Entomology* **4**: 455–473.

(1975b). Pollination of pyrrolizidine alkaloid-containing plants by male Lepidoptera. *Environmental Entomology* **4**: 474–479.

Pliske, T. E. and Eisner, T. (1969). Sex pheromones of the queen butterfly: biology. *Science* **164**: 1170–1172.

Ritchey, G. E. and McKee, R. (1941). *Crotalaria for Forage, Bulletin 361*. Gainesville, FL: University of Florida Agricultural Experimental Station.

Roeder, K. D. (1974). Acoustic responses and possible bat-evasion tactics of certain moths. In *Proceedings of the Annual Meeting of the Canadian Society of Zoologists*, University of New Brunswick, Fredericton, June 1974, pp. 74–78.

Roque-Albelo, L. (2000). The tiger moths (Arctiidae) of the Galápagos Islands, their biogeography and life history. In *Proceedings of the Annual Meeting of the Lepidopterists' Society*, Wake Forest University, Winston-Salem, July 2000.

Roque-Albelo, L., Schroeder, F. C., Conner, W. E. *et al.* (2002). Chemical defense and aposematism: the case for *Utetheisa galapagensis*. *Chemoecology* **12**: 153–157.

Rossini, C., Hoebeke, E. R., Iyengar, V. K., Conner, W. E., Eisner, M. and Eisner, T. (2000). Alkaloid content of the pupal parasitoids of an alkaloid sequestering arctiid moth (*Utetheisa ornatrix*). *Entomological News* **111**: 287–290.

Rothschild, M. (1961). Defensive odours and Mullerian mimicry among insects. *Transactions of the Royal Entomological Society of London* **113**: 101–113.

(1963). Is the buff ermine (*Spilosoma lutea* (Huf.)) a mimic of the white ermine (*Spilosoma lubricipeda* (L.))? *Proceedings of the Royal Entomological Society of London* **38**: 159–164.

(1972a). Colour and poisons in insect protection. *New Scientist*, 11 May, 318–320.

(1972b). Secondary plant substances and warning colouration in insects. In *Insect/Plant Relationships: Symposium of the Royal Entomological Society of London 6*, ed. H. F. van Emden, pp. 59–83. Oxford: Blackwell Scientific.

(1972c). Some observations on the relationship between plants, toxic insects and birds. In *Phytochemical Ecology: Proceedings of the Phytochemical Society 8*, ed. J. B. Harborne, pp. 1–12. New York: Academic Press.

(1985). British aposematic Lepidoptera. In *The Moths and Butterflies of Great Britain and Ireland*, Part 2, eds. J. Heath and A. M. Emmet, pp. 9–62. Colchester: Harley Books.

Rothschild, M. and Aplin, R. T. (1971). Toxins in tiger moths (Arctiidae: Lepidoptera). *Pesticide Chemistry* **3**: 177–182.

Rothschild, M., Reichstein, T., von Euw, J., Aplin, R. and Harman, R. R. M. (1970). Toxic Lepidoptera. *Toxicon* **8**: 293–299.

Rothschild, M., von Euw, J. and Reichstein, T. (1973). Cardiac glycosides (heart poisons) in the polka-dot moth *Syntomeida epilais* Walk. (Ctenuchidae: Lep.) with some observations on the toxic qualities of *Amata* (= *Syntomis*) *phegea* (L.). *Proceedings of the Royal Society of London, Series B* **183**: 227–247.

Rothschild, M., Aplin, R. T., Cockrum, P. A., Edgar, J. A., Fairweather, P. and Lees, R. (1979). Pyrrolizidine alkaloids in arctiid moths (Lep.) with a discussion on host plant relationships and the role of these secondary plant substances in the Arctiidae. *Biological Journal of the Linnaean Society* **12**: 305–326.

Rothschild, M., Moore, B. P. and Vance Brown, W. (1984). Pyrazines as warning odour components in the Monarch butterfly, *Danaus plexippus*, and in moths of the genera *Zygaena* and *Amata* (Lepidoptera). *Biological Journal of the Linnaean Society* **23**: 375–380.

Sargent, T. D. (1995). On the relative acceptabilities of local butterflies and moths to local birds. *Journal of the Lepidopterists' Society* **49**: 148–162.

Schmeller, T., El-Shazley, A. and Wink, M. (1997). Allochemical activities of pyrrolizidne alkaloids: interactions with neuroreceptors and acetylcholine related enzymes. *Journal of Chemical Ecology* **23**: 399–416.

Schmitz, B., Buck, M., Egelhaaf, A. and Schneider, D. (1989). Ecdysone and a dietary alkaloid interact in the development of the pheromone gland of a male moth (*Creatonotos*, Lepidoptera: Arctiidae). *Roux's Archives of Developmental Biology* **198**: 1–7.

Schneider, D. (1986). The strange fate of pyrrolizidine alkaloids. In *Perspectives in Chemoreception and Bihavior*, eds. R. F. Chapman, E. A. Bernays and J. G. Stoffolano, pp. 123–142. New York: Springer-Verlag.

Schneider, D., Boppré, M., Zweig, J. *et al.* (1982). Scent organ development in *Creatonotos* moths: regulation by pyrrolizidine alkaloids. *Science* **215**: 1264–1265.

Schneider, D., Schultz, S., Kittmann, R. and Kanagaratnam, P. (1992). Pheromones and glandular structures of both sexes of the weed defoliator moth *Pareuchaetes pseudoinsulata* Rego Barros (Lep., Arctiidae). *Journal of Applied Entomology* **113**: 280–294.

Schneider, D., Schulz, S., Priesner, E., Ziesmann, J. and Franke, W. (1998). Autodetection and chemistry of female and male pheromone in both sexes of the tiger moth *Panaxia quadripunctaria*. *Journal of Comparative Physiology A* **182**: 153–161.

Schultz, S., Franke, W., Boppré, M., Eisner, T. and Meinwald, J. (1993). Insect pheromone biosynthesis: stereochemical pathway of hydroxydanaidal production from alkaloid precursors in *Creatonotos transiens* (Lepidoptera, Arctiidae). *Proceedings of the National Academy of Sciences, USA* **90**: 6834–6838.

Simmons, R. B. (2001). Phylogenetic studies of mimetic tiger moths based on morphological and molecular data (Lepidoptera: Arctiidae: Euchromiini). Ph.D. Thesis. University of Minnesota, Minneapolis.

Simmons, R. B. and Weller, S. J. (2002). What kind of signals do mimetic tiger moths send? A phylogenetic test of wasp mimicry systems (Lepidoptera: Arctiidae: Euchromiini). *Proceedings of the Royal Society of London, Series B* **269**: 983–990.

Singer, M. S. (2000). Ecological maintenance of food-mixing in the woolly bear caterpillar *Grammia geneura* (Strecker)(Lepidoptera: Arctiidae). Ph.D. Thesis. University of Arizona, Tucson.

Stoneman, M. G. and Fenton, M. B. (1988). Disrupting foraging bats: the clicks of arctiid moth. In *NATO ASI Series A, Life Sciences*, vol. 156, *Animal Sonar: Processes and Performance*, eds. P. E. Nachtigall and P. W. B. Moore, pp. 635–638. Brussels: NATO.

Stretch, R. H. (1882). Larva of *Gnophaela hopfferi*. *Papilio* **2**: 82–83.

Teas, H. J., Dyson, J. G. and Whisenant, B. R. (1966). Cycasin metabolism in *Seirarctia echo* Abbot and Smith (Lepidoptera: Arctiidae). *Journal of the Georgia Entomological Society* **1**: 21–22.

Tietz, H. M. (1972). *An Index to the Life Histories, Early Stages and Hosts of the Macrolepidoptera of the Continental United States and Canada*, vols. I and II. Sarasota, FL: Allyn Museum of Entomology.

Tougaard, J., Casseday, J. H. and Covey, E. (1998). Arctiid moths and bat echolocation: broad band clicks interfere with neural responses to auditory stimuli in the nuclei of the lateral lemniscus of the big brown bat. *Journal of Comparative Physiology A* **182**: 203–215.

J. R., Witte, L., Brown, K. S., Jr, Trigo, Hartmann, T. and Barata, L. E. S. (1993). Pyrrolizidine alkaloids in the arctiid moth, *Hyalurga syma*. *Journal of Chemical Ecology* **19**: 669–679.

Trigo, J. R., Brown, K. S., Jr, Witte, L., Hartmann, T., Ernst, L. and Barata, L. E. S. (1996). Pyrrolizidine alkaloids: differential acquisition and use patterns in Apocynaceae and Solanaceae feeding ithomiine butterflies (Lepidoptera: Nymphalidae). *Biological Journal of the Linnaean Society* **58**: 99–123.

Vasoconcellos-Neto, J. and Lewinsohn, T. M. (1984). Discrimination and release of unpalatable butterflies by *Nephila clavipes*, a neotropical orb-weaving spider. *Ecological Entomology* **9**: 337–344.

von Nickisch-Rosenegk, E. and Wink, M. (1993). Sequestration of pyrrolizidine alkaloids in several arctiid moths (Lepidoptera: Arctiidae). *Journal of Chemical Ecology* **19**: 1889–1903.

Watson, A. and Goodger, D. T. (1986). Catalogue of the neotropical tiger-moths. *Occasional Papers on Systematic Entomology* No. 1. London: British Museum of Natural History.

Weller, S. J., Jacobsen, N. L. and Conner, W. E. (2000a). The evolution of chemical defences and mating systems in tiger moths (Lepidoptera: Arctiidae). *Biological Journal of the Linnaean Society* **68**: 557–578.

Weller, S. J., Simmons, R. B., Boada, R. and Conner, W. E. (2000b). Abdominal modifications occurring in wasp mimics of the Ctenuchine–Euchromiine clade (Lepidoptera: Arctiidae). *Annals of the Entomological Society of America* **93**: 920–928.

Willis, M. A. and Birch, M. C. (1982). Male lek formation and female calling in a population of the arctiid moth, *Estigmene acrea*. *Science* **218**: 168–170.

Wink, M. L. D. and Schneider, D. (1988). Carrier-mediated uptake of pyrrolizidine alkaloids in larvae of the aposematic and alkaloid-exploiting moth *Creatonotus*. *Naturwissenschaften* **75**: 524–525.

Wink, M. and von Nickisch-Rosenegk, E. (1997). Sequence data of mitochondrial 16S rDNA of arctiidae and nymphalidae: evidence for a convergent evolution of pyrrolizidine alkaloid and cardiac glycoside sequestration. *Journal of Chemical Ecology* **23**: 1549–1568.

Wink, M. L., Schneider, D. and Witte, L. (1988). Biosynthesis of pyrrolizidine alkaloid-derived pheromones in the arctiid moth, *Creatonotos transiens*: stereochemical conversion of heliotrine. *Zeitschrift für Naturforschung* **43c**: 737–741.

Wunderer, H., Hansen, K., Bell, T. W., Schneider, D. and Meinwald, J. (1986). Sex pheromones of two Asian moths (*Creatonotus transiens, C. gangis*; Lepidoptera, Arctiidae): behavior, morphology, chemistry and electrophysiology. *Experimental Biology* **46**: 11–27.

Yosef, R., Carrel, J. E. and Eisner, T. (1996). Contrasting reactions of loggerhead shrikes to two types of chemically defended insect prey. *Journal of Chemical Ecology* **22**: 173–181.

8

Structure of the pheromone communication channel in moths

Ring T. Cardé

Department of Entomology, University of California, Riverside, USA

Kenneth F. Haynes

Department of Entomology, University of Kentucky, Lexington, USA

Introduction

Moths are among the most speciose of insect groups, comprising perhaps 140 000 species, despite their remarkably undiversified and almost exclusively phytophagous larval lifestyle. Among the factors that are likely to have promoted such speciose success is their ability to persist at relatively low densities, facilitated by a pheromone communication system that allows males to locate conspecific females over distances of tens and in some species perhaps thousands of meters. The pheromone communication system also serves a primary basis of premating (prezygotic) reproductive isolation among species. Chemical communication channels that are distinctive at the species level permit the co-existence of many species in the same habitat or region. Although there is no firm evidence yet that the process of speciation itself has been fostered by splitting of the pheromone channel in sympatry, there are saltational mechanisms that could account for a rapid shift in the compounds produced by the emitter and a parallel tracking shift in the responder. There also are a few cases which suggest that either reinforcement or communication interference has caused divergence of chemical channels. Nonetheless, how changes in these communication systems evolve remains largely speculative. We also have no explanation for why in some species the female's production of a pheromone blend is variable and in others highly canalized.

Long-distance mate location that is mediated by pheromones is true "communication" as defined by Burghardt (1970), in that there are selective constraints on both the females' production of the signal and the males' response – in other words, selection favors some individuals over others in finding a mate. In a provocative argument, Williams (1992) contended that the female odor used by males for mate location is *not* a pheromone, because in moths there is no special "machinery"

Advances in Insect Chemical Ecology, ed. R. T. Cardé and J. G. Millar. Published by Cambridge University

or "signaling device" for the females' production of odor. Furthermore, Williams noted that the females' odor is produced in minuscule amounts, which he inferred to be additional evidence that production is not under active selection. In this view, odor-mediated attraction of the male by the female in moths fails to meet the strict definition of communication, because selection has not acted on signal production. However, pheromone production in moths and many other insect taxa involves specialized glandular structures at the abdominal tip for biosynthesis and dissemination of pheromone to the external environment. Furthermore, biosynthesis is under hormonal and/or neural regulation, and it often involves unique enzymatic reactions. In some cases, pheromone is transported in the hemolymph via proteins called lipophorins from the site of synthesis to specialized storage reservoirs (Schal *et al.*, 1998). Pheromone release itself is a specialized behavior (termed "calling"), which involves protrusion of the pheromone gland, sometimes rhythmically, and usually in a stereotypical stance. Some species even atomize their pheromone from reservoirs through specialized pores. The amount of pheromone emitted clearly is sufficient to attract males over long distances. Calling occurs at a set daily time and its periodicity and that of pheromone biosynthesis have, when examined, been found to have a circadian basis. The machineries for production and release of pheromone by female moths are thus clearly specialized. Williams' hypothesis may be a useful explanation of the origin of female signals and male response, which may have initially relied on inadvertent emission of chemicals by females, followed by males using these cues to find females, but it is demonstrably inconsistent with the indisputable evidence that selection has shaped the mechanisms of signaling.

This review considers evidence for selective forces that mold the chemical signal and behavioral response, including the characteristics of pheromone dispersal and the plume-tracking maneuvers that influence the success of mate finding. Recent reviews that consider mechanisms of evolutionary change in moth pheromones include Phelan (1992, 1996, 1997), Löfstedt (1993), Linn and Roelofs (1995), and Löfstedt and Kozlov (1996).

Why do females signal and males respond?

Emission of pheromone by females, causing males to fly to them, appears to be the basal condition in moths. In the Trichoptera or caddisflies, a sister group to the Lepidoptera, mate finding also is mediated by a female-released pheromone (e.g., Resh and Wood, 1985; Löfstedt *et al.*, 1994); this system also predominates among the basal and derived lineages of moths. Among the primitive Lepidoptera, however, the Hepialidae have many species in which this system is reversed, that is, males attract females, often by evident release of pheromone while flying in groups. However, such lekking formations are found in the more-derived hepialids, and, therefore, male attraction to females is also presumed to be the basal state of hepialids (Wagner and Rosovsky, 1991). Among the approximately 28 ditrysian ("advanced") superfamilies of moths, male signaling with female attraction over a long distance is a rare derived condition, so far documented only in a few species of Pyralidae, Arctiidae, and one species of Noctuidae (Phelan, 1997).

The basal motif in moths of female production of pheromone and male response may be a consequence of the relative disparity in parental investment by females and males in reproduction (Trivers, 1972; Thornhill, 1979). In general, female production of offspring is limited by the number of eggs produced, whereas production of offspring by males is determined by the number of matings secured. Once a female has mated, her reproductive success is maximized by finding appropriate oviposition sites, whereas males can only benefit by finding additional mates. This leads to a skewing of the operational sex ratio toward a male bias. Females are a resource to be competed for and males that are proficient at locating females at great distances and in rapidly navigating a course along her plume are likely to be favored (Cardé and Baker, 1984).

In many moth species, a single mating is sufficient to fertilize all or nearly all of the females' eggs; even so, females of many species have multiple partners and, because of sperm precedence, a later-mating male typically will fertilize the preponderance of subsequently laid eggs (Drummond, 1984). In some species, a female's fecundity (Delisle and Hardy, 1997), and the genetic diversity of her offspring (Drummond, 1984), is increased by multiple partners, but even in these cases, a male's reproductive success is more tightly coupled to the number of matings than is that of the female. In addition, because of the disparity in parental investment and the skewing of the operational sex ratio, males often have the riskier role in mate signaling and response (Greenfield, 2002). Signaling from a hidden location with minute quantities of pheromone presents little risk to females, but navigating through open spaces exposes males to predators (e.g., bats). The influence of the level of parental investment and the type of mate recruitment system employed has been noted in other insect groups (Thornhill and Alcock, 1983).

Selection for rapid mate finding

A widely held assumption is that there should be substantial competition among males to be the first to locate a calling female, and, therefore, there should be strong selection for behavioral traits in males that promote rapid location of calling females. These behaviors include: (i) efficient ranging movements, that is, flight patterns (timed to coordinate with the females' rhythm of calling) that sweep a wide area for the presence of a pheromone plume; (ii) a low threshold for detection of pheromone; and (iii) orientation maneuvers that enhance a male's success in tracking a pheromone plume to its source.

There are few field data to bear on the contention that there is strong selective pressure for rapidity of mate finding, which is not surprising in view of the difficulty of making such observations, particularly because most moths are nocturnal. In the case of the gypsy moth (*Lymantria dispar*), which mates around midday (Cardé *et al.*, 1996; Charlton *et al.*, 1999) and, therefore, can be observed in the field readily, some important features of its mate-finding strategies are documented. Field observations indicate that ranging gypsy moth males fly randomly with respect to the direction of current wind flow (Elkinton and Cardé, 1983). Males *do not* direct their flight with a strong bias toward wind headings that might enhance their probability of contact with the plume, even though flight is energetically expensive, and plume-finding behaviors with a generally crosswind directional bias would appear to improve efficiency of plume acquisition (Li *et al.*, 2001).

The first male to reach the female is in nearly all cases successful in mating (Cardé and Hagaman, 1984; Charlton and Cardé, 1990), and a single mating is sufficient to fertilize all of a female's eggs (Doane, 1968). A second mating by females under natural circumstances is very rare. The female gypsy moth has few discernible behavioral reactions before coupling and it is rare that a calling female rejects a courting male (unless he fails to perform the requisite copulatory movements). For the gypsy moth, the selective value for a male being the first to locate the female is clear.

Females of other moths, however, may choose among several simultaneously courting males. For example, females of *Ephestia elutella* tend to select the larger of two competing males (Phelan and Baker, 1986). *Utethesia ornatrix* females also choose larger males over smaller ones; large males transfer relatively more nutrients and defensive pyrrolizidine alkaloids during mating. Females that have mated with large males have increased chemical protection and also lay more eggs, which also have enhanced levels of chemical protection (Iyengar *et al.*, 2002). These issues are considered further on pp. 248–282 in Ch. 7.

Males may compete for females. For example, *Grapholita molesta* males can display variability in their sequence of courtship behaviors (see p. 290), such that a later-arriving male may truncate his courtship sequence, bypassing extrusion of his hairpencils and dissemination of courtship pheromone, and proceed directly to a copulatory attempt, at the expense of the "honest," hairpencil-extruding signaler (Baker, 1983).

These three examples suggest that a male's mating success does not always hinge on whether or not he is the first male to reach the female. It remains unclear, nonetheless, how frequently under *field circumstances* males arrive in rapid enough succession to permit either female choice or male–male competition. The known examples of female choice are from laboratory assays where the simultaneous arrival or presence of males was an arranged event. Some of the complications of measuring female mating preference by using only simultaneous presentations of potential male partners were reviewed by Wagner (1998). Despite these two kinds of sexual selection (female choice and male–male competition), in the field, a first-arriving male ought to have some advantage, because it is likely that he will be the only courting male.

Assuming that a female has normal characteristics of pheromone release, how likely is it that she will *not* attract a suitor? Few field experiments have addressed this question. Sharov *et al.* (1995), working with tethered female gypsy moths in forests with a very low population (near the leading edge of a range expansion), found that many females remained unmated, even though it was known from catches in nearby pheromone-baited traps that males were available. Failure to attract a mate is not the only issue. In the gypsy moth, a delay in mating of 2 days greatly reduces fecundity (Doane, 1968); similar effects are widely reported in other moth species. What can be inferred from such trials is that there is strong selective pressure for females to have effective advertisement strategies in sparse gypsy moth populations. Females may do so by optimizing the synchrony of their time of calling with male ranging flights, maximizing their emission of pheromone so as to increase its downwind projection, and selecting calling sites (such as height on a tree that matches that of the males' ranging flight) that enhance the probability that their plumes will be detected.

The same selective pressures for efficient advertisement also apply to moth species that mate two or more times. Such subsequent matings can improve fertility (Delisle and Hardy, 1997) and the genetic diversity of a female's progeny (Drummond, 1984). Multiple matings ought not to be "superfluous," because they also can incur disadvantages such as increasing the risk of predation (Drummond, 1984).

Selection for male threshold of response and female rate of emission

The lower a male's threshold of response to pheromone, the further downwind a male ought to be able to *detect* a plume from a calling female. A logical corollary is that selection will favor males that have the lowest threshold of pheromone detection. However, a male's ability to follow a pheromone plume successfully to its source is limited by two features of plume dispersal that are not directly related to concentration of pheromone in the plume (Murlis *et al.*, 1992). First, as pheromone is transported downwind, turbulence fragments the plume into discontinuous pockets; far downwind, these gaps can expand to a scale of meters in length, and they are an impediment to plume following. Second, because of changes in wind direction and speed, the direction due upwind within the plume often is not along the plume's long axis (Elkinton *et al.*, 1987). Consequently, males face two meteorological constraints while heading upwind within the pheromone plume: they can encounter within-plume gaps or they may exit the plume's overall boundaries, after which they may have difficulty in reestablishing contact with the plume. Together, these two meteorological effects render tracking a plume to its source a substantial challenge.

Field experiments with male gypsy moths in a forest have shown that, although pheromone can be sensed by almost all males at distance of 120 m, few males (8%) released while in the plume successfully navigated a course to the plume's origin. Moreover, their median time in transit was about 10 min (Elkinton *et al.*, 1987), an interval far longer than what would be achieved by continuous direct flight to the source of pheromone (gypsy moth males have a typical flight speed in plumes of approximately 1 m/s). These findings suggest that the principal constraints on the range of pheromones are of two sorts. First, the male's threshold of response and the female's rate of release of pheromone together dictate how far downwind the plume's active space will project. Second, fragmentation of the plume by turbulence and the unreliability of wind direction as a directional guide place strong constraints on the feasibility of males tracking plumes distances of ten or more meters. Variability in the wind field will cause considerable minute-to-minute changes in a female's distance of communication. Nonetheless, selection should act to match the male's threshold to the lowest concentrations that typically occur near the maximum distances that males can track the plume. Active spaces for detection, in other words, ought not to be appreciably larger than the active space for successful plume tracking.

Does selection always favor females maximizing their release to achieve a maximum downwind projection of pheromone? Greenfield (1981) proposed a seemingly counterintuitive argument, in which females should favor males with low thresholds of pheromone responsiveness by emission of minimal quantities of pheromone, a "barely detectable signal," which would constitute a "passive mechanism of female choice" (Greenfield, 2002). Release of low amounts of pheromone (titer in the gland is not a reliable correlate of release rate) when the female is fully competent for reproduction would be supportive of Greenfield's passive filtering hypothesis. However, the relationships between the timing of calling and the rate of emission are complex and difficult to quantify (McNeil, 1991). Factors that need to be considered include: when calling begins and ends, whether it is continuous or occurs in distinct bouts, and the effect of the age of the female. Females of some moth species may release lower amounts of pheromone on the first night of calling than on subsequent nights, but this could be an incidental consequence of a recently eclosed female not being reproductively fully mature, rather than being an overt strategy for selection of sensitive males.

Lundberg and Löfstedt (1987) pointed out that releasing more pheromone and thereby attracting a mate over a longer distance potentially could select for a male with good abilities to follow a plume to its source and, indirectly, for other fitness traits correlated with strong flight. There is little evidence to substantiate such supposed variability in mate-finding abilities of male moths. The speed of flight of gypsy moth males flying along pheromone plumes in a wind tunnel is directly correlated with size: the larger the male the more rapid his flight (Kuenen and Cardé, 1993). It is unclear, however, whether male size in this species is heritable, or whether rapid flight along a plume of pheromone increases (or decreases) the chance of losing contact with the plume. In the field, small gypsy moth males are most apt to be indicative of crowded larval conditions or disease, both of which are environmental rather than heritable factors.

An important point is that the desirability of males with high sensitivity to pheromone and good plume-tracking abilities and females with a high rate of pheromone release would be most valuable when population density is low. Greenfield's hypothesis (1981) for females releasing a low "discriminating" amount of pheromone remains intriguing, but its validity hinges on the possible benefit of a female attracting a sensitive male versus the increased risk of not attracting any mate. This proposal also rests on demographic issues that are difficult to assess: how closely is rate of emission tied to a female's probability of mating, and how does this vary with density? Variation in rate of release of individual females can be measured, and it is clear that rates vary widely among taxa. At one extreme, females of the arctiid *Holomelina lamae* atomize pheromone, with individuals emitting up to 350 ng/min in the first 10 min of calling and 835 ng in 1 h (Schal *et al.*, 1987). Most moths, however, have much lower rates of emission. The pink bollworm, *Pectinophora gossypiella*, releases about 0.16 ng/min (Haynes and Baker, 1988), a rate typical of many moths.

Male-produced pheromones

In addition to female signaling that results in long-range attraction by males, male moths often release sex pheromones. In a few species, these male-produced signals attract females, and sometimes other males as well. More frequently, male signals are involved in short-range interactions that are more immediately associated with copulatory success or failure. In such cases, the male releases scent by very rapidly extruding a tuft of hairs (modified scales) with a large surface area (Birch *et al.*, 1990). These scent-disseminating structures are morphologically diverse, and in different species may be found on the abdomen, thorax, legs, or wings. Although such tufts are often displayed overtly in courtship, their effects can range from no apparent influence on female behavior to being a key stimulus in determining female mate preference. The morphological characteristics of these eversible structures have been studied in some taxa (Birch, 1972), but the detailed behavioral studies needed to characterize a pheromonal role for the compounds emitted have only been conducted on a few species. To illustrate the diversity of structures and functions of scent brush compounds, we will focus on five species: the oriental fruit moth, *G. molesta* (Tortricidae), *U. ornatrix* (Arctiidae), the salt-marsh caterpillar, *Estigmene acrea* (Arctiidae), the African sugarcane borer, *Eldana sacharina* (Pyralidae), and the cabbage looper, *Trichoplusia ni* (Noctuidae).

After flying upwind in response to female-produced pheromone and landing near a calling female, a male *G. molesta* everts a pair of hairpencils from a pocket located between the 7th and 8th abdominal segments (Baker and Cardé, 1979). The scent-disseminating brushes are infused with a blend of ethyl *trans*-cinnamate, methyl jasmonate, methyl 2-epijasmonate, and (*R*)-mellein (Baker *et al.*, 1981). A blend of ethyl *trans*-cinnamate and methyl 2-epijasmonate does indeed attract females from several centimeters downwind. The attracted female contacts the signaling male, and he whirls around to make a copulatory attempt. Adult males that were fed sugar water laced with ethyl *trans*-cinnamate attracted females from greater distances and more rapidly, thus garnering greater mating success than males that were fed only sugar water (Löfstedt *et al.*, 1989a). Adult males from larvae that have been reared on artificial diets apparently lack the pheromone components (Baker *et al.*, 1981; Nishida *et al.*, 1982). Therefore, the evidence suggests that sequestration of ethyl *trans*-cinnamate either by adult males or by male larvae influences their mating success.

Larval diet also influences the reproductive success of adult male *U. ornatrix*. Larvae of this arctiid moth consume seeds containing pyrrolizidine alkaloids (Conner *et al.*, 1981; see Ch. 7). Males and females gain protection from potential predators by sequestration of these toxic chemicals. In the adult male, some of the alkaloid is converted to hydroxydanaidal, which is released from inflatable, hair-covered structures (coremata) during courtship. Males that were denied access to pyrrolizidine alkaloids during the larval stage often were rejected by pheromone-releasing females (Conner *et al.*, 1981). There is a strong positive relationship between the quantity of alkaloid consumed by the larvae and the males' prenuptial content of alkaloid. Furthermore, the male's prenuptial content of alkaloid was strongly correlated with his pheromone titer (Dussourd *et al.*, 1991). Females that were reared on alkaloid-free diet were protected from spider predators by alkaloids that were transferred to them during copulation (González *et al.*, 1999). The benefits of mating with an alkaloid-rich male were also passed on to the eggs, as they were, in turn, protected by the defensive chemicals that can come from both mother and father (Dussourd *et al.*, 1988). In addition, if male alkaloid content reflects a heritable ability to forage and accumulate alkaloids, then the choosy female will increase the probability that her offspring will be good foragers and, therefore, well defended. These intricate relationships are reviewed by Conner and Weller in Ch. 7.

G. molesta and *U. ornatrix* are, therefore, two species in which a role for scent brushes and specific pheromone compounds in courtship has been established. The cabbage looper, *T. ni*, stands in stark contrast, because despite considerable research effort, no roles for scent brushes in *courtship* have been established. Males have a variety of modified scales associated with glands (Grant, 1971). The large scent brushes of the 8th abdominal segment are everted by male *T. ni* when they are close to a female (Gothilf and Shorey, 1976) and they produce an odor that is detectable to the human nose (K. F. Haynes, personal observation). Amputation of these scent brushes, however, does not influence the males' mating success (Gothilf and Shorey, 1976; K. F. Haynes, personal observation). In a different context, these scent brushes release *d*-linalool, *m*-cresol, and *p*-cresol, which attract females from long range (Landolt and Heath, 1990). The role for a female-produced sex attractant is well established in *T. ni* (review: Haynes, 1996), but it is not known whether male attraction or female attraction predominates in nature, or if there is a facultative change between these two signaling systems. Nonetheless, the overt use of scent brushes in *T. ni* courtship remains enigmatic. Perhaps the impact of released scents on females' acceptance of mates is so subtle that more sophisticated assays are required. Another possibility is that perception of the male's scent could alter postcopulatory events such as sperm precedence. Alternatively, a courtship function of scent brushes in this species could be vestigial (see below), with long-distance attraction of females by signaling males playing an increasingly important role.

Attraction of females mediated by compounds released from scent brushes or coremata of males has been documented in other moths, including *E. acrea* and *E. saccharina* (Atkinson, 1981; Willis and Birch, 1982), but many unanswered questions remain about the communication system of these species. Male saltmarsh caterpillar moths, *E. acrea*, have paired hair-lined coremata, which when inflated are filled with air and remain inflated for extended periods (Willis and Birch, 1982). Early in the evening, males displayed their coremata (and presumably released pheromone), stimulating both males and females to fly upwind into the area. Males that joined the aggregation typically inflated their coremata. Females that enter the aggregation mate with displaying males. *E. acrea* females release (*Z,Z*)-9,12-octadecadienal, (*Z,Z,Z*)-9,12,15-octadecatrienal, and (*Z,Z*)-3,6-*cis*-9,10-epoxyheneicosadiene (Hill and Roelofs, 1981). A blend of these compounds stimulated males to fly upwind in a wind tunnel. Males attempt to mate with calling females without inflating their coremata (Willis and Birch, 1982), which contrasts with coremata use in *U. ornatrix*. The mean time of male-signal-induced mating was about 1.5 h earlier than when females signaled (Willis and Birch, 1982). The chemical composition of the male signal has not been definitively determined.

Hydroxydanaidal has been extracted from coremata (Krasnoff and Roelofs, 1989). Like the cabbage looper moth, the salt marsh caterpillar moth has a dual signaling system, but in the case of *E. acrea* both males and females are attracted, leading to aggregations.

E. saccharina, the African sugarcane borer, has a distinctive communication system in which the male signals using wing glands, abdominal hairpencils, and high-frequency sound (Bennett *et al.*, 1991). The wing glands are situated at the base of each forewing, and the hairpencils are extruded from the 8th abdominal segment (Atkinson, 1982). Homologous wing glands in males of closely related moths (e.g., *Galleria mellonella*, the greater wax moth, and *Achroia grisella*, the lesser wax moth) have been shown to release a sex pheromone, and ultrasonic signals act in conjunction with the pheromone to attract females (Spangler, 1987; Spangler *et al.*, 1984). At night, *E. saccharina* males assume a posture in which they vibrate their wings and evert their hairpencils (Atkinson, 1981; Zagatti, 1981). Simultaneously, males produce high-frequency sonic communication signals from tymbal organs associated with the tegulae near the base of the forewing. Apparently, the ultrasonic signal, wing gland pheromone, and hairpencil pheromone act in concert to mediate attraction and/or courtship (Bennett *et al.*, 1991; Burger *et al.*, 1993; Zagatti, 1981).

In *G. molesta* and *U. ornatrix*, the pheromone signal and response have likely evolved through sexual selection. Mating does not occur at random but instead is influenced by mating preferences. Such selectivity means that some individuals have much greater reproductive success than others. Darwin (1871) recognized sexual selection as a principal factor influencing the evolution of mating systems (Bonduriansky, 2001). Signals that are the basis for mate selection may have evolved through a process of runaway selection, if signal and preference are genetically correlated (runaway selection: Fisher, 1958). Birch *et al.* (1990) point out that the initial preference of females for a male scent may be related to some fitness advantage, including avoiding interspecific mating mistakes. The signals may be a direct reflection of fitness in the context of natural selection (as would seem to be the case with *U. ornatrix*). Fisher's model of runaway sexual selection emphasized the arbitrariness of which set of characteristics would become more and more extreme. The diversity and spotty distribution of eversible scent-releasing structures in moths are consistent with the fickle nature of sexual selection. In addition, scent brushes have been lost (become vestigial). In some species, all that remain are non-functional brushes without glands or even only the levers (used for extrusion of the scent brush) (Birch, 1972, 1974). Adaptation to a new host with different plant chemistry may deprive insects of access to pheromone precursors. In geographic areas where such a switch has occurred, females could no longer be selective based on that signal. In this case, there should be selection against a female preference, and perhaps selection against the morphological adaptations for releasing scents in males.

Male-produced long-range signals may have evolved from courtship pheromones or the reverse, as was suggested by Birch (1974). The use of homologous glands for courtship and long-range attraction in related species suggests an evolutionary linkage between these two characteristics of sexual communication. The broad distribution of female-produced long-range sexual attractants and their occurrence in the sister order Trichoptera suggest that ancestral female Lepidoptera also used attractive pheromones. Courtship pheromones used by males, and male-produced long-range attractants, are more patchily distributed in Lepidoptera, suggesting that these have evolved independently many times. Baker and Cardé (1979) proposed that these signals could have evolved via sexual selection with a Fisherian runaway process. Phelan and Baker (1987) suggested that the initial evolution of male scent structures also was related to their value as reproductive isolating mechanisms. This view is further debated by Krasnoff (1996) and Phelan (1996).

Predators and parasitoids

Predators and parasitoids that exploit communication systems have the potential to influence the evolution of signal and/or response characteristics, but there is no good evidence to suggest that the evolution of sex pheromone communication systems of moths have been influenced by either illicit signalers or responders. Predatory "eavesdropping" on bark beetle aggregation pheromones (Wood, 1982), parasitoid and predator use of sex pheromones of stink bugs (Aldrich, 1995; Aldrich *et al.*, 1989; Millar *et al.*, 2001), and exploitation of other types of pheromone and modalities of communication in diverse taxa are well established (Haynes and Yeargan, 1999). Evidence for exploitation of female-produced pheromones of moths by illicit receivers, however, is limited. Egg parasitoids in the genus *Trichogramma* may remain in habitats in which female-produced pheromone of their host is present (Lewis *et al.*, 1982; Noldus, 1988; Noldus *et al.*, 1991). A direct response of these parasitoids to the moth pheromone would not lead them to ovipositional resources, but the presence of pheromone might be correlated with moth eggs in the habitat. Even such an indirect response to their hosts suggests an exceptional sensitivity of the parasitoids to sex pheromone. Two general characteristics of the sensory and central nervous systems of moths are their extraordinary sensitivity and their high specificity to pheromone blends. If sensory sensitivity and specificity are correlated in moths, how can generalist egg parasitoids, such as *Trichogramma* spp., show sensitivity without the same degree of specialization? Studies of the sensory physiology of these parasitoids could help us to understand the apparent trade-offs between sensitivity and specificity.

Responses of specialist egg parasitoids to sex pheromones of their host species illustrate a remarkable adaptation for host-finding. *Telemonus euproctidis* parasitizes eggs of the lymantrid moth, *Euproctis taiwana* (Arakaki *et al.*, 1996). The adult female parasitoid is phoretic on virgin female moths. After mating when the female moth oviposits, the wasp jumps off and parasitizes the moth eggs. Production of sex pheromone by the moth is a clear indicator of a female that has the potential for future egg laying. The response of the wasp to (*Z*)-16-methyl-9-heptadecenyl isobutyrate, a sex pheromone component of this moth, leads the wasp indirectly to resources for her offspring (Arakaki *et al.*, 1996). Because the parasitoid is a specialist, the evolution of a sensitive and highly specialized sensory system for detecting the kairomonal signal would be advantageous: this does not pose the theoretical conflicts involved in a generalist responding to specialized cues.

There are no known examples of parasitoids that attack adult moths, although there are parasitoids that attack adults of many other orders of insects. Perhaps the low-amplitude sex pheromone signals of moths represent an adaptation to predators: an evolutionary escape from predators that might use higher-amplitude signals as beacons to locate food. Greenfield (1981) suggested that low-amplitude signals might be an adaptation to avoid predation (he also proposed the hypothesis that the low-amplitude signal may have arisen through sexual selection favoring males that were the most sensitive to pheromone). Similar evolutionary arguments have been advanced to explain the evolution of substrate-borne vibrational signals, which would presumably be more difficult for a predator to detect than airborne acoustical signals (see Haynes and Yeargan, 1999).

Evidence supporting direct exploitation of sex pheromone communication in moths by illicit receivers is limited, but with the notable exception of bolas spiders, there are also few examples of illicit signalers that take advantage of the predictable responses of male moths to sex pheromones. Adult female bolas spiders mimic the female-produced sex pheromones of moths and catch the attracted males with a specialized weapon called a bolas (Eberhard, 1977, 1980; Stowe *et al.*, 1987; Yeargan, 1988, 1994). The bolas consists of a single strand of silk with a terminal ball of liquid adhesive. The female spider holds the bolas with either of her forelegs, and when she detects wing vibrations of an approaching moth, she swings the bolas in the downwind direction (Haynes *et al.*, 2001). If the sticky globule comes in contact with the moth, the specialized and highly effective glue penetrates through the scales. The spider then reels in and envenomates the moth, paralyzing it before she wraps it in restraining silk. For aggressive chemical mimicry to be effective, the spider has to control precisely the biosynthesis of the illicit signal, because the blend ratio is critical to species specificity of the normal pheromone signal. The bolas spider, *Mastophora hutchinsoni*, releases an allomonal blend that includes a blend ratio of (*Z*)-9-tetradecenyl acetate and (*Z,E*)-9,12-tetradecadienyl acetate (Gemeno *et al.*, 2000a) that is similar to the pheromonal blend of one of its victims, the bristly cutworm moth, *Lacinipolia renigera* (Haynes, 1990). Does predation of this kind impose selection on characteristics of the moths' chemical communication system? The spider is not common, and, therefore, its impact on the prey population densities may be negligible. From the perspective of the male moth, it seems unlikely that the risk of predation exceeds the risks of going unmated, which could be associated with avoidance of the false signal. However, this conclusion only applies if the illicit signal is identical to the functional signal emitted by the female moth. Selection favoring male discrimination between spider allomone and moth pheromone may have little or no cost and an obvious benefit. At relatively high population densities of prey compared with population densities of predators, it is unlikely that females will go unmated, and, therefore, there would be little or no selection on female signaling. These arguments leave us with unresolved questions. First, given an apparently efficient mechanism of hunting, why are predator populations so limited? Second, have these predators always had relatively low population densities? If not, could the predator have influenced the prey's chemical communication system in the past?

Phylogeny and chemical diversity of structure and biosynthetic pathways

Closely related species of moths tend to share pheromone components, which is a reflection of their common evolutionary history. Not surprisingly, closely related species also have similarities in the biosynthetic pathways that produce pheromone components. Enzymatic changes can result in modifications of the placement of double bonds (e.g., Δ^9, versus Δ^{11}), isomerization (e.g., (*Z*)- versus (*E*)-isomers), chain shortening (β-oxidation), or functional groups, amongst others. The noctuid subfamily Plusiinae illustrates how pheromone diversity can be generated around a core of biochemically similar enzymatic reactions. (Z)-7-Dodecenyl acetate is a component of many species within this subfamily (see Arn *et al.*, 1992, 2003). The basic enzymatic reactions used in pheromone biosynthesis in this subfamily have been determined in the cabbage looper, *T. ni* (Fig. 8.1) (Bjostad *et al.*, 1984). Some species include unesterified (Z)-7-dodecenol as a pheromone component. The soybean looper, *Pseudoplusia includens*, pheromone includes (Z)-7-dodecenyl proprionate and (Z)-7-dodecenyl butyrate as minor components, indicative of esterification with short-chain substrates other than acetate. For some species, (Z)-11-hexadecenyl acetate, (Z)-9-tetradecenyl acetate, and (Z)-5-decenyl acetate are used as pheromone components. These compounds are related to (Z)-7-dodecenyl acetate in that they are zero, one, and three cycles of chain shortening (losing two carbon atoms per cycle) removed from a common precursor, (Z)-11-hexadecenyl fatty acid ((*Z*)-7-dodecenyl acetate is two cycles downstream of this precursor). Because chain shortening is a common enzymatic step in pheromone biosynthesis for many species, compounds that are either two carbons shorter or two carbons longer are

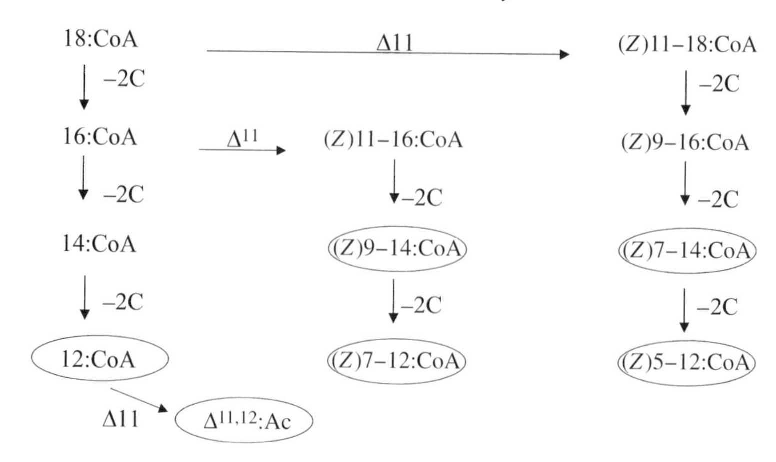

Fig. 8.1. Pheromone biosynthetic pathway in the cabbage looper, *Trichoplusia ni.* Key steps are chain shortening (−2C) and Δ^{11} desaturation ($\Delta 11$). Circled acyl-coenzyme A (CoA) derivatives are reduced and acetylated to form the pheromone components. (Modified from Jurenka *et al.*, 1994.)

easily produced via the same general pathways. Mutations that affect one element of the biosynthetic pathway can alter the presence, absence, or relative abundance of key pheromone precursors.

The biochemical machinery for producing pheromones has similarities across broadly divergent taxa. Chain shortening, Δ^{11} desaturation, reduction, and acetylation are common enzymatically driven reactions in the Noctuidae, Pyralidae, Tortricidae, Gelechiidae, and other families of Lepidoptera. Qualitative and quantitative variation in substrate specificity or other aspects of reaction kinetics are apparently sufficient to generate a tremendous diversity of compounds that could serve as pheromone components. Because the underlying biosynthetic reactions are alike, similar pheromone components are often found in distantly related taxa. For example, similar patterns of use of (*Z*)-7-dodecenyl acetate and biosynthetically closely related compounds noted above for Noctuidae: Plusiinae are repeated in the noctuid subfamilies Hadeninae and Noctuinae, and even in Tortricidae: Olethreutinae (Arn *et al.*, 2003). Similarly, (*Z*)-11-hexadecenal is an identified pheromone component in species within six families of moths (Hieroxestidae (1 sp.), Plutellidae (1 sp.), Sphingidae (1 sp.), Pterophoridae (2 spp.), Pyralidae (9 spp.), and Noctuidae (19 spp.); Arn *et al.*, 2003). The use of (*Z*)-11-hexadecanal as a pheromone component has evolved more than once. By plotting the occurrence of pheromone components on a phylogeny of two tortricid genera (*Ctenopseustis* and *Panotortrix*), Newcomb and Gleeson (1998) found evidence for the independent gains and losses of pheromone components. They concluded that the use of (*Z*)-5-tetradecenyl acetate has evolved twice within this complex.

Compounds that are characteristic of a genus or even subfamily level do not always distinguish one genus from a distantly related genus. A single gene mutation in the cabbage looper moth (*T. ni*, Noctuidae: Plusiinae) leads to production of a pheromonal blend that is more attractive to black cutworm moths (*Agrotis ipsilon*, Noctuidae: Noctuinae), than to wild-type conspecifics. In natural habitats, the potential for these types of interspecific (even interfamilial) interaction may be rare because of other factors such as diel periodicity, seasonal occurrence, and geographic distribution, but this potential suggests that not only reinforcement of isolation mechanisms between incipient species but also communication interference between distantly related taxa may be important.

Two general classes of pheromone compound have been identified in moths, and these have some broad, although not uniform, associations with certain taxa. The polyene hydrocarbons and epoxides of various chain lengths are pheromones found in some subfamilies of the Geometridae and Noctuidae, and in the Arctiidae and Lymantridae (Millar, 2000). These compounds are probably derived from dietary linoleic and linolenic acids. The other major class of pheromone compounds includes acetate, alcohols, and aldehydes, which are found in the Tortricidae, Pyralidae, Gelechiidae, Sessiidae, and Noctuidae. This class of compounds is derived from the insect's fatty acid synthesis pathway, with enzymatic modifications discussed above. Both classes of pheromone are broadly represented in the Noctuidae but are typically found in different subfamilies (Arn *et al.*, 1992, 2003).

Genetic architecture of communications systems and mutation

The genetic architecture of communication systems provides clues to the evolutionary history of the system, as well as an indication of the potential for selection to influence signaling and response characteristics. These traits may be influenced by single genes with major impacts or by many genes with additive influences on communication. Genetic coupling or linkage disequilibrium between signal and response characteristics may have been favored by selection, and such linkage could then influence the potential for genetically coordinated changes in both sides of communication. However, some of the best-studied systems suggest that linkage disequilibrium has not played a major role in the evolution of species-specific pheromone blends.

Polygenic influences on pheromone signals have been documented in several species. Hunt *et al.* (1990) found that more than one gene influences the subtle variation in the pheromone blend that is observed in field populations of the cabbage looper moth, *T. ni*. At a minimum, one autosomal gene and one sex-linked gene were indicated. Estimates of heritabilities of the quantities of components of the pheromone blend were moderate (Gemeno *et al.*, 2001), but genetic correlations between components (both positive and negative correlations) suggest that there are genetic limits on the overall structure of the pheromone blend. These limits may be indicative of the underlying biosynthetic pathways (i.e., shunting a precursor preferentially down one branch of a biosynthetic pathway may preclude it from going down an alternative pathway). Collins and Cardé (1985) documented heritable variation in both pheromone component ratios and pheromone quantity in the pink bollworm moth, *P. gossypiella.* Selection for altered pheromone blend ratios (Collins and Cardé, 1989) and increased pheromone quantity (Collins *et al.*, 1990) resulted in predictable changes in these characteristics of the chemical communication system. Selection for increased pheromone titer resulted in no change in component ratios, but selection for altered pheromone-component ratios caused a decrease in pheromone quantity (Collins *et al.*, 1990). Genetic correlations between such characteristics could lead to multiple effects of selection on the communication system.

Single genes can have a major influence on production of pheromone blend, and on the response to pheromones, suggesting that these genes may have a central role in reproductive isolation between species. The best-documented case is the genetic basis of differences between two strains of European corn borer, *Ostrinia nubilalis.* Crossing studies between the strains have documented that a single autosomal gene is responsible for the reversal of the pheromone blend ratio from 3:97 to 99:1 for (*E*)-11-tetradecenyl acetate to (*Z*)-11-tetradecenyl acetate (Klun and Maini, 1979; Roelofs *et al.*, 1987). Zhu *et al.* (1996a) found that the substrate specificity of the reductase enzyme (converting fatty acids to alcohols) was responsible for the pheromone differences between the two strains. The behavioral response of males is influenced by an independent gene that is sex linked (in Lepidoptera, females are the heterogametic sex). Another independent autosomal gene affects which of the pheromone components stimulates high-amplitude action potentials in receptor neurons (Löfstedt *et al.*, 1989b).

Although major genes clearly explain much of the pheromone differences between the (*E*)- and (*Z*)-strains, modifier genes also are involved. Crossing studies revealed that additional alleles and different genes have the potential to affect the pheromone blend ratio (Zhu *et al.*, 1996b). This genetically based variation is normally hidden within the (*E*)- and (*Z*)-containing strains but is exposed in F_1 crosses and backcrosses between these strains. Because crosses occur in nature, this variation may occasionally be exposed and, therefore, has the potential to influence the evolution of the communication system.

Single autosomal gene differences appear to control the pheromonal blend differences between *Spodoptera latifascia* and *Spodoptera descoinsi.* Chemical communication in these two species can be distinguished not only by the pheromone blend but also by the periodicity of calling behavior (Monti *et al.*, 1997). The ratio of (*Z,E*)-9,12-tetradecadienyl acetate and (*Z*)-9-tetradecenyl acetate is a critical difference between the two species. These two species can produce fertile offspring in the laboratory. Crossing experiments revealed that a single autosomal gene was primarily responsible for the pheromonal differences, but modifier genes also may control more subtle differences (including the possibility of sex-linked modifiers). In addition, the mean time of onset of calling of *S. latifascia* is substantially later at night than that of *S. descoinsi*. Hybrid females show an intermediate calling time. Monti *et al.* (1997) provisionally concluded that calling time was under polygenic control.

The genetic basis of chemical communication differences between two species of *Ctenopseustis* involves both single gene and polygenic differences (Foster *et al.*, 1996). *Ctenopseustis obliquana* females produce both (*Z*)-8-tetradecenyl acetate and (*Z*)-5-tetradecenyl acetate, whereas *Ctenopseustis herana* females produce only the latter. A single sex-linked gene influences the difference between the two species in sensory receptors for pheromone components, but variability in the responses of hybrids suggested that other genes may be involved as well. Another sex-linked gene controls the difference in behavioral responses to pheromone blends. Because the two genes are sex linked they are genetically coupled, but the linkage distance is not known. The pheromone blend differences did not fit a simple single gene inheritance model.

The impact that a point mutation can have on a complex pheromone blend is illustrated by the cabbage looper moth, *T. ni* (Haynes and Hunt, 1990). The mutation affects a chain-shortening step in the biosynthetic pathway (Jurenka *et al.*, 1994). The result is that females expressing the mutation, which is a single autosomal recessive gene, produce much less (*Z*)-5-dodecenyl acetate, which is three cycles of chain shortening downstream of the Δ^{11} desaturation step. These females also produce much more (*Z*)-9-tetradecenyl acetate, which is only one step removed from the desaturation step. The other pheromone components are affected by the mutation as well, but to a lesser extent. Mutant females are much less effective than normal females in attracting male *T. ni* in the field. In fact, they are more effective in attracting male black cutworm moths, *A. ipsilon*, than males of their own species. Male *T. ni* that carry the mutant gene initially responded preferentially to the normal pheromone blend, as did the normal males. However, within colonies in which all females expressed the mutant phenotype, a gradual evolutionary change in the male response phenotype occurred, so that males responded equally well to mutant and normal pheromone blends (Liu and Haynes, 1994). Males that continued to discriminate in favor of the normal pheromone blend would be at a reproductive disadvantage in these colonies. Evenden *et al.* (2002) found that there was heritable variation in the breadth of responsiveness to the mutant and normal pheromone blend even in the normal colony. The response to selection observed in the mutant colony is consistent with Phelan's (1992) asymmetric tracking hypothesis, which predicts that males are more likely to track changes in the female pheromone blend than the reverse. The occurrence and impact of a mutation in this laboratory colony of *T. ni*, and the response of males to selection, suggest a role for both major gene and additive genetic effects in the evolutionary diversification of pheromone blends.

There are very few studies of the genetic basis of signal and response characteristics in pheromone communication. The studies that have been conducted indicate that additive genetic characteristics explain the relatively subtle variation in signal characteristics within populations and may help to explain geographical variation (e.g., the black cutworm, *A. ipsilon* (Gemeno *et al.*, 2000b) and the turnip moth, *Agrotis segetum* (LaForest *et al.*, 1997)). However, the case of the European corn borer suggests that major gene effects are important determinants of differences between field strains. Based on this limited dataset, it seems likely that both major gene and polygenic effects will play an important role in the evolution of species-specific communication systems. To understand variation among individuals, it will be especially useful to have quantitative trait locus analyses.

Signal specificity at the species level

Female-emitted pheromones form a major element of the mate-finding and species-recognition system for nearly all moths. Distinctive communication channels among closely related species sharing the same habitat and with overlapping seasonal phenologies are achieved by a number of strategies. The pheromones of each species may differ in structure or ratio, and the presence of behavioral antagonists may confer additional specificity. Non-overlapping windows of reproductive activity provide a mechanism for species to share the same chemical channel. Additional species specificity in mate recognition after attraction can be achieved by differences in mating behaviors (e.g., Phelan and Baker, 1990) and by male-produced pheromones (Phelan and Baker, 1987). Often more than one strategy is employed to achieve unique channels.

There are numerous examples of species pairs and species complexes isolated by distinctive female-emitted pheromones and, in many cases, antagonistic, inter-species interactions. For example, four limacodid (nettle caterpillar) moths from Borneo occurring in oil palm plantations differ in their pheromone blends, with further isolation being conferred by some separation in times of sexual activity, and even the preferential height at which males seem to locate pheromone sources. The pheromones of some species act antagonistically when added to the pheromone of another limacodid, but the extent to which such antagonism augments the blend exclusivity is not clear, given that cross-attractivity of blends without antagonists added is relatively infrequent (Sasaerila *et al.*, 2000).

Among the tortricine moths, the chemistry of pheromone communication is especially well known (Roelofs and Brown, 1982; Arn *et al.*, 1992, 2003). There are many examples of closely related species that share components of their pheromone while achieving species-specific attractants with unique components, unique ratios of components, or interspecific antagonistic effects. For example, among the leafroller moths of eastern North America, *Platynota idaeusalis* and *Platynota flavedana* emerge on the same hosts at the same time of year. Although these congeners share one component in common, signal specificity is based on attraction to unique blends and reciprocal antagonism. *P. idaeusalis* uses an approximately equal mix of (*E*)-11-tetradecenyl acetate and (*E*)-11-tetradecenol (Hill *et al.*, 1974). The pheromone of *P. flavedana* is an 85:15 blend of (*E*)- and (*Z*)-11-tetradecenol (Hill *et al.*, 1977). If a synthetic lure consisting of the natural blend of either species is augmented by addition of the unique component of the other species' blend, the lure is no longer effective.

There are many other examples of species isolation by distinctive blends in tortricids and other moth groups. One of the more remarkable cases is found in the two sibling species of the dingy cutworm moth, *Feltia jaculifera*, complex (Byers and Struble, 1990). These two species were uncovered during attempts to define a field-attractive blend for this noctuid. Subsequent studies of the pheromone composition of isofemale lines revealed that there were two pheromone forms. These have not been separable by either morphological differences or distinctive mitochondrial lineages (Sperling *et al.*, 1996). They are common, polyphagous, and broadly sympatric across the prairies of western Canada. Both species produce a blend of (*Z*)-7-dodecenyl, (*Z*)-9-tetradecenyl, and (*Z*)-11-hexadecenyl acetates, in a 100:13:3 ratio for the A species and 0.3:0.5:100 ratio for the B species. Optimum attraction for A occurs with an 8 μg lure in a 10:1:0 ratio; for B the optimum is a 500 μg lure in a 1:1:2000 ratio. The role of pheromonal differences in the initial divergence of these two species is unknown.

The existence of two pheromone races of the European corn borer, *O. nubilalis*, has been considered in the context of the genetic architecture regulating pheromone production and response. These two strains were transported from Europe to new areas (such as North America) by human activity and much of the current sympatry of the two strains in eastern North America and some regions of Europe may have been fostered by both strains shifting to maize as a larval food source. It is uncertain how these strains diverged, but this may have occurred allopatrically as it appears that hybridization in the field in areas of current sympatry is infrequent (Klun and Huettel, 1988; Bengtsson and Löfstedt, 1990).

Another example of pheromone diversification occurs in the larch budmoth, *Zeiraphera diniana*, complex. The two species use (*E*)-9- and (*E*)-11-tetradecenyl acetates as a pheromone, the larch-feeding type in a 100:1 ratio and the pine-feeding type in the inverse ratio (Priesner and Baltensweiler, 1987). Although earlier studies suggested that hybridization and introgression between these two forms was likely (Guerin *et al.*, 1984), allozyme studies by Emelianov *et al.* (1995) confirmed that populations of the larch and pine forms are not panmictic and instead are "good" species. Certainly differences in the pheromone channel alone would account for nearly complete reproductive isolation between the two species, and phenological differences in host acceptability of larch and pine afford an additional degree of isolation because these result in somewhat differing times of seasonal flight activity in areas of sympatry. The initial divergence of these two species in sympatry via a host shift was proposed to be as probable as conventional allopatric differentiation (Emelianov *et al.*, 1995).

Geographic variation

Löfstedt and colleagues have provided the most thoroughly documented example of geographic variation in a sex pheromone in the turnip moth, *A. segetum*. This noctuid occurs throughout much of Africa and most of Europe and Asia and it is now recognized to have four pheromone components: (*Z*)-5-decenyl, (*Z*)-5-dodecenyl, (*Z*)-7-dodecenyl, and (*Z*)-9-tetradecenyl acetates (Löfstedt *et al.*, 1985; Wu *et al.*, 1995, 1999). Most of the studies have examined the proportions across Europe of three of these components ((*Z*)-5-decenyl, (*Z*)-7-dodecenyl, and (*Z*)-9-tetradecenyl acetates) in the blend of individual females, and the attraction of males in the field to a range of blend ratios (Tòth *et al.*, 1992) (Fig. 8.2). There is substantial variability within populations and between geographic regions in the ratio produced by the female and the male attraction (reviewed: Löfstedt, 1993; see also Hansson *et al.*, 1990). Generally, female blend production and the proportions of male antennal receptor attuned to each component co-vary geographically (Löfstedt, 1993) (Fig. 8.3). Some of the European populations, such as those from France and those from Armenia and Bulgaria, differ sufficiently in their pheromone systems such that they would appear to be "good" species from this viewpoint. In Hungry, two differing parapatric populations are present.

The evolutionary basis for such populational divergence is unclear (Löfstedt, 1993; Hansson *et al.*, 1990; Wu *et al.*, 1999), and *A. segetum* may actually comprise a complex of sibling species that originated in allopatry, with communication interference being the main force causing the pheromone channels to diverge. In areas of current sympatry of strains, intermediate types occur, indicative of some hybridization. Over time, strain differences could be magnified through reinforcement if hybrids have reduced fitness (Butlin, 1987). An alternative path for divergence of the pheromone channel in either sympatry or parapatry supposes differing selective pressures from ecological factors, for example differences in host plant

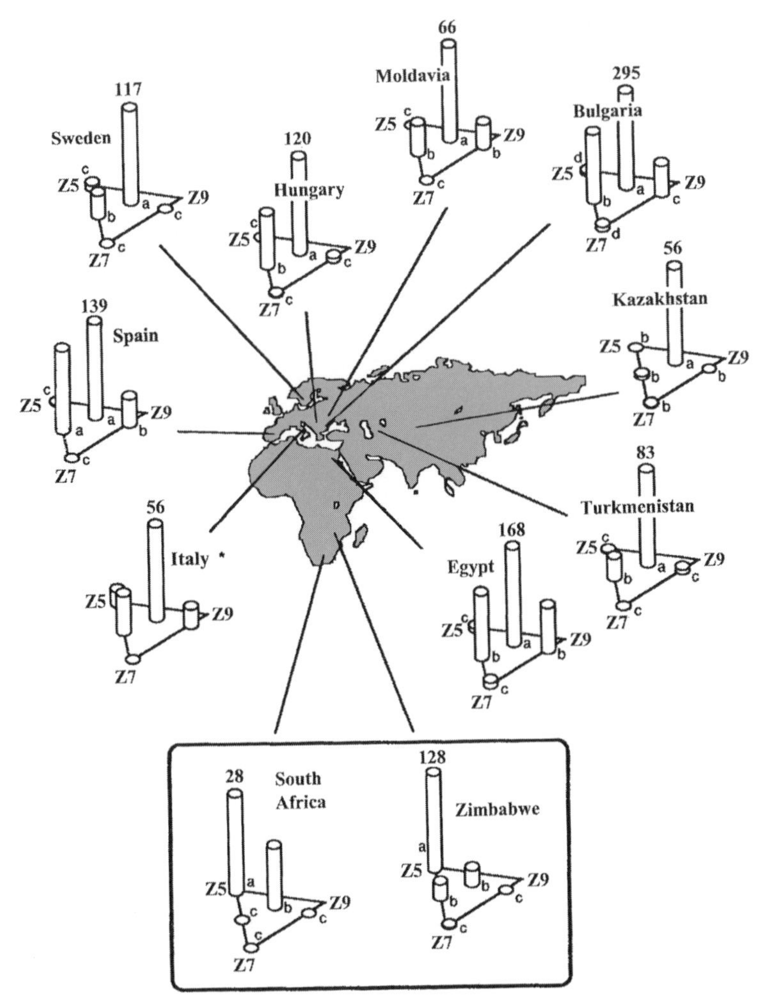

Fig. 8.2. Geographic variation in male attraction to pheromone in the turnip moth, *Agrotis segetum* as percentage capture at 11 localities in Eurasia and Africa. The numbers above each column indicate the highest catch of the best lure. Z5, (Z)-5-decenyl acetate; Z7, (Z)-7-dodecenyl acetate; and Z9, (Z)-9-tetradecenyl acetate. Within each diagram, differing letters indicate $P = 0.05$. (From Tòth *et al.*, 1992.)

utilization. Insight into which of these alternatives may have fostered geographical divergence in the pheromone system of *A. segetum* requires further investigation of the genetic control of differences in these strains and their probable evolutionary history.

A second example of geographically based differentiation is seen in the New Zealand torticid *Planotortrix exessana* (Foster *et al.*, 1989). This species employs a mix of (Z)-5- and (Z)-7-tetradecenyl acetates. Initially, it was thought that there were two discrete, geographically based groups of females (Foster *et al.*, 1986), one with

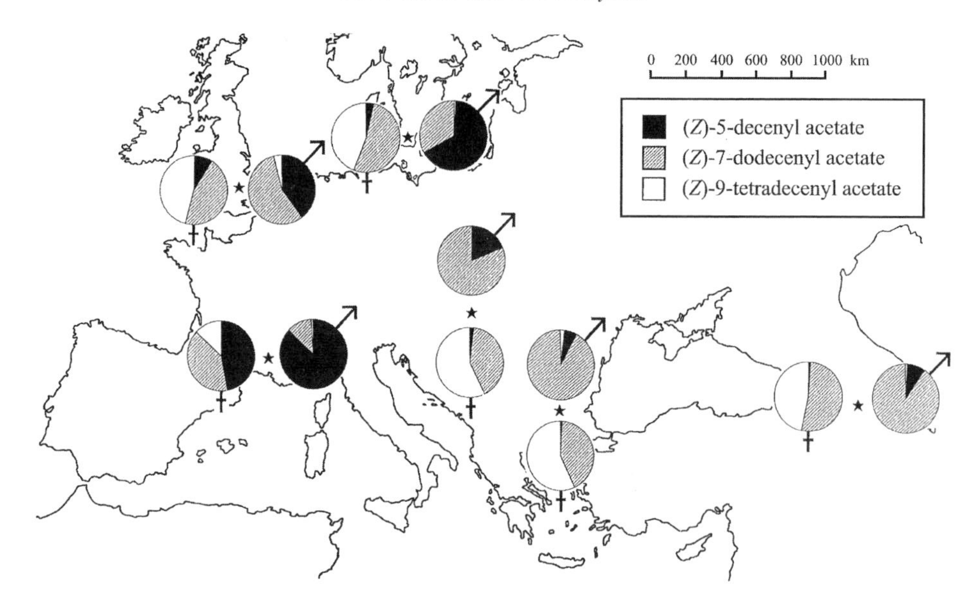

Fig. 8.3. Geographic variation in the ratio of pheromone components in the female sex pheromone and the relative abundance of the male antennal sensilla attuned to these components in the turnip moth, *Agrotis segetum*. (From Löfstedt, 1993.)

a ratio of 46:54 to 70:30 of (*Z*)-5 to (*Z*)-7 with a mean of 3.5 ng/female and another population with ratios ranging from 12:88 to 33:67 and a mean of 22 ng/female. Sampling of a greater number of females from additional geographic locations, however, revealed that the North Island and South Island populations generally did not differ in either component ratio or titer, with ratios ranging continuously from 3:97 to 71:29, and with titer extending nearly continuously up to 36 ng/female. Foster *et al.* (1989) also determined the extent to which tethered females from either the Waikato (North Island) or Dunedin (South Island) populations, which differed in mean ratios and titers, were likely to mate in field cages with either Waikato or Dunedin males. These studies established substantial variation within and between localities for *P. exessana*. Some of this variation could be the result of geographical isolation of South and North Island populations, but it is also clear that this tortricid has a high within-population variability in pheromone production and an evidently catholic response by males.

Among the mechanisms that have been identified (Cardé, 1987) as candidates for promoting distinctive pheromone communication channels for mate finding are: (i) reduction or elimination of response to and production of pheromone components among co-existing taxa that hybridize (reinforcement *sensu* Butlin); (ii) reduction or elimination of response to and production of pheromone components among co-existing taxa that cross-attract but do not interbreed (communication interference); (iii) narrow tuning of the response and production, reducing variance in the channel (stabilizing selection); and (iv) selection of mates that enhance the fitness of offspring (sexual selection). Support for these mechanisms in the evolution of communication by pheromones is surprisingly limited, given that our knowledge of the chemistry of the attractants used by hundreds of species of moths is rather extensive.

Stabilizing selection and sexual selection

Stabilizing selection is the force that maintains stasis in a pheromone channel. Given that the pheromone comprises a blend of components, females at or tending toward the mean blend should have the greatest probability of attracting mates: either more mates for those species in which females engage in multiple matings or an enhanced probability of securing at least one mating for those species in which the female mates once. Similarly, males whose optimal response is centered near the most prevalent female blend – by having, for example, the lowest threshold of response for that blend – should have an increased probability of finding a mate (or mates), because the median blend is the most common phenotype. Also, although males should favor the normative blend by having the lowest threshold of response for this blend, males also should be willing to accept females that deviate from the mean. The kind of information needed to assess the reproductive or fitness "cost" of a male mating with a female that deviates somewhat from the norm is difficult to acquire. A potential cost is that the female progeny of a male mating with an off-blend female might be somewhat less fit (attractive), provided that the production of the off-blend is a heritable characteristic.

De Jong (1988) used a population genetics model to determine the effects of the number of matings on male and female traits for moth mating systems. He proposed that at evolutionary equilibrium there should be a larger (genetically based) variance among female traits for pheromone attractiveness than among male traits for response to pheromone. This difference was attributed to asymmetric sexual selection, with males having opportunities to mate more often than females. De Jong argued that, in the pool of unmated females, the most attractive ones would be the first to mate and the remaining, less-attractive ones (termed "wallflowers") would over the season increase in proportion. Some of these females might not mate at all, and even a delay in mating can reduce fecundity. It remains unclear, however, why the model predicts that selection would not act against such heritable variants, the wallflower females, because they are by definition at a reproductive disadvantage in luring a mate. Furthermore, the outcome of the wallflower effect model may be very biased by the assumption that males are much more likely to be attracted to females with the most common pheromone blend than to other potential mates (De Jong, 1988; De Jong and Sabelis, 1991).

Phelan (1992) also has emphasized that the strength of stabilizing selection on males and females should be asymmetrical because of differential parental investment. Females are viewed as being under "weak" stabilizing selection to emit the most attractive blend, causing "significant" genetically based pheromonal variation in blend production to arise (De Jong, 1988; Phelan, 1992). Males, generally having little investment beyond sperm, were expected to show "maximum sensitivity to the most common blend" while not having so narrow a response as to exclude "other blends that signal viable mates" (Phelan, 1997). The point that males would be attracted to most or all females in the population differs from the De Jong model, which postulates wallflower females. Phelan's (1992) prediction that the non-limited sex, the male, follows changes in the communication system of the limited sex, the female, was termed "asymmetric tracking."

How much evidence is available regarding the contention that males have less genetically set variation in response than females have in signal production? The extent of variability seen in responses of a population of males to blend variants could be mainly the consequence of the variability in response of *individual* males, rather than signifying the presence of males with differing, genetically based preferences for particular ratios. Some evidence of individual variability lacking a clear genetic component is seen in the results of field experiments that allowed males to be attracted more than once to a range of blend ratios. On their visit to one or more artificial lures on the first day, males marked themselves with a powder that was color coded for the ratio in the lures; their fidelity to that particular ratio would be evident in their capture on subsequent days in traps baited with lures with the same range of ratios. If the response to a particular ratio was somewhat fixed, then we should expect males to tend to be reattracted to the lure that they visited on the first day. Such experiments with *G. molesta* (Cardé *et al.*, 1976) and *P. gossypiella* (Haynes and Baker, 1988) have shown that variability in attraction to a spectrum of ratios seems attributable to the capacity of individuals to respond to a range of ratios (albeit with a clear preference on average), rather than the presence of distinct response types.

In some species, male variation in response to component ratio offset from the natural blend is somewhat modulated by ambient temperature (Linn *et al.*, 1988). The response specificity of *G. molesta* and *P. gossypiella* to off-ratios of pheromone acetate components in a wind tunnel assay was narrower at 20 °C than at 26 °C. In the field, sexual activity in both species occurs at both of these temperatures, depending on time of year. Some field evidence of this phenomenon with *P. gossypiella* appears in the distribution of catch in traps baited with a range of ratios measured at various times of the flight season. Flint *et al.* (1977) found an evidently narrower response breadth early in the season (when temperatures were cool) compared with late-season responses. In the omnivorous leafroller *Platynota stultana*, the optimum ratio of its two components for attraction seems to shift with temperature in the field, whereas the ratio of components produced by the female does not (Baker *et al.*, 1978). The extent to which elevated temperatures broaden the acceptability of ratios away from the mean, or even shift the optimum ratio, has not been widely studied in other moths. To provide unambiguous data, it will be necessary to have dispenser systems that emit a constant ratio of components at the test temperatures, and traps that remain equally effective at retaining lured moths as they capture more individuals. Temperature-induced plasticity in response to ratio is consistent with the prediction that response variability in the tracking sex should be largely non-genetic (Phelan, 1992).

Although we have provided cases where control of component ratio by the females is precise, there are many other species in which the ratio of components varies widely. Perhaps the most extreme case is the potato tuberworm, *Phthorimaea operculella*, which uses (*E,Z,Z*)-4,7,10-tridecatrienyl and (*E,Z*)-4,7-tridecadienyl acetates. The percentage of the triene is influenced by the rearing temperature, with the ratio of triene to diene ranging from about 4:1 at 15 °C to an equal mix at 30 °C (Ono, 1993). However, even at a constant rearing temperature, female moths showed wide variability in ratios. Males, not surprisingly, show a wide latitude in responsiveness to ratio, with good attraction occurring to ratios ranging from 9:1 to 1:9 (Ono and Orita, 1986).

Female variability in ratio production has not been determined for *G. molesta*, but it is established for *P. gossypiella* (Collins and Cardé, 1985). The mean ratio of the (*Z,Z*)- and (*Z,E*)-isomers of the 7,11-hexadecadienyl acetate components is 56:44, with little variability (coefficient of variation, 5.3%). Based on field tests with a range of lures, the response of males broadly overlaps the ratios produced by females (Haynes and Baker, 1988) and also the males' threshold for response is equivalent for the range of ratios produced by females (Linn and Roelofs, 1985). One consequence of this in *P. gossypiella* is that a male would not select a female based on the range of ratios that the females emit.

The genetic basis of pheromone production and response suggests that these traits could be changed by the disruptive selection imposed by using synthetic pheromone for mating disruption. Heritability of blend ratio in female *P. gossypiella* was 0.34 (Collins and Cardé, 1985). Directional selection experiments intended to alter the ratio of components produced a change only for increased (*Z,E*)-isomer in the ratio, which was elevated from 42.9 to 48.2% over 12 generations, although in the selected line the amount produced was depressed by about half (Collins and Cardé, 1989). Such a minor change in ratio in female production, however, would be transparent for *P. gossypiella* males, given their catholic reaction to all ratios within this range. In the male, the duration of wing-fanning and the propensity to initiate upwind flight to a range of ratios are well correlated, and so heritabilities of wing-fanning to a range of ratios have been used as a guide to the probable heritabilities of attraction to the same ratios (Collins and Cardé, 1989). Heritability to the natural blend was

0.385; heritability to a blend with 25% (*Z,E*)-isomer was 0.377, and that to a 65% blend was −0.145. This asymmetry suggests that directional selection would be much more likely to shift male response in the direction of low rather than high ratios of the (*Z,E*)-isomer. However, in *P. gossypiella* there is no clear evidence that there is more genetically based variation in female production of blend ratio than in male response to blend ratio.

The asymmetric tracking hypothesis does not explicitly set forth conditions that cause males to have a genetically based breadth of response to blend variants that is more narrowly governed than the genetically based female's production of pheromone, other than the supposition that the cost to males of making the mistake of not responding to a given variant female is assumed to be higher – females are less likely to remain unmated than males. In general, heritabilites of communication systems for mate finding by males would be predicted to be comparatively low because males are assumed to be under strong selective pressure to find females (Falconer, 1981), and characters closely related to fitness should have low additive genetic variance (Maynard Smith, 1978). This is the paradox characterized by Lewontin (1979) as "natural selection destroys the genetic variance on which it feeds."

Reinforcement and reproductive character displacement

Natural selection can enhance prezyogotic barriers to mating in zones of overlap of species that are already isolated by effective postzygotic barriers or in zones of hybridization between incipient species. One of the potential products of such reinforcement is reproductive character displacement, which Howard (1993) categorized as a "pattern of greater divergence of an isolating trait in areas of sympatry between closely related taxa than in areas of allopatry." Reproductive character displacement is, therefore, an observed pattern, whereas reinforcement is a process that contributes to the divergence. Butlin (1987), however, has argued to limit reinforcement to cases where fertile hybrids are produced and selected against. Butlin viewed reproductive character displacement to be a process of enhancing prezygotic barriers among taxa that are already fully isolated by postzygotic barriers. Reinforcement in Butlin's view is a component of some forms of speciation, whereas reproductive character displacement of prezygotic barriers can only occur after postzygotic barriers are effective and speciation is already complete.

Communication interference

The selective action of communication interference supposes that the disadvantage in responding to the pheromone of another species is not reduced hybrid viability with pairing of closely related taxa, as would be the case with reinforcement in the sense of Butlin (1997), or wastage of gametes. Instead, it is the expenditure of energy and the possible loss of mating opportunities for the male searcher that serves as the selective force. For the pheromone emitter, it may be the arrival of a heterospecific that disrupts calling and thereby diminishes her probability of mating with a conspecific. Such postulated costs need to be measured in the field, and they likely will be contingent on the density of emitters and responders. Communication interference has been invoked to explain why the pheromone channels of co-existing species that share some components differ in other ways, for example in the ratio of components that are optimal or their use of other exclusive constituents. Communication interference also has been used to explain the selective value of behavioral antagonists (also called "inhibitors"); these are components emitted by one species that interfere with attraction of another species. Antagonists may be pheromone components of the emitting species or they may have no pheromonal activity but function to lessen or eliminate attraction of other species.

In most cases, it will be difficult or impossible to know if the past divergence in the chemical channels occurred during the speciation process by the selective force of hybrid disadvantage (reinforcement *sensu* Butlin (1987)) or as a product of communication interference among taxa that were already separated by effective postzygotic barriers. Exceptions to this difficulty of interpretation may be found in incipient species pairs that can be shown to hybridize. It is even feasible that communication interference could arise among long-diverged lineages (e.g., Cardé, 1987; Löfstedt *et al.*, 1991).

The mechanisms that enhance differences in the chemical channel, whether by selection against hybrids or by selection against cross-attraction of species that already possess good postzygotic barriers, can lead to the pattern of reproductive character displacement. Reproductive character displacement is viewed as a pattern of distribution distinguished by increased divergence in the signal channel in zones of sympatry (this is also consistent with the original definition of this concept by Brown and Wilson (1957)). Reproductive character displacement can occur either with hybridizing taxa or with those already possessing effective postzygotic barriers that prevent the formation of hybrids.

Perhaps the most well-established cases for reproductive character displacement have been documented by McElfresh and Millar in saturniid (hemileucine) moths. In one example, the pheromone has diverged in one set of populations of *Hemileuca eglanterina* (McElfresh and Millar, 2001). Over much of the western United States into southern Canada, *H. eglanterina* is broadly sympatric with *Hemileuca nuttalli*, which employs a 2:1 blend of the (*E,Z*)-10,12- and (*E,E*)-10,12-hexadecadienyl acetates (McElfresh and Millar, 1999a). In populations of *H. eglanterina* in the San Gabriel Mountains of southern California where this species is found without

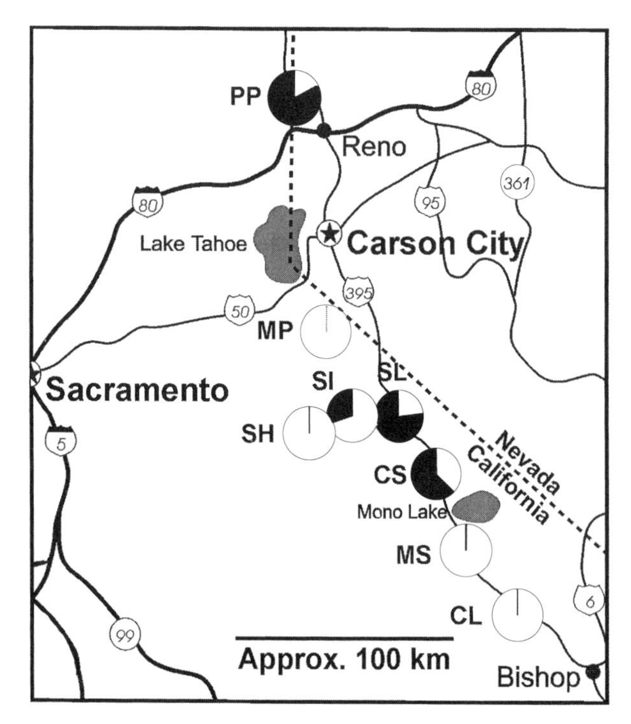

Fig. 8.4. Proportions of male *Hemileuca eglanterina* attracted to lures baited with either 100:10:1 (white area) or 0:100:10 (black area) acetate–alcohol–aldehyde blends at various locations along the eastern slope of the Sierra Nevada Mountains. (From McElfresh and Millar (2001), where designations for localities can be found.)

H. nuttalli, optimal attraction of *H. eglanterina* occurs with a blend of three components: (*E,Z*)-10,12-hexadecadienal, (*E,Z*)-10,12-hexadecadienol, and (*E,Z*)-10,12-hexadecadienyl acetate. In the Robinson Summit population near Ely, Nevada, where both species co-occur, males of *H. eglanterina* are attracted to a two-component blend of the alcohol and aldehyde, and addition of the (*E,Z*)-10,12-hexadecadienyl acetate does not augment attraction.

In a series of populations studied along the eastern Sierra Nevada Mountains of California, three types of male exist (based on electrophysiological evidence): those most responsive to the alcohol, aldehyde, and acetate mix; those responsive to the alcohol and aldehyde mix; and an intermediate form attracted to either mixture. The distribution of attraction of *H. eglanterina* males to the alcohol–aldehyde–acetate blend and to the alcohol–aldehyde blend was documented along a transect of several hundred kilometers along the eastern Sierras (Fig. 8.4). Two *H. eglanterina* populations in the south and two in the west that were outside the range of *H. nuttalli* were attracted only to the three-component blend. Where *H. eglanterina* and *H. nuttalli* overlapped in other sites to the north, there was a mixture of males, usually with the most common type attracted to the two-component alcohol–aldehyde mix, but with a "subpopulation" of males attracted to the three-component mixture. McElfresh and Millar (2001) hypothesized that the ancestral blend of *H. eglanterina* contained the acetate component and that the loss of response to (*E,Z*)-10,12-hexadecadienyl acetate in areas of sympatry with *H. nuttalli* may have occurred because *H. nuttalli* shares this pheromone component. Deletion of the acetate component might have reduced attraction of *H. eglanterina* males to *H. nuttalli* females. Deletion of a pheromone component may be a relatively simple evolutionary occurrence (Löfstedt *et al.*, 1991), in contrast to the addition of novel components.

In addition to divergence in their pheromone blends, *H. nuttalli* and *H. eglanterina* are further isolated by differing times of sexual activity, with the former mainly being attracted between the hours of 10:30 and 13:30 and the latter between 13:30 and 16:30. Notwithstanding this difference, some males of *H. nuttalli* are attracted to *H. eglanterina* females, while the reverse situation does not occur (Collins and Tuskes, 1979). There is little evidence of natural hybridization of these two species, although a few putative natural hybrid moths have been identified (Tuskes *et al.*, 1996); consequently, explanations for differences in their pheromone channels based on reproductive character displacement more logically rest on communication interference rather than hybrid disadvantage as the selective force. Because addition of the (*E,E*)-10,12-hexadecadienyl acetate component of *H. nuttalli* to synthetic blends for *H. eglanterina* eliminates their attractiveness (McElfresh and Millar, 2001), this antagonistic effect is likely one factor in the lack of attraction of *H. eglanterina* males to females of *H. nuttalli*. However, the attraction of *H. nuttalli* males to *H. eglanterina* females could interfere with their calling and, thus, select for females that produce less of their acetate component. Males of *H. eglanterina* are not especially sensitive to the proportion of this component in the blend, and so male dependence on this compound could be reduced gradually instead of in a single step or, in other terms, the male change would track female change.

McElfresh and Millar (2001) also suggested an alternative evolutionary path: that the loss of response to the (*E,Z*)-10,12-hexadecadienyl acetate might have been achieved by *H. eglanterina* males arising that avoid attraction to females of *H. nuttalli* by eliminating their response to the common (*E,Z*)-10,12-hexadecadienyl acetate component, presumably followed by females of *H. eglanterina* producing less of this component or eliminating it entirely. Given that the (*E,E*)-10,12-hexadecadienyl acetate component of *H. nuttalli* is already antagonistic to *H. eglanterina*, this second scenario seems the less likely path. A compelling argument for reproductive character displacement is the geographic distribution of male response types along the eastern Sierras, with the distribution of the alcohol–aldehyde responders of *H. eglanterina* concordant with the distribution of *H. nuttalli*.

A second example of reproductive character displacement in hemileucine moths appears to be represented by geographic variation in the pheromone of *Hemileuca electra* from southern California, described by McElfresh and Millar (1999b). Two subspecies, *H. electra electra* and *H. electra mojavensis*, differ in the proportions of (*E*,*Z*)-10,12-hexadecenadienal in their attractive blends. Females of *H. e. mojavensis* produce eight-fold more of the aldehyde component and one half as much of the acetate component as females of *H. e. electra.* Males of each subspecies are optimally attracted to three-component blends that reflect those produced by their females. Another congener, *Hemileuca burnsi*, overlaps broadly in distribution with *H. e. mojavensis.* These two species are sufficiently distinct that (forced) hybridization yields no viable offspring. Females of *H. e. electra* transported to the range of *H. burnsi* and *H. e. mojavensis* attract males of both species, but cross-attraction between females of *H. e. mojavensis* and males of *H. burnsi* is exceedingly rare. Some isolation of *H. e. mojavensis* and *H. burnsi* is attributable to somewhat differing times of activity: the former species calls from mid morning to afternoon, while the latter calls from afternoon to dusk. Males of *H. burnsi* are much more attracted to the pheromone blend of *H. e. electra* than to that of *H. e. mojavensis.* Together these findings are consistent with a divergence in the pheromone of the two populations of *H. e. electra* and *H. e. mojavensis* being the result of past communication interference.

Other patterns consistent with reproductive character displacement have been documented. Among leafroller (tortricine) moths, there is field evidence that *Archips argyrospila* males from New York have a more acutely tuned response to one of their minor pheromone components than is the case for populations in British Columbia. In the Hudson Valley region of New York, *A. argyrospila* "competes" with a sibling species, *Archips mortuana*, for a distinctive channel. Both species are attracted to mixtures of (*Z*)-11, (*E*)-11-, and (*Z*)-9-tetradecenyl acetates and it seems that the narrowed response in *A. argyrospila* males from New York to the proportion of the (*Z*)-9-component may be attributable to reproductive character displacement (Cardé *et al.*, 1977). A second case may be the obliquebanded leafroller, *Christoneura rosaceana*, which has differing pheromone response in populations from Quebec and British Columbia (Thompson *et al.*, 1991). The presence of differing species of sympatric leafroller species in these two regions may have promoted these population differences. Before these two cases can be substantiated as reproductive character replacement, further documentation of specificity of production and response in additional populations is needed.

Still another case of reproductive character displacement has been proposed for the nun moth, *Lymantria monacha*, by Gries *et al.* (2001). In Honshu, Japan, *L. monacha* is sympatric with five other congeners, including *Lymantria fumida*, which has (7*R*,8*S*)-*cis*-7,8-epoxy-2-methyloctadecane (called "(+)-disparlure") as a major constituent of its pheromone. In Bohemia, Czech Republic, addition of (+)-disparlure to the blend of *L. monacha* increased attraction some 20-fold. In Honshu, however, the addition of (+)-disparlure to the pheromone blend of *L. monacha* increased trap catch only 1.2-fold. Additional differences in the chemistry of their attractants impart specificity to the channels separating *L. monacha* and *L. fumida*, and in Honshu the time of sexual activity of these two species also may have been altered by reproductive character displacement. Greis *et al.* (2001) suggested that the loss of response to (+)-disparlure by *L. monacha* in Honshu would have reduced communication interference between these two congeners. As Greis *et al.* (2001) pointed out, to verify this hypothesis, information on the blends used by other sympatric and allopatric populations in Asia would be useful.

Since reproductive character displacement was proposed in 1957 by Brown and Wilson, its role in fostering further divergence of preexisting but insufficiently robust prezygotic isolating mechanisms between newly evolved species has been much debated. Its general importance in diversification of the pheromone communication channels of moths remains to be established. To establish these cases fully as examples of reproductive character displacement (*sensu* Howard, 1993) would require that the genetics of male variation in response be established and that we learn the extent of blend variation in individual females and its heritability. It also would be useful to quantify the extent to which communication interference lowers reproductive success of males and females. Geographic variation in response and emission is the kind of information that would be especially useful in delineating reproductive character displacement, but strictly comparable studies are difficult to perform. To our knowledge, selection on reproductive success of moths caused by communication interference remains unmeasured in the field, despite the fact that this mechanism, rather than selection against hybridization, is frequently invoked as the selective force (e.g., Cardé, 1987; Löfstedt, 1993).

Change in the number of pheromone components

In general, if the pheromone consists of multiple components, deletion of response to a component of the blend could be advantageous if it lowers the threshold of a male's response while still allowing the male to find the "correct" female. Some blends have behaviorally active components that form a very small proportion of the blend, several percent or less, and although these components augment attraction, they are not crucial to it. Such constituents also may not contribute to creating a private communication channel. A male that could detect females at some distance downwind without requiring a minor constituent could, therefore, have an advantage in mate finding in that such a male presumably could do so at a lower concentration of pheromone; this would be particularly advantageous if an assessment of the ratio of the minor component was not particularly important to threshold (Cardé and Charlton, 1984). This mechanism differs from asymmetric tracking in that males would initiate change in the channel.

The mechanisms promoting a decrease in the number of components in the ermine moth *Yponomeuta rorellus* have been considered by Löfstedt *et al.* (1986, 1990, 1991). *Y. rorellus* is, judging from its nearly monomorphic allozyme variation and reduced chromosome number, a product of a population bottleneck. Unlike other European ermine moth species, it uses tetradecyl acetate as its primary pheromone component, rather than achieving specificity by use of specific blends of (*Z*)-11- and (*E*)-11-tetradecenyl acetates. The model proposed by Löfstedt assumes that a "genetic revolution" was closely followed by a shift of the communication channel. Females of the new *Yponomeuta* population would not produce (*Z*)-11- and (*E*)-11-tetradecenyl acetates and, therefore, they would be unattractive to males of the ancestral species. Derived males might still be attracted to females of the progenitor species, but hybrids might be at a strong disadvantage because of differing number of chromosomes. This explanation seems far more plausible than the alternative hypothesis, that Δ^{11} desaturase evolved independently in the *Yponomeuta* spp. related to *Y. rorellus* (Löfstedt *et al.*, 1986). Although theoretical models generally assume that the likelihood of speciation via genetic revolution is very low, this case seems well substantiated. It is unlikely, however, that similar genetic revolution generally accounts for the simplification of pheromone blends.

Increasing the number of components is not as straightforward as their deletion, because the addition of constituents in most cases necessitates the simultaneous modification of female production and male response, and these traits are not genetically linked. One simplified scenario for adding components would have males acquire the ability to recognize an already emitted component (either one that is emitted "inadvertently" and is not recognized by conspecific males or perhaps one that serves as an antagonist for another species). This could be done via "forgiving receptors" that respond to two compounds, as has been shown for pheromone receptors of *Helicoverpa zea* (Vickers *et al.*, 1991) and *T. ni* (Todd *et al.*, 1992). A possible example of this is the response of the codling moth, *Cydia pomonella*, to (*E,Z*)-8,10-dodecadienol, which has been termed a minor pheromone component. This compound augments attraction somewhat when added to (*E,E*)-8,10-dodecadienol in certain proportions and dosages, but the males do not seem to have any receptors specifically attuned to (*E,Z*)-8,10-dodecadienol. Instead, its perception seems to rely on simultaneous stimulation of receptors that are maximally sensitive to (*E,E*)-8,10-dodecadienol (Bäckman *et al.*, 2000). Eventually there could be selection in such a system for receptors that are specific to the added constituent, and interpretation of its presence at the more central processing level. Avoidance of communication interference or reinforcement (*sensu* Butlin, 1987) could select for such a change.

A second and perhaps more common possibility is suggested by the occasional attraction of male moths to compounds very closely related to actual constituents of a pheromone. For example, in wind tunnel trials, about 5% of the males of the strain of the European corn borer (*O. nubilalis*) that uses (*Z*)- and (*E*)-11-tetradecenyl acetate in a 97:3 ratio are attracted to a 2:1 mix of (*Z*)- and (*E*)-12-tetradecenyl acetates, which is the mix of the Asian corn borer, *Ostrinia furnacalis* (Roelofs *et al.*, 2002). Such occasional "mistakes" also are found in many species when the attractancy of pheromone analogs are evaluated in the field: small proportions of males are attracted to the "wrong" lures. The presumption is that there is a heritable basis for such deviance in at least some of these males. Again, these compounds could be recognized by forgiving receptors (Vickers *et al.*, 1991; Baker, 2002). Change in the communication system then supposes that males track females producing a new component or an additional component.

The presence in the pheromone gland of analogs of the pheromone has been documented in innumerable moths, and many of these constituents have been shown to be emitted with the pheromone (Arn *et al.*, 1992, 2003). These non-pheromonal compounds are thus available for males to detect. (It is also possible in some cases that these additional constituents are true pheromones, but that their behavioral activity remains to be documented.)

Spread of novel signals

Shifts in the communication system over time may be inferred from the evolutionary history of the species and its relatives. Roelofs *et al.* (2002) documented such a case in the pheromone of *O. furnacalis*, a species which seems to have diverged about 1 million years ago from the *O. nubilalis*-containing lineage. *O. furnacalis* differs from other *Ostrinia* spp. in using a mixture of (*Z*)- and (*E*)-12-tetradecenyl acetates as a pheromone, rather than the (*Z*)- and (*E*)-11-components that are common to all other *Ostrinia* spp. investigated to date. Biosynthesis of pheromone in *O. furnacalis* from hexadecanoic acid involves Δ^{14}-desaturation, followed by β-oxidation, reduction, and acetylation (Zhao *et al.*, 1990). In *O. nubilalis*, biosynthesis from hexadecanoic acid involves β-oxidation, followed by Δ^{11}-desaturation, reduction, and acetylation. Based on the phylogeny of desaturase multigene families in insects, Roelofs *et al.* (2002) proposed that the Δ^{14} desaturase gene in *Ostrinia* is a resurrected pseudogene that dates to a duplication event predating the divergence of the Lepidoptera from the Diptera. Roelofs *et al.* (2002) have speculated that the switch in the pheromone would appear to be the consequence of a mutation that activated synthesis of unsaturated fatty acids from the Δ^{14} desaturase transcript and eliminated either the presence or the function of the Δ^{11} desaturase in a single saltational event. Evolution of males responsive to the new Δ^{12}-tetradecenyl acetate pheromone would be facilitated by a small proportion of males already being responsive to this new pheromone (Roelofs *et al.*, 2002). The mechanism explaining the abandonment of the Δ^{11} blend at the population level is not fully specified in this model, but Baker (2002) speculated that if the new blend types become common enough, assortative mating between the new blend types would promote linkage disequilibrium, that is, a non-random association between loci for the signal and its reception. As Butlin and Trickett (1996) pointed out, then selection on one trait would cause a correlated response in the other character, producing a Fisherian runaway process.

Notwithstanding, it is difficult to envisage, even in small populations, just how the new types would become frequent enough for this process to begin. The cost of producing the new pheromone at the expense of losing the old one would seem substantial – a female with the mutant pheromone might well remain unmated – and so drift in a small population should not allow the new blend to proliferate. However, if such an increase did occur, then the new preference allele would be at an initial advantage, because those few males attracted to the mutant blend would have access to more, albeit mutant, females (Butlin and Trickett, 1996). The conundrum is addressed in part by the observation that in some cases there could be a small proportion of males in the population *already* responsive to the blend switch (Roelofs *et al.*, 2002).

An alternative, plausible scenario for the switch from Δ^{11}- to Δ^{12}-tetradecenyl acetates in *O. furnacalis* that cannot be excluded is a stepwise change. First, Δ^{12}-tetradecenyl acetates would be added to the Δ^{11}-tetradecenyl acetates, followed by the elimination of the Δ^{11}-tetradecenyl acetates. The initial addition of the Δ^{12}-components and the subsequent loss of the Δ^{11}-components could have been favored by either communication interference or reinforcement (*sensu* Butlin) in contact with other *Ostrinia* spp. using the ancestral Δ^{11}-components.

Simulation modeling offers a way to examine how changes in the blend production and preference of the responder might influence the fate of mutants. Butlin and Trickett (1996) created a two-locus simulation model that considered "internal" selection pressures and explored how a variety of inheritance patterns (sex linked, autosomal recessive, autosomal additive, and autosomal dominant) would dictate the spread of a new signal system. The model can be set to take into account the cost of these mutations in terms of the success of mate procurement compared for populations of infinite size and of 100 individuals. Not surprisingly, sex linkage would be more likely to promote replacement. Butlin and Trickett's simulation runs suggest many limitations to change in the signaling system, such as the need for simultaneous presence of males with the "correct" mutation preferring the mutant blend in a small population wherein the appropriate blend mutant has already segregated. Although such a concordance of events is expected to be exceedingly rare, there is ample opportunity over time and space for such improbable events to occur. Once set in a small population, however, a further difficulty is the seeding of the new variant into other population demes.

The concept of "forgiving receptors" (Vickers *et al.*, 1991; Baker, 2002) implies that some kinds of variant female will be mated by males that *do not* possess a preference mutation, similar to the model of Roelofs *et al.* (2002). This would obviate the requirement of the Butlin and Trickett (1996) model that the "correct" mutant for male preference be present.

Redundancy

The complexity of pheromone blends is a contributor to reproductive isolation amongst species in the same habitat, but this complexity may also reflect redundancy in the chemical communication system. Linn *et al.* (1984) determined that the six-component pheromone blend of the cabbage looper moth included redundant information. In a wind tunnel assay, progressive subtraction of individual minor components demonstrated that each compound as part of a blend could play a role in stimulating or sustaining upwind flight. However, each of the minor components could be deleted singly from the full blend without a deleterious effect on the response. The neurophysiological basis for this apparent redundancy appears to be that redundant pairs of compounds send information to the same macroglomerular compartment of the brain (Todd *et al.*, 1995). The redundancy may have arisen as a byproduct of a biosynthetic pathway that produced related compounds (Bjostad *et al.*, 1984). Alternatively, redundancy may allow the signal-to-noise ratio to be boosted where species overlap in their geographical distribution, and their pheromone blends overlap in their composition. In this sense, the pheromone components of sympatric species may represent noise. Redundancy may buffer against the potential adverse effects of noise on the effectiveness of communication.

Detecting redundancy in a chemical communication system requires more rigorous testing of blends than is common in pheromone identifications. Typically, such identifications focus on the minimum number of compounds to duplicate the behavioral response stimulated by a female moth. A subtraction bioassay that documented no adverse effect of deletion of a compound could lead to the conclusion that the compound is not a pheromone component. Clearly redundant components could be overlooked in this way. Perhaps for this reason, very few cases of redundancy have been documented (but see Linn *et al.*, 1984; Rhainds *et al.*, 1994; King *et al.*, 1995).

Evolution of antagonists

Evidence from field experiments with many moth species, including cases cited above, shows that addition of very small amounts of pheromone components from a closely related species can abolish the attractiveness of a pheromone. For example, addition of as little as 1% (*Z*)-11-tetradecenyl acetate to the synthetic pheromone of the ermine moth *Y. rorellus* greatly reduces its attractiveness (Löfstedt *et al.*, 1990, 1991). Males of *Y. rorellus* have receptors that are specifically attuned to this antagonist and to other unsaturated acetates, all of which can serve as pheromone components for other sympatric *Yponomeuta* spp. This pattern of antagonists enhancing the discreteness of the communication channel of each species is common among the nine sibling species of *Yponomeuta* of Europe. As Löfstedt (1993) pointed out, it is difficult to envisage how such specialized receptors and behavioral avoidance of their own pheromone tainted with antagonist could have evolved (or be maintained) in these *Yponomeuta* spp. without communication interference as the principal selective force. Löfstedt *et al.* (1991) have also noted that subtraction of one constituent of some of the blends of *Yponomeuta* spp. produces mixtures attractive to sympatric tortricid species that also rely on 14-carbon chain pheromone mixtures. This observation raises the intriguing possibility that communication interference between species in different families could influence blend composition.

Following peripheral reception, the neural circuitry allowing the discrimination of behavioral antagonists by males is understood in only a few moth species. In *H. zea* and *Heliothis virescens*, for example, the behavioral antagonists are targeted by a unique compartment within the macroglomerular complex separate from those attuned to the pheromone components (Vickers *et al.*, 1998). Similarly, in another noctuid, *Spodoptera littoralis*, the behavioral antagonist (*Z*)-9-tetradecenyl acetate is detected by a particular glomerulus (Carlsson *et al.*, 2002). These cases imply that the original evolution of interruption of attraction to pheromone may have required either acquisition of a new glomerulus linked to this response (Hansson and Christensen, 1999), or altering the behavioral response linked to the output of an existing compartment. In *H. zea*, some of the peripheral receptors attuned to one of the pheromone components, (*Z*)-11-hexadecenal, also can detect a behavioral antagonist, (*Z*)-9-tetradecenal, emitted by females of *H. virescens* (Vickers *et al.*, 1998). Whether attraction or antagonism of attraction in *H. zea* males occurs is governed by the proportion of (*Z*)-9-tetradecenal that is detected. As Vickers *et al.* (1998) suggested, such a system could have arisen by changes in the subcircuits of glomeruli in the macroglomerular complex "that reflect differing sensitivities of glomerular output pathways to the same odorant." Given that in most moths attraction of males to a pheromone blend is abolished by presentation of unnatural ratios of components, these types of change may provide a ready avenue for evolution of male response to antagonists.

The presence of behavioral antagonists also is problematic to Paterson's (1985) concept of specific mate recognition systems (SMRS). In Paterson's view, a SMRS is adapted to function well in the species' typical habitat and it is maintained by the forces of stabilizing selection. Change in the SMRS is thought to come about when either the habitat changes or the species invades a new habitat (rather than by reinforcement or communication interference). In these new circumstances, the SMRS is assumed to have increased additive variance, and this allows directional selection to adapt the SMRS to the new environment. Change in the SMRS is thought not to be governed by interaction with other taxa with similar SMRSs, as in reinforcement *sensu* Butlin or in communication interference. The existence of behavioral antagonists is not predicted by the SMRS. The difficulties of applying the SMRS concept to change in moth pheromone systems are summarized by Linn and Roelofs (1995). They emphasized among several concerns the need to arrive at a precise definition of factors in the habitat that would promote such changes and the setting forth of testable hypotheses.

Potential of evolution of resistance to mating disruption

The diversity of signaling and response characteristics in moths indicates that these communication systems are adaptable to both biotic and abiotic factors. The use of synthetic pheromones to disrupt mating has the potential to select for changes in the communication systems that parallel those that have occurred over evolutionary history. It seems prudent to understand the evolutionary processes that could lead to that resistance and to avoid using pheromones in a way that increases the probability of the loss of a valuable control tactic. The factors that need to be understood are the natural heritable variation in characteristics both of signal and of response, the hidden genetically based versatility in pheromone biosynthesis, the potential for genetic linkage between signaling and response, the importance of normalizing selection as a constraint on directional selection imposed by mating disruptants, and the potential for selection for other characteristics such as leaving the treatment area to mate. The potential for selection will be further influenced by the relationship between the synthetic blend used to disrupt mating and the natural blend. The males' ability to pick out or discriminate against pheromone blends that differ from the females' blends could strongly influence both the potential for successful mating disruption and the strength of the imposed selection.

Evolution of resistance to pheromones by a shift in the pheromone to a new blend is fundamentally different from resistance to other control tactics in at least two ways. First, because both signaler and responder are involved in communication, resistance would have to involve both sides of signaling. Changes in pheromone production that are not paralleled by changes in response specificity would not lead to restoration of effective mating. Without linkage between signal and response characteristics, communication would be resistant to change (owing to normalizing or stabilizing selection countering directional selection). This contrasts with resistance to an insecticide, where an outlying resistant individual has a distinct fitness advantage in insecticide-treated habitats independent of other individuals in the population. Second, if changes in the pheromone blend ratio do occur, we should be able to mimic those changes and modify the pheromone blend used to optimize control. In contrast, resistance to an insecticide may translate not only into loss of effectiveness of one compound, but cross-resistance with other compounds that have similar modes of action.

In a field cage experiment, Evenden and Haynes (2001) showed that selection imposed by mating disruption with the normal pheromone blend of *T. ni* resulted in an increased (compared with control cages) frequency of females with a mutant pheromone blend. In this study, the mutant phenotype started at a high frequency and, therefore, did not duplicate the expected scenario where mutant phenotypes are very rare. Nonetheless, the study suggested that, under the right conditions, selection with mating disruptants could lead to changes in the communication system.

The only conclusively documented instance of loss of efficacy of mating disruption over time has occurred in Japan with the smaller tea tortrix, *Adoxophyes honmai* (Mochizuki *et al.*, 2002). The natural blend of this species comprises a 63:31:4:2 blend of (*Z*)-9-, (*Z*)-11-, and (*E*)-11-tetradecenyl acetates, and 10-methyldodecyl acetate (Tamaki *et al.*, 1979). Mating disruption by application of formulated (*Z*)-11-tetradecenyl acetate has been used for control of this species in Japan since 1983, but efficacy of control in one locality (Shimada) declined markedly after some 10 years of application. Control was restored by using the complete, four-component blend as a disruptant. In this case, resistance could have evolved, for example, either by males becoming more discriminating to the natural blend amongst a disruptant background of (*Z*)-11-tetradecenyl acetate or by the original formulation becoming less effective in diminishing male response via sensory adaptation/habituation (Cardé and Minks, 1995). Without knowing how the original formulation of (*Z*)-11-tetradecenyl acetate once interfered with mate finding, it would be difficult to identify how resistance evolved. Because the resistant population occurs only in one region, it may be feasible to establish the mechanism of resistance by behavioral comparison of populations. The rapidity with which resistance evolved in the tea tortrix may have been fostered by using only one component of the natural blend of this species as a disruptant, potentially allowing resistance to evolve by changes in only the male's behavior. Alternatively, females might have become more apparent by increasing the proportions of components in their blend relative to (*Z*)-11-tetradecenyl acetate. In either case, resistance undoubtedly was fostered by the skewed selection imposed by use as a disruptant of only one component of the four-component pheromone. Resistance would have been much less likely to have evolved had the disruptant been formed with the natural blend.

An example of the potential of males to orient to off-ratios of a blend when a single component of a two-component blend is used as a disruptant was shown by Flint and Merkle (1984) in field trials with the pink bollworm, *P. gossypiella*. Attraction to traps baited with a range of component ratios in plots treated with only the (*Z,Z*)-isomer showed a marked shift in male response to a blend high in the (*Z,Z*)-component. This marked alteration in male response, however, probably was the result of *P. gossypiella* males undergoing some sort of sensory impairment that altered the males' perception of ratio; in this case, no evolutionary modification of response was involved. This comparison illustrates the kind of variation which can exist that could be subject to selection, but in this case effective female-to-male communication under pressure from application of a disruptant would necessitate an evolutionary shift in the female's ratio, possibly accompanied by an increased favoring by the male for this blend.

Future directions

Since the first identification of an insect pheromone in 1961, the female-emitted attractant of the silkworm moth, *Bombyx mori*, there has been remarkable progress in deciphering the attractant blends used in all the major lineages of moths. We now know that female-emitted attractants typically have two or more components, and that component ratio can be important in evoking a response. There are now methods to verify biosynthetic pathways and to identify the genes that are responsible (e.g., Roelofs *et al.*, 2002). New brain imaging and recording techniques permit examination of how messages perceived by antennal receptors are interpreted and integrated in the olfactory lobe (e.g., Vickers *et al.*, 1991; Carlsson *et al.*, 2002); such approaches will help us to understand the basis for variation in male behavior and ultimately will provide avenues for dissecting the genetics of response. These approaches also will be helpful in establishing major-gene effects for pheromone synthesis and regulation of response, but quantitative trait locus analyses will be needed to understand many features of individual variation. Stabilizing selection, various forms of sexual selection, asymmetric tracking, reinforcement, and communication interference all have been proposed to be important selective forces in the evolution of sex pheromone communication in moths. To establish how these mechanisms promote change in the communication channel in any given species, it will be essential to have an understanding of the genetic architecture underlying pheromone production and response, and how individual variation in these traits expressed in the field alters the success and rapidity of mate finding. To date, the importance of variation in the communication channel remains incompletely documented and poorly understood.

Acknowledgement

We are grateful to Christer Löfstedt for his comments on this review.

References

Aldrich, J. R. (1995). Chemical communication in the true bugs and parasitoid exploitation. In *Chemical Ecology of Insects 2*, eds. R. T. Cardé and W. J. Bell, pp. 318–363. New York: Chapman & Hall.

Aldrich, J. R., Lusby, W. R., Marron, B. E., Nicolaou, K. C., Hoffmann, M. P. and Wilson, L. T. (1989). Pheromone blends of green stink bugs and possible parasitoid selection. *Naturwissenschaften* **76**: 173–175.

Arakaki, N., Wakamura, S. and Yasuda, T. (1996). Phoretic egg parasitoid, *Telenomus euproctidis* (Hymenoptera: Scelionidae), uses sex pheromone of tussock moth *Euproctis taiwana* (Lepidoptera: Lymantriidae) as a kairomone. *Journal of Chemical Ecology* **22**: 1079–1085.

Arn, H, Tòth, M. and Priesner, E. (1992). *List of Sex Pheromones of Lepidoptera and Related Attractants*, 2nd edn. Montfavet, France: International Organization of Biological Control.

(2003). *The Pherolist.* http://www.nysaes.cornell.edu/pheronet.

Atkinson, P. R. (1981). Mating behaviour and activity patterns of *Eldana saccharina* Walker (Lepidoptera: Pyralidae). *Journal of the Entomological Society of Southern Africa* **44**: 265–280.

(1982). Structure of the putative pheromone glands of *Eldana saccharina* Walker (Lepidoptera: Pyralidae). *Journal of the Entomological Society of Southern Africa* **45**: 93–104.

Bäckman, A.-C., Anderson, P., Bengtsson, M., Löfqvist, J., Unelius, C. R. and Witzgall, P. (2000). Antennal response of codling moth males, *Cydia pomonella* L. (Lepidoptera: Tortricidae), to the geometrical isomers of codlemone and codlemone acetate. *Journal of Comparative Physiology A* **186**: 513–519.

Baker, J. L., Hill, A. S. and Roelofs, W. L. (1978). Seasonal variations in the pheromone trap catches of male omnivorous leafroller moths, *Platynota stultana. Environmental Entomology* **7**: 399–401.

Baker, T. C. (1983). Variations in male oriental fruit moth courtship patterns due to male competition. *Experientia* **39**: 112–114.

(2002). Mechanism for saltational shifts in pheromone communication system. *Proceedings of the National Academy of Sciences, USA* **99**: 13368–13370.

Baker, T. C. and Cardé, R. T. (1979). Courtship behavior of the oriental fruit moth (*Grapholitha molesta*): experimental analysis and consideration of the role of sexual selection in the evolution of courtship pheromones in the Lepidoptera. *Annals of the Entomological Society of America* **72**: 173–188.

Baker, T. C., Nishida, R. and Roelofs, W. L. (1981). Close-range attraction of female oriental fruit moths to herbal scent of male hairpencils. *Science* **214**: 1359–1361.
Bengtsson, B. O. and Löfstedt, C. (1990). No evidence for selection in a pheromonally polymorphic population. *American Naturalist* **138**: 722–726.
Bennett, A. L., Atkinson, P. R. and La Croix, N. J. S. (1991). On communication in the African sugarcane borer, *Eldana saccharina* Walker (Lepidoptera: Pyralidae). *Journal of the Entomological Society of Southern Africa* **54**: 243–259.
Birch, M. C. (1972). Male abdominal brush-organs in British noctuid moths and their value as a taxonomic character. Part II. *The Entomologist* **105**: 233–244.
(1974). Aphrodisiac pheromones in insects. In *Pheromones*, ed. M.C. Birch, pp. 115–134. Amsterdam: North-Holland.
Birch, M. C., Poppy, G. M. and Baker, T. C. (1990). Scents and eversible scent structures of male moths. *Annual Review of Entomology* **35**: 25–58.
Bjostad, L. B., Linn, C. E., Du, J.-W. and Roelofs, W. L. (1984). Identification of new sex pheromone components in *Trichoplusia ni*, predicted from biosynthetic precursors. *Journal of Chemical Ecology* **10**: 1309–1323.
Bonduriansky, R. (2001). The evolution of male mate choice in insects: a synthesis of ideas and evidence. *Biological Reviews* **76**: 305–339.
Brown, W. L., Jr and Wilson, E. O. (1957). Character displacement. *Systematic Zoology* **5**: 49–64.
Burger, B. V., Nell, A. E., Smit, D. *et al.* (1993). Constituents of wing gland and abdominal hair pencil secretions of male African sugarcane borer, *Eldana saccharina* Walker (Lepidoptera: Pyralidae). *Journal of Chemical Ecology* **19**: 2255–2277.
Burghardt, G. M. (1970). Defining communication. In *Advances in Chemoreception*, vol. 1, eds. J. W. Johnson and A. Turk. pp. 5–18. New York: Appleton, Century-Crofts.
Butlin, R. K. (1987). Speciation by reinforcement. *Trends in Ecology and Evolution* **2**: 8–13.
Butlin, R. K. and Trickett, A. J. (1996). Can population genetics simulations help to interpret pheromone evolution? In *Insect Pheromone Research: New Directions*, eds. R. T. Cardé and A. K. Minks, pp. 548–562. New York: Chapman & Hall.
Byers, J. A. and Struble, D. L. (1990). Identification of sex pheromones of two sibling species in dingy cutworm complex, *Feltia jaculifera* (Gn.) (Lepidoptera: Noctuidae). *Journal of Chemical Ecology* **16**: 2981–2992.
Cardé, R. T. (1987). The role of pheromones in reproductive isolation and speciation of insects. In *Evolutionary Genetics of Invertebrate Behavior*, ed. M. D. Huettel. pp. 303–317. New York: Plenum.
Cardé, R. T. and Baker, T. C. (1984). Sexual communication with pheromones. In *Chemical Ecology of Insects*, eds. W. J. Bell and R.T. Cardé, pp. 355–384. London: Chapman & Hall.
Cardé, R. T. and Charlton, R. E. (1984). Olfactory sexual communication in Lepidoptera: strategy, sensitivity and selectivity. In *Insect Communication*, ed. T. Lewis, pp. 241–265. London: Academic Press.
Cardé, R. T. and Hagaman, T. E. (1984). Mate location strategies of gypsy moths in dense populations. *Journal of Chemical Ecology* **10**: 25–31.
Cardé, R. T. and Minks, A. K. (1995). Control of moth pests by mating disruption: successes and constraints. *Annual Review of Entomology* **40**: 559–585.
Cardé, R. T., Baker, T. C. and Roelofs, W. L. (1976). Sex attractant responses of male oriental fruit moths to a range of component ratios: pheromone polymorphism? *Experientia* **32**: 1406–1407.

Cardé, R. T., Cardé, A. M., Hill, A. S. and Roelofs, W. L. (1977). Sex pheromone specificity as a reproductive isolating mechanism among the sibling species *Archips argyrospilus* and *A. mortuanus* and other sympatric tortricine moths (Lepidoptera: Tortricidae). *Journal of Chemical Ecology* **3**: 71–84.

Cardé, R. T., Charlton, R. E., Wallner, W. E. and Baranchikov, Y. N. (1996). Pheromone-mediated diel activity rhythms of male Asian gypsy moth (Lepidoptera: Lymantriidae) in relation to female eclosion and temperature. *Annals of the Entomological Society of America* **89**: 745–753.

Carlsson, M. A., Galizia, C. G. and Hansson, B. S. (2002). Spatial representation of odours in the antennal lobe of the moth *Spodoptera littoralis* (Lepidoptera: Noctuidae). *Chemical Senses* **27**: 231–244.

Charlton, R. E. and Cardé, R. T. (1990). Behavioral interactions in the courtship of the gypsy moth, *Lymantria dispar* (Lepidoptera: Lymantriidae). *Annals of the Entomological Society of America* **83**: 89–96.

Charlton, R. E., Cardé, R. T. and Wallner, W. E. (1999). Synchronous crepuscular flight of female Asian gypsy moths: relationships of light intensity and ambient and body temperatures. *Journal of Insect Behavior* **12**: 517–531.

Collins, M. M. and Tuskes, P. M. (1979). Reproductive isolation in sympatric species of dayflying moths (*Hemileuca*: Saturniidae). *Evolution* **33**: 728–733.

Collins, R. D. and Cardé, R. T. (1985). Variation in and heritability of aspects of pheromone production in the pink bollworm moth, *Pectinophora gossypiella* (Lepidoptera: Gelechiidae). *Annals of the Entomological Society of America* **78**: 229–234.

(1989). Selection for altered pheromone-component ratios in the pink bollworm moth, *Pectinophora gossypiella* (Lepidoptera: Gelechiidae). *Journal of Insect Behavior* **2**: 609–621.

Collins, R. D., Rosenblum, S. L. and Cardé, R. T. (1990). Selection for increased pheromone titre in the pink bollworm moth, *Pectinophora gossypiella* (Lepidoptera: Gelechiidae). *Physiological Entomology* **15**: 141–147.

Conner, W. E., Eisner, T., Vander Meer, R. K., Guerrero, A. and Meinwald, J. (1981). Precopulatory sexual interaction in an arctiid moth (*Utetheisa ornatrix*): role of a pheromone derived from dietary alkaloids. *Behavioral Ecology and Sociobiology* **9**: 227–235.

Darwin, C. (1871). *The Descent of Man, and Selection in Relation to Sex.* London: J. Murray.

De Jong, M.C.M. (1988). *Evolutionary Approaches to Insect Communication Systems. Bark Beetle Host Colonization and Mate Finding in small ermine moths.* Ph.D. Thesis, Leiden University.

De Jong, M. C. M. and Sabelis, M. W. (1991). Limits to runaway sexual selection: the wallflower paradox. *Journal of Evolutionary Biology* **4**: 637–655.

Delisle, J. and Hardy, M. (1997). Male larval nutrition influences the reproductive success of both sexes of the spruce budworm, *Christoneura fumiferana* (Lepidoptera: Tortricidae). *Functional Ecology* **11**: 451–463.

Doane, C. C. (1968). Aspects of mating behavior of the gypsy moth. *Annals of the Entomological Society of America* **68**: 768–773.

Drummond, B. A., III (1984). Multiple mating and sperm competition in the Lepidoptera. In *Sperm Competition and the Evolution of Animal Mating Systems*, ed. R. L. Smith, pp. 291–370. San Diego, CA: Academic Press.

Dussourd, D. E., Ubik, K., Harvis, C., Resch, J., Meinwald, J. and Eisner, T. (1988). Biparental defensive endowment of eggs with acquired plant alkaloid in the moth

Utetheisa ornatrix. *Proceedings of the National Academy of Sciences, USA* **85**: 5992–5996.

Dussourd, D. E., Harvis, C. A., Meinwald, J. and Eisner, T. (1991). Pheromonal advertisement of a nuptial gift by a male moth (*Utetheisa ornatrix*). *Proceedings of the National Academy of Sciences, USA* **88**: 9224–9227.

Eberhard, W. G. (1977). Aggressive chemical mimicry by a bolas spider. *Science* **198**: 1173–1175.

(1980). The natural history and behavior of the bolas spider *Mastophora dizzydeani* sp. N. (Araneidae). *Psyche* **87**: 143–169.

Elkinton, J. S. and Cardé, R. T. (1983). Appetitive flight behavior of male gypsy moths (Lepidoptera: Lymantriidae). *Environmental Entomology* **12**: 1702–1707.

Elkinton, J. S., Schal, C., Ono, T. and Cardé, R. T. (1987). Pheromone puff trajectory and upwind flight of male gypsy moths in a forest. *Physiological Entomology* **12**: 399–406.

Emelianov, I., Mallet, J. and Baltensweiler, W. (1995). Genetic differentiation in *Zeiraphera diniana* (Lepidoptera: Tortricidae, the larch budmoth): polymorphism, host races, or sibling species. *Heredity* **75**: 416–424.

Evenden, M. L. and Haynes, K. F. (2001). Potential for the evolution of resistance to pheromone-based mating disruption tested using two pheromone strains of the cabbage looper, *Trichoplusia ni*. *Entomologia Experimentalis et Applicata* **100**: 131–134.

Evenden, M. L., Spohn, B. G., Moore, A. J., Preziosi, R. F. and Haynes, K. F. (2002). Inheritance and evolution of male response to sex pheromone in *Trichoplusia ni* (Lepidoptera: Noctuidae). *Chemoecology* **12**: 53–59.

Falconer, D.S. (1981). *Introduction to Quantitative Genetics*, 2nd edn. Harlow, UK: Longmans.

Fisher, R. A. (1958). *The Genetical Theory of Natural Selection*, 2nd edn. New York: Dover.

Flint, H. S. and Merkle, J. R. (1984). The pink bollworm (Lepidoptera: Gelechiidae): alteration of male response to gossyplure by release of its component *Z,Z*-isomer. *Journal of Economic Entomology* **77**: 1099–1104.

Flint, H. S., Smith, R. L., Forey, D. E. and Horn, B. R. (1977). Pink bollworm: response of males to (*Z,Z*-) and (*Z,E*-) isomers of gossyplure. *Journal of Economic Entomology* **70**: 274–257.

Foster, S. P., Clearwater, J. R., Muggleston, S. J., Dugdale, J. S. and Roelofs, W. L. (1986). Probable sibling species complexes within two described New Zealand leafroller moths. *Naturwissenschaften* **73**: 156–158.

Foster, S. P., Clearwater, J. R., and Muggleston, S. J. (1989). Intraspecific variation of two components in sex pheromone gland of *Planotortrix excessana* sibling species. *Journal of Chemical Ecology* **15**: 457–465.

Foster, S. P., Muggleston, S. J., Löfstedt, C. and Hansson, B. (1996). A genetic study on pheromonal communication in two *Ctenopseustis* moths. In *Insect Pheromone Research: New Directions*, eds. R. T. Cardé and A. K. Minks, pp. 514–524. New York: Chapman & Hall.

Gemeno, C., Yeargan, K. V. and Haynes, K. F. (2000a). Aggressive chemical mimicry by the bolas spider *Mastophora hutchinsoni*: identification and quantification of a major prey's sex pheromone components in the spider's volatile emissions. *Journal of Chemical Ecology* **26**: 1235–1243.

Gemeno, C., Lutfallah, A. F. and Haynes, K. F. (2000b). Pheromone blend variation and cross-attraction among populations of the black cutworm moth (Lepidoptera: Noctuidae). *Annals of the Entomological Society of America* **93**: 1322–1328.

Gemeno, C., Moore, A. J., Preziosi, R. F. and Haynes, K. F. (2001). Quantitative genetics of signal evolution: a comparison of the pheromonal signal in two populations of the cabbage looper, *Trichoplusia ni. Behavior Genetics* **31**: 157–165.
González, A., Rossini, C., Eisner, M. and Eisner, T. (1999). Sexually transmitted chemical defense in a moth (*Utetheisa ornatrix*). *Proceedings of the National Academy of Sciences, USA* **96**: 5570–5574.
Gothilf, S. and Shorey, H. H. (1976). Sex pheromones of Lepidoptera: examination of the role of male scent brushes in courtship behavior in *Trichoplusia ni. Environmental Entomology* **5**: 115–119.
Grant, G. G. (1971). Scent apparatus of the male cabbage looper, *Trichoplusia ni. Annals of the Entomological Society of America* **64**: 347–352.
Greenfield, M. D. (1981). Moth sex pheromones: an evolutionary perspective. *Florida Entomologist* **64**: 4–17.
(2002). *Signalers and Receivers. Mechanisms and Evolution of Arthropod Communication.* Oxford: Oxford University Press.
Gries, G., Schaeffer, P. W., Gries, R., Liška, J. and Gotoh, T. (2001). Reproductive character displacement in *Lymantria monacha* from Northern Japan? *Journal of Chemical Ecology* **27**: 1163–1176.
Guerin, P. M., Baltensweiler, W., Arn, H. and Buser, H.-R. (1984). Host race pheromone polymorphism in the larch budmoth. *Experientia* **40**: 892–894.
Hansson, B. S. and Christensen, T. A. (1999). Functional characteristics of the antennal lobe. In *Insect Olfaction*, ed. B. S. Hansson, pp. 125–161. Berlin: Springer-Verlag.
Hansson, B. S., Tòth, M., Löfstedt, C., Szöcs, G., Subchev, M. and Löfqvist, J. (1990). Pheromone variation among eastern European and western Asian populations of the turnip moth *Agrotis segetum. Journal of Chemical Ecology* **16**: 1611–1622.
Haynes, K. F. (1990). Identification of sex pheromone of bristly cutworm, *Lacinipolia renigera* (Stephens). *Journal of Chemical Ecology* **16**: 2615–2621.
(1996). Genetics of pheromone communication in the cabbage looper moth, *Trichoplusia ni.* In *Insect Pheromone Research: New Directions*, eds. R. T. Cardé and A. K. Minks, pp. 525–534. New York: Chapman & Hall.
Haynes, K. F. and Baker, T. C. (1988). Potential for evolution of resistance to pheromones: worldwide and local variation in chemical communication system of the pink bollworm moth, *Pectinophora gossypiella. Journal of Chemical Ecology* **14**: 1547–1560.
Haynes, K. F. and Hunt, R. E. (1990). A mutation in pheromonal communication system of cabbage looper moth, *Trichoplusia ni. Journal of Chemical Ecology* **16**: 1249–1257.
Haynes, K. F and Yeargan, K.V. (1999). Exploitation of intraspecific communication systems: illicit signalers and receivers. *Annals of the Entomological Society of America* **92**: 960–970.
Haynes, K. F., Yeargan, K. V. and Gemeno, C. (2001). Detection of prey by a spider that aggressively mimics pheromone blends. *Journal of Insect Behavior* **14**: 535–544.
Howard, D. J. (1993). Reinforcement: origin, dynamics, and fate of an evolutionary hypothesis. In *Hybrid Zones and the Evolutionary Process*, ed. R. G. Harrison, pp. 46–69. New York: Oxford University Press.
Hill, A. S. and Roelofs, W. L. (1981). Sex pheromone of the saltmarsh caterpillar moth, *Estigmene acrea. Journal of Chemical Ecology* **7**: 655–668.
Hill, A., Cardé, R., Comeau, A., Bode, W. and Roelofs, W. (1974). Sex pheromones of the tufted apple bud moth (*Platynota ideausalis*). *Environmental Entomology* **3**: 249–252.

Hill, A. S., Cardé, R. T, Bode, W. M. and Roelofs, W. L. (1977). Sex pheromone components of the variegated leafroller moth, *Platynota flavedana*. *Journal of Chemical Ecology* **3**: 369–376.

Hunt, R. E., Zhao, B.-G. and Haynes, K. F. (1990). Genetic aspects of interpopulational differences in pheromone blend of cabbage looper moth, *Trichoplusia ni*. *Journal of Chemical Ecology* **16**: 2935–2946.

Iyengar, V. K., Reeve, H. K. and Eisner, T. (2002). Paternal inheritance of a female moth's mating preference. *Nature* **419**: 830–832.

Jurenka, R. A., Haynes, K. F., Adlof, R. O., Bengtsson, M. and Roelofs, W. L. (1994). Sex pheromone component ratio in the cabbage looper moth altered by a mutation affecting the fatty acid chain-shortening reactions in the pheromone biosynthetic pathway. *Insect Biochemistry and Molecular Biology* **24**: 373–381.

King, G. G. S., Gries, R., Gries, G. and Slessor, K. N. (1995). Optical isomers of 3,13-dimethylheptadecane: sex pheromone components of the western false hemlock looper, *Nepytia freemani* (Lepidoptera: Geometridae). *Journal of Chemical Ecology* **21**: 2027–2045.

Klun, J. A. and Huettell, M. D. (1988). Genetic regulation of sex pheromone production and response: interaction of sympatric pheromonal types of the European corn borer, *Ostrinia nubilalis* (Lepidoptera: Pyralidae). *Journal of Chemical Ecology* **14**: 2047–2061.

Klun, J. A. and Maini, S. (1979). Genetic basis of an insect chemical communication system: the European corn borer. *Environmental Entomology* **8**: 423–426.

Krasnoff, S. B. (1996). Evolution of male lepidopteran pheromones: a phylogenetic perspective. In *Insect Pheromone Research: New Directions*, eds. R. T. Cardé and A. K. Minks, pp. 490–504. New York: Chapman & Hall.

Krasnoff, S. B. and Roelofs, W. L. (1989). Quantitative and qualitative effects of larval diet on male scent secretions of *Estigmene acrea, Phragmatobia fuliginosa*, and *Pyrrharctia isabella* (Lepidoptera: Arctiidae). *Journal of Chemical Ecology* **15**: 1077–1093.

Kuenen, L. P. S. and Cardé, R. T. (1993) Effects of moth size on velocity and steering during upwind flight toward a sex pheromone source by *Lymantria dispar* (Lepidoptera: Lymantriidae). *Journal of Insect Behavior* **6**: 177–193.

LaForest, S., Wu, W. and Löfstedt, C. (1997). A genetic analysis of population differences in pheromone production and response between two populations of the turnip moth, *Agrotis segetum*. *Journal of Chemical Ecology* **23**: 1487–1503.

Landolt, P. J. and Heath, R. R. (1990). Sexual role reversal in mate-finding strategies of the cabbage looper moth. *Science* **249**: 1026–1028.

Lewontin, R. C. (1979). Sociobiology as an adaptationist program. *Behavioral Science* **24**: 5–14.

Lewis, W. J., Nordlund, D. A., Gueldner, R. C., Teal, P. E. A. and Tumlinson, J. H. (1982). Kairomones and their use for management of entomaphagous insects. XIII. Kairomonal activity for *Trichogramma* spp. of abdominal tips, excretion, and a synthetic sex pheromone blend of *Heliothis zea* (Boddie) moths. *Journal of Chemical Ecology* **8**: 1323–1331.

Li, W., Farrell, J. A. and Cardé, R. T. (2001). Tracking of fluid-advected odor plumes: strategies inspired by insect orientation to pheromone. *Adaptive Behavior* **9**: 143–167.

Linn, C. E., Jr and Roelofs, W. L. (1985). Response specificity of male pink bollworm moths to different blends and dosages of sex pheromone. *Journal of Chemical Ecology* **11**: 1583–1590.

(1995). Pheromone communication in moths and its role in the speciation process. In *Speciation and the Recognition Concept*, eds. D. M. Lambert and H. G. Spencer, pp. 263–300. Baltimore: Johns Hopkins University Press.

Linn, C. E., Bjostad, L. B., Du, J. W. and Roelofs, W. L. (1984). Redundancy in a chemical signal: behavioral responses of male *Trichoplusia ni* to a 6-component sex pheromone blend. *Journal of Chemical Ecology* **10**: 1635–1658.

Linn, C. E., Campbell, M. G., and Roelofs, W. L. (1988). Temperature modulation of behavioural thresholds controlling male moth sex pheromone response specificity. *Physiological Entomology* **13**: 59–67.

Liu, Y.-B. and Haynes, K. F. (1994). Evolution of behavioral responses to sex pheromone in mutant laboratory colonies of *Trichoplusia ni. Journal of Chemical Ecology* **20**: 231–238.

Löfstedt, C. (1993). Moth pheromone genetics and evolution. *Philosophical Transactions of the Royal Society London, Series B* **340**: 167–177.

Löfstedt, C. and Kozlov, M. (1996). A phylogenetic analysis of pheromone communication in primitive moths. In *Insect Pheromone Research: New Directions*, eds. R. T. Cardé and A. K. Minks, pp. 473–489. New York: Chapman & Hall.

Löfstedt, C., Lanne, B. S., Löfquist, J. and Bergström, G. (1985). Individual variation in the pheromone of the turnip moth *Agrotis segetum. Journal of Chemical Ecology* **11**: 1181–1196.

Löfstedt, C., Herrebout, W. M. and Du, J.-W. (1986). Evolution of the ermine moth pheromone tetradecyl acetate. *Nature* **323**: 621–623.

Löfstedt, C., Vickers, N. J., Roelofs, W. L. and Baker, T. C. (1989a). Diet related courtship success in the oriental fruit moth, *Grapholita molesta* (Tortricidae). *Oikos* **55**: 402–408.

Löfstedt, C., Hansson, B. S., Roelofs, W. L. and Bengtsson, B. O. (1989b). No linkage between genes controlling female pheromone production and male pheromone response in the European corn borer, *Ostrinia nubilalis* Hübner (Lepidoptera: Pyralidae). *Genetics* **123**: 553–556.

Löfstedt, C., Hannson, B. S., Dijkerman, H. J. and Herrebout, W. M. (1990). Behavioural and electrophysiological activity of unsaturated analogues of the pheromone tetradecyl acetate in the small ermine moth *Yponomeuta rorellus. Physiological Entomology* **15**: 47–54.

Löfstedt, C., Herrebout, W. M. and Menken, S. B. J. (1991). Sex pheromones and their potential role in the evolution of reproductive isolation in small ermine moths (Yponomeutidae). *Chemoecology* **2**: 20–28.

Löfstedt, C., Hansson, B. S., Petersson, E., Valeur, P. and Richards, A. (1994). Pheromonal secretions from gland on the 5th abdominal sternite of hydrodpsychid and rhyacophilid caddisflies (Trichoptera). *Journal of Chemical Ecology* **20**: 153–170.

Lundberg, S. and Löfstedt, C. (1987). Intra-specific competition in the sex communication channel: a selective force in the evolution of moth pheromones. *Journal of Theoretical Biology* **125**: 15–24.

Maynard Smith, J. (1978). *The Evolution of Sex.* Cambridge: Cambridge University Press.

McElfresh, J. S. and Millar, J. G. (1999a). Sex pheromone of Nuttall's sheep moth, *Hemileuca nuttalli*, from the eastern Sierra Mountains of California. *Journal of Chemical Ecology* **25**: 711–726.

(1999b). Geographic variation in sex pheromone blend of *Hemileuca electra* from southern California. *Journal of Chemical Ecology* **25**: 2505–2525.

(2001). Geographic variation in the pheromone system of the saturniid moth *Hemileuca eglanterina. Ecology* **82**: 3505–3518.

McNeil, J. N. (1991). Behavioral ecology of pheromone-mediated communication in moths and its importance in the use of pheromone traps. *Annual Review of Entomology* **36**: 407–430.
Millar, J. G. (2000). Polyene hydrocarbons and epoxides: a second major class of lepidopteran sex attractant pheromones. *Annual Review of Entomology* **45**: 575–604.
Millar, J. G., Rice, R. E., Steffan, S. A., Daane, K. M., Cullen, E. and Zalom, F. G. (2001). Attraction of female digger wasps, *Astata occidentalis* Cresson (Hymenoptera: Sphecidae) to the sex pheromone of the stink bug *Thyanta pallidovirens* (Hemiptera: Pentatomidae). *Pan-Pacific Entomologist* **77**:244–248.
Mochizuki, F., Fukumoto, T., Noguchi, H., Sugie, H., Morimoto, T. and Ohtani, K. (2002). Resistance to a mating disruptant composed of (Z)-11-tetradecenyl acetate in the smaller tea tortrix, *Adoxophyes honmai* (Yasuda) (Lepidoptera: Tortricidae). *Applied Entomology and Zoology* **37**: 299–304.
Monti, L., Génermont, J., Malosse, C. and Lalanne-Cassou, B. (1997). A genetic analysis of some components of reproductive isolation between two closely related species, *Spodoptera latifascia* (Walker) and *S. descoinsi* (Lalanne-Cassou and Silvain) (Lepidoptera: Noctuidae). *Journal of Evolutionary Biology* **10**: 121–134.
Murlis, J., Elkinton, J. S. and Cardé, R. T. (1992). Odor plumes and how insects use them. *Annual Review of Entomology* **37**: 505–532.
Newcomb, R. D. and Gleeson, D. M. (1998). Pheromone evolution within the genera *Ctenopseustis* and *Planotortrix* (Lepidoptera:Tortricidae) inferred from a phylogeny based on cytochrome oxidase I gene variation. *Biochemical Systematics and Ecology* **26**: 473–484.
Nishida, R., Baker, T. C. and Roelofs, W. L. (1982). Hairpencil pheromone components of male oriental fruit moths, *Grapholitha molesta. Journal of Chemical Ecology* **8**: 947–959.
Noldus, L. P. J. J. (1988). Response of the egg parasitoid *Trichogramma pretiosum* to the sex pheromone of its host *Heliothis zea. Entomologia Experimentalis et Applicata* **48**: 293–300.
Noldus, L. P. J. J., van Lenteren, J. C. and Lewis, W. J. (1991). How *Trichogramma* parasitoids use moth sex pheromones as kairomones: orientation behaviour in a wind tunnel. *Physiological Entomology* **16**: 313–327.
Ono, T. (1993). Effect of rearing temperature on pheromone component ratio in potato tuberworm moth, *Phthorimaea operculella*, (Lepidoptera: Gelechiidae). *Journal of Chemical Ecology* **19**: 71–81.
Ono, T. and Orita, S. (1986). Field trapping of the potato tuber moth, *Phthorimaea operculella* (Lepidoptera: Gelechiidae), with the sex pheromone. *Applied Entomology and Zoology* **21**: 632–634.
Paterson, H. E. H. (1985). The recognition concept of species. In *Museum Monograph 4: Species and Speciation*, ed. E. S. Virba, pp. 21–29. Pretoria: Transvaal Museum.
Phelan, P. L. (1992). Evolution of sex pheromones and the role of asymmetric tracking. In *Insect Chemical Ecology. An Evolutionary Approach*, eds. B. D. Roitberg and M. B. Isman, pp. 265–314. New York: Chapman & Hall.
(1996). Genetics and phylogenetics in the evolution of sex pheromones. In *Insect Pheromone Research: New Directions*, eds. R. T. Cardé and A. K. Minks, pp. 563–579. New York: Chapman & Hall.
(1997). Evolution of mate-signaling in moths: phylogenetic consideration and predictions from the asymmetric tracking hypothesis. In *Evolution of Insect Mating Systems in Insects and Arachnids*, eds. J. C. Choe and B. J. Crespi, pp. 240–256. Cambridge: Cambridge University Press.

Phelan, P. L. and Baker, T. C. (1986). Male-size-related courtship success and intersexual selection in the tobacco moth, *Ephestia elutella*. *Experientia* **42**: 1291–1293.

(1987). Evolution of male pheromones in moths: reproductive isolation through sexual selection. *Science* **235**: 205–207.

(1990). Comparative study of courtship in 12 phycitine moths (Lepidoptera: Pyralidae). *Journal of Insect Behavior* **3**: 303–326.

Priesner, E. and Baltensweiler, W. (1987). Studien zum Pheromon-Polymorphismus von *Zeiraphera diniana* Gn. (Lep., Tortricidae). 1. Pheromon-Reacktionstypen mannlicher falter in europaischen Wildpopulationen, 1978–85. *Journal of Applied Entomology* **104**: 234–256.

Resh, V. H. and Wood, J. R. (1985). The site of pheromone production in three species of Trichoptera. *Aquatic Insects* **7**: 65–71.

Rhainds, M., Gries, G., Li, J. X. *et al.* (1994). Chiral esters: sex pheromone of the bagworm, *Oiketicus kirbyi* (Lepidoptera: Psychidae). *Journal of Chemical Ecology* **20**: 3083–3096.

Roelofs, W. L. and Brown, R. L. (1982). Pheromones and the evolutionary relationships of Tortricidae. *Annual Review of Ecology and Systematics* **13**: 395–422.

Roelofs, W. L., Glover, T., Tang, X. H. *et al.* (1987). Sex pheromone production and perception in European corn borer moths is determined by both autosomal and sex-linked genes. *Proceedings of the National Academy of Sciences, USA* **84**: 7585–7589.

Roelofs, W. L., Liu, W., Hao, G., Jiao, H., Rooney, A. P. and Linn, C. E., Jr (2002). Evolution of moth sex pheromones via ancestral genes. *Proceedings of the National Academy of Sciences, USA* **99**: 13621–13626.

Sasaerila, Y., Greis, G., Greis, R. and Boo, T. C. (2000). Specificity of communication channels in four limacodid moths: *Darna bradleyi, Darna trima, Setothosea asigna,* and *Setore nitens* (Lepidoptera: Limacodidae). *Chemoecology* **10**: 193–199.

Schal, C., Charlton, R. E. and Cardé, R. T. (1987). Temporal patterns of sex pheromone titers and release rates in *Holomelina lamae* (Lepidoptera: Arctiidae). *Journal of Chemical Ecology* **13**: 1115–1129.

Schal, C., Sevala, V. and Cardé, R. T. (1998). Novel and highly specific transport of a volatile sex pheromone by hemolymph lipophorin in moths. *Naturwissenschaften* **85**: 339–342.

Sharov, A. A., Liebhold, A. M. and Ravlin, F. W. (1995). Prediction of gypsy moth (Lepidoptera: Lymantriidae) mating success from pheromone trap counts. *Environmental Entomology* **24**: 1239–1244.

Spangler, H. G. (1987). Acoustically mediated pheromone release in *Galleria mellonella* (Lepidoptera: Pyralidae). *Journal of Insect Physiology* **33**: 465–468.

Spangler, H. G., Greenfield, M. D. and Takessian, A. (1984). Ultrasonic mate calling in the lesser wax moth. *Physiological Entomology* **9**: 87–95.

Sperling, F., Byers, R. and Hickey, D. (1996). Mitochondrial DNA sequence variation among pheromotypes of the dingy cutworm, *Feltia jaculifera* (Gn.) Lepidotera: Noctuidae). *Canadian Journal of Zoology* **74**: 2109–2117.

Stowe, M. K., Tumlinson, J. H. and Heath, R. R. (1987). Chemical mimicry: bolas spiders emit components of moth prey species sex pheromones. *Science* **236**: 964–967.

Tamaki, Y., Noguchi, H., Sugie, H., Sato, R. and Kariya, A. (1979). Minor components of the female sex-attractant pheromone of the smaller tea tortrix (Lepidoptera: Tortricidae): isolation and identification. *Applied Entomology and Zoology* **14**: 101–113.

Thompson, D. R., Angerilli, N. P. D., Vincent, C. and Gaunce, A. P. (1991). Evidence for regional differences in the response of the obliquebanded leafroller (Lepidoptera: Tortricidae) to sex pheromone blends. *Environmental Entomology* **20**: 935–938.
Thornhill, R. (1979). Male and female sexual selection and the evolution of mating strategies in insects. In *Sexual Selection and Reproductive Competition in Insects*, eds. M. S. Blum and N. A. Blum, pp. 81–121. New York: Academic Press.
Thornhill, R. and Alcock, J. (1983). *The Evolution of Insect Mating Systems.* Cambridge, MA: Harvard University Press.
Todd, J. L., Haynes, K. F. and Baker, T. C. (1992). Antennal neurones specific for redundant pheromone components in normal and mutant *Trichoplusia ni* males. *Physiological Entomology* **17**: 183–192.
Todd, J. L., Anton, S., Hansson, B. S. and Baker, T. C. (1995). Functional organization of the macroglomerular complex related to behaviourally expressed olfactory redundancy in male cabbage looper moths. *Physiological Entomology* **20**: 349–361.
Tòth, M., Löfstedt, C., Blair, B. W. *et al.* (1992). Attraction of male turnip moths *Agrotis segetum* (Lepidoptera: Noctuidae) to sex pheromone components and their mixtures at 11 sites in Europe, Asia, and Africa. *Journal of Chemical Ecology* **18**: 1337–1347.
Trivers, R. L. (1972). Parental investment and sexual selection. In *Sexual Selection and the Descent of Man, 1871–1971*, ed. B. Campbell, pp. 136–179. Chicago, IL: Aldine.
Tuskes, P.M., Tuttle, J.P. and Collins, M.M. (1996). *The Wild Silk Moths of North America.* Ithaca, NY: Cornell University Press.
Vickers, N. J., Christensen, T. A., Mustaparta, H. and Baker, T. C. (1991). Chemical communication in heliothine moths. III. Flight behavior of male *Heliocoverpa zea* and *Heliothis virescens* in response to varying ratios of intra- and interspecific sex pheromone components. *Journal of Comparative Physiology A* **169**: 275–280.
Vickers, N. J., Christensen, T. J. and Hildebrand, J. G. (1998). Combinatorial odor discrimination in the brain: attractive and antagonistic odor blends are represented in distinct combinations of uniquely identifiable glomeruli. *Journal of Comparative Neurology* **400**: 35–56.
Wagner, D. L. and Rosovsky, J. (1991). Mating systems in primitive Lepidoptera, with emphasis on the reproductive behaviour of *Korscheltellus gracilis* (Hepialidae). *Zoological Journal of the Linnaean Society* **102**: 227–303.
Wagner, W. E., Jr (1998). Measuring female mating preferences. *Animal Behaviour* **55**: 1029–1042.
Williams, G. C. (1992). *Natural Selection.* Princeton, NJ: Princeton University Press.
Willis, M. A. and Birch, M. C. (1982). Male lek formation and female calling in a population of the arctiid moth *Estigmene acrea. Science* **218**: 168–170.
Wood, D. L. (1982). The role of pheromones, kairomones, and allomones in the host selection and colonization behavior of bark beetles. *Annual Review of Entomology* **27**: 411–446.
Wu, W.-Q., Hansson, B. S. and Löfstedt, C. (1995). Electrophysiological and behavioural evidence for a fourth sex pheromone component in the turnip moth, *Agrotis segetum. Physiological Entomology* **20**: 81–92.
Wu, W., Cottrell, C. B., Hansson, B. S. and Löfstedt, C. (1999). Comparative study of pheromone production and response in Swedish and Zimbabwean populations of turnip moth, *Agrotis segetum. Journal of Chemical Ecology* **25**: 177–196.
Yeargan, K. V. (1988). Ecology of a bolas spider, *Mastophora hutchinsoni*: phenology, hunting tactics, and evidence of aggressive chemical mimicry. *Oecologia (Berlin)* **74**: 524–530.
(1994). Biology of bolas spiders. *Annual Review of Entomology* **39**: 81–99.

Zagatti, P. (1981). Comportement sexuel de la pyrale de la canne à sucre *Eldana saccharina* (Wlk.) lié à deux phéromones émises par le male. *Behaviour* **78**: 81–98.

Zhao, C., Löfstedt, C. and Xuying, W. (1990). Sex pheromone biosynthesis in the Asian corn borer *Ostrinia furnacalis* (II): Biosynthesis of (*E*)- and (*Z*)-12-tetradecenyl acetate involves *delta*-14 desaturation. *Archives of Insect Biochemistry and Physiology* **15**: 57–65.

Zhu, J. W., Zhao, C. H., Lu, F., Bengtsson, M. and Löfstedt, C. (1996a). Reductase specificity and the ratio regulation of E/Z isomers in pheromone biosynthesis of the European corn borer, *Ostrinia nubilalis* (Lepidoptera: Pyralidae). *Insect Biochemistry and Molecular Biology* **26**: 171–176.

Zhu, J., Löfstedt, C. and Bengtsson, B. O. (1996b). Genetic variation in the strongly canalized sex pheromone communication system of the European corn borer, *Ostrinia nubilalis* Hübner (Lepidoptera; Pyralidae). *Genetics* **144**: 757–766.

Index